MARINE BIOLOGY

SEAWEEDS

AGRICULTURAL USES, BIOLOGICAL AND ANTIOXIDANT AGENTS

MARINE BIOLOGY

Additional books in this series can be found on Nova's website under the Series tab.

Additional e-books in this series can be found on Nova's website under the e-book tab.

MARINE BIOLOGY

SEAWEEDS

AGRICULTURAL USES, BIOLOGICAL AND ANTIOXIDANT AGENTS

VITOR H. POMIN
EDITOR

nova publishers
New York

For permission to use material from this book please contact us:
Telephone 631-231-7269; Fax 631-231-8175
Web Site: http://www.novapublishers.com

Library of Congress Cataloging-in-Publication Data

ISBN: 978-1-63117-571-8

Published by Nova Science Publishers, Inc. † New York

CONTENTS

PREFACE

Seaweeds are well-known to be a natural resource rich of ingredients with healthy and therapeutic properties. For this reason, cultivation of seaweeds is highly demanded to afford the international market. This book aims to report possible applications of seaweeds in agriculture as well as the composing ingredients of nutritional value and of therapeutic properties against oxidative stress. The information set provided in this book comes from very recent scientific results obtained by research groups of several countries. This information set is useful not solely for the academic community (undergraduate or graduate students, staff and faculty personnel), but also to any of those individuals involved somewhat with the agricultural, commercial, biological and medical business of seaweeds. The contributing authors are renowned scientific leaders in the field. This feature associated with the fresh type of information provided herein collaborates significantly to the high-quality of this publication.

Chapter 1- Seaweeds form an integral part of marine coastal ecosystems. They include the multicellular marine algae that commonly inhabit the coastal regions of the world's oceans, where suitable substrata exist. These marine seaweed species are often regarded as an underutilized bioresource, and many of them have been widely used as a source of food, industrial raw materials, and in therapeutic, cosmetic and botanical applications for centuries. Seaweeds likewise have been used for millennia as fodder supplements to improve animal nutrition and productivity. Moreover, some seaweeds have the capacity to purify water by removing heavy metals and can be used in the bioremediation of such pollutants. In agriculture, there is a growing concern that environmental pollution caused by nutrient imbalance and overuse of chemical fertilizers and pesticides is directly or indirectly related to human health problems. Consequently, worldwide, there is a tendency towards a shift from chemical-based conventional farming towards to low input sustainable, organic, or alternative agriculture. With increasing demand, availability of organic fertilizers from limited sources was not adequate. To meet the increasing demand many viable options should be explored and one such option is the use of seaweed extracts as fertilizers. More specifically, plants grown in soils treated with seaweed manures, or extracts applied either to the soil or to foliage, exhibit a wide range of responses that have been well documented in a number of reports. Positive responses include improved seed germination, root development, leaf quality, plant vigor and resistance to pathogens. Also, benefits in flower set, fruit production and marketable qualities of fruit are recorded for a large number of both herbaceous and woody crop species. Many reports indicate that seaweeds have a high content of macro- and microelements that are necessary for the development and growth of plants, amino acids, antibiotics, and vitamins. It is now well recognized that many of the common plant hormones (abscisic acid, auxins, cytokinins, gibberellic acid) occur in algae. Moreover, seaweed synthesize polymers such as agars, alginates, phlorotannins and others that show a variety of biological activities in both plants and animals and involved in host defense mechanisms. This short summary indicates the significance of algae as sources of various chemicals in agriculture and their potential use in the near future.

Chapter 2 - Seaplants (a better alternative to the misnomer "Seaweeds"), by all means, are "future plants"; they have been projected as the future viand for ever-increasing human populations, viable and sustainable source for biofuel without disturbing global food scenario, as potential candidates for carbon capture and sequestration that is considered as a practical remedy for global warming, and they have a number of pharmaceutical, industrial and biotechnological applications. However, information on its cultivation methods or life history remains obscure to a majority of marine botanists. While life histories of seaweeds have traditionally been an exotic topic for specialists-language of which is ciphered with scientific jargons incomprehensible to general scientific audience, its agronomy had been a trade secret for coastal communities in East Asian countries, especially Japan, the Philippines and Indonesia. In this up-to-date illustrated review, current scientific understanding on the life-histories of agronomically pertinent seaweeds are presented in a fashion akin to popular science journalism with an overview of major coastal and offshore seaweed mariculture techniques, presented with the aid of clear-to-understand illustrations. Also discussed in this report are recent advances in the algal natural products; including uses in hydrocolloid and pharmaceutical industries, Integrated Multi Trophic Aquaculture, energy production, environmental impacts of the seafarming and its counter measures, before concluding with an overview of future research avenues.

Chapter 3 - Seaweeds are major coastal resources which are valuable to human consumption and environment. Also, it is a foodstuff that has been historically consumed around the globe but is only consumed in appreciable amounts in certain areas of the world today and it is also considered as a renewable natural resource with extensive distribution along the ocean coast. Biologically, seaweeds are categorized by their pigmentation, morphology, anatomy, and nutritional composition as brown (*Phaeophyta*), red (*Rhodophyta*) or green algae (*Chlorophyta*). The chemical composition of seaweeds varies with species, habitat, maturity and environmental conditions. Seaweeds are valuable sources of protein, fiber, vitamins, polyunsaturated fatty acids, macro and trace elements, as well as important bioactive compounds. Thus, they have been recognized as being beneficial for human and animal health and may be used to the food industry as a source of ingredients with high nutritional value. Seaweeds also can be provided as a dietary alternative due to its nutritional value and its commercial value can be enhanced by improving the quality and expanding the range of seaweed – based products. In recent years, seaweeds have caused emerging interest in biomedicine and the food area, because they possess a wealth of bioactive compounds (such as sulfated polysaccharides, carotenoids, dietary fiber, protein, essential fatty acids, vitamins, minerals, terpenoids, oxylipins, phlorotannins and steroids) with potential industrial and agricultural applications.

Chapter 4 - There is a long tradition in East-Asian countries of consuming seaweed, although its intake has recently increased in Western countries. Nutritional aspects of seaweeds as foods are well established, as it is considered a valuable food source. In addition, an extensive body of evidence of the associated health effects of seaweeds measured *in vitro* is described. However, knowledge of the health effects of seaweed based upon intervention human studies is very limited. The purpose of this chapter is to describe the scarce substation of scientific of the health-promoting effects attributed to brown and red edible seaweeds. Different health aspects will be described, e.g. protection of postmenopausal breast cancer, modulation of postpandrial plasma glucose and insulin levels, or modulation of serum estrogen hormone levels.

Chapter 5 - Several studies have demonstrated that seaweeds besides their nutritional properties such as high content in dietary fibre, proteins, polyunsaturated fatty acids and low content in saturated fats are an excellent source of compounds with biological activity. Among the bioactive substances, antioxidants have attracted the attention of the scientific community due to the positive effects against serious disorders. Several compounds, of different nature, with antioxidant activity have been identified in macroalgae including, carotenoids (i.e. fucoxanthin), polyphenols (i.e. phlorotannins) or vitamins such as α-tocopherol (Vitamin E) and ascorbic acid (Vitamin C). This study reports an up-to-date overview on the analysis of bioactive compounds in seaweeds. Relevant features of chromatographic methods are commented on. In addition methods to evaluate the antioxidant activity are also reviewed.

Chapter 6 - Eight carrageenophytes, representing seven genera and three families of the Gigartinales (Florideophyceae, Rhodophyta), were studied for 15 months. Reproductive status, dry weight and carrageenan content have been monitored d by monthly random sampling. The highest carrageenan yields were found in *Chondracanthus acicularis* (61.1%), *Gigartina pistillata* (59.7%) and *Chondracanthus teedei* var. *lusitanicus* (58.0%). Cultivated *Chondrus crispus* from an integrated multi-trophic aquaculture (IMTA) system and wild *C. crispus* were compared regarding carrageenan yield and quality. Species from Cystocloniaceae family predominantly produce iota-carrageenans; Gigartinaceae family produces hybrid kappa-iota carrageenans (gametophytic plants) and lambda-family carrageenans (sporophytic plants); Phyllophoraceae family produces kappa-iota hybrid-carrageenans. Quadrate (10x10 cm) destructive sampling method was used to determine the carrageenophytes biomass in two periods: autumn/winter and spring/summer. The highest biomass was found in *C. crispus* (570 g/m^2), *C. acicularis* (99 g/m^2), *C. teedei* var. *lusitanicus* (207.5 g/m^2) and *Mastocarpus stellatus* (520 g/m^2). Quadrate (50x50 cm) non-destructive sampling method, applied along perpendicular transects to the shoreline, was also used to determine the carrageenophytes coverage during spring/summer and along exposed and shelters *facies*. Considering the overall coverage at both exposed and sheltered areas, *C. teedei* var. *lusitanicus* (26.39%) is the most abundant species.

Chapter 7 - The composition of free amino acids is responsible for the flavor and nutritional value of seaweeds such as *Pyropia yezoensis*, *Saccharina japonica*, *Undaria pinnatifida*, and *Sargassum fusiforme*, which are edible foods that are popular in Japan. Although only L-amino acids were thought to occur naturally, recent advances in analytical techniques have led to the identification of various free D-amino acids in eukaryotic organisms, including seaweeds. We detected high levels of D-aspartate in *S. fusiforme*, and the functional significance of this amino acid is supported by its specific localization in the medullary layer of blades and quantitative variation due to changes in environmental conditions. Despite the presence in various kinds of seaweeds, the biological function of D-alanine remains unclear. Based on the genome-wide survey, we have found gene candidates in seaweeds that encode putative serine racemase [EC 5.1.1.18] involving in the racemization of D- and L-serine, while alanine racemase [EC 5.1.1.1] and aspartate racemase [EC 5.1.1.13] were not found. Seaweed serine racemases are homologous to the human enzyme, suggesting the functional enzymatic activity of seaweed proteins. However, D-serine was not detected in any seaweed analyzed. Thus, the analysis of the enzymatic activity of putative serine racemases encoded by the genes discussed herein provides new insight on the biological significance of D-amino acids in seaweeds.

Chapter 8 - Recent developments in medicinal field have reported a number of diseases linked to free radicals that have been allied to an unhealthy lifestyle, exposure of chemicals, pollution, cigarette smoking, drugs, illness, and stress etc. In the past, coastal populations across the globe consumed dietary foods such as edible seaweeds as a part of their habitual diet including many several Asian countries, that protected them from the free radical induced damages. Their anti-oxidant composition alone provides a useful alternative to non-natural substances, while at the same time providing worthwhile nutritional benefits as foods, food supplements, nutraceuticals and medicine. Previous studies have shown that phenolic compounds are the main contributors to the anti-oxidant activity of various seaweeds. In fact, a positive correlation has been well documented between total phenol content (TPC) and anti-oxidant activity. Many types of seaweed are already in use as alternative and complementary medicine, and obtainable in numerous ways, such as, teas, syrups, oils, liquid extracts, tinctures, and dry extracts (pills or capsules) for therapeutic purposes. However, economic utilization of seaweeds from the perspective of anti-oxidant benefits to human health still remains in its infancy. This chapter will highlight current lack of data in this area and help outline seaweed extracts and seaweed derived isolated with major health benefits that merit consequent analytical studies and possible utilization as natural anti-oxidants.

Chapter 9 - Seaweeds belong to a group of marine organisms that are frequently subjected to ultraviolet light, high oxygen content, and saline and hydric stress. These conditions produce continuous damage to important biomolecules in these organisms, such as proteins, lipids and DNA, in addition to the high production of free radicals that can destroy the cells. Due to the consequences of this damage, seaweeds have many endogenous mechanisms to prevent the destruction of cells, especially the production of antioxidant molecules. As a result, seaweeds have been studied as a potential source for the extraction of antioxidant compounds, and these molecules might be used in various antioxidant therapies. Among the antioxidant compounds present in the seaweeds, sulfated polysaccharides are the most studied. Each seaweed species produces its own sulfated polysaccharide, which makes sulfated polysaccharide research in algae a vast field of study. Brown seaweeds synthesize polysaccharides rich in sulfated α-L-fucose and are called fucans; red seaweeds synthesize polysaccharides rich in galactose and are called named galactans; while green seaweeds synthesize more heterogeneous sulfated polysaccharides. Many studies have shown that the antioxidant capacity of these polysaccharides depend on structural characteristic, such as sulfation and molecular weight. Moreover, studies have shown that sulfated polysaccharides can act to prevent the production of free radicals or neutralize already existing free radicals. The present study reports the progress of research in antioxidant activity from sulfated polysaccharides obtained from seaweeds, the principal methods used to evaluate this activity, and the principal applications of antioxidant sulfated polysaccharides in several human activities.

In: Seaweeds
Editor: Vitor H. Pomin
ISBN: 978-1-63117-571-8

Chapter 1

ROLE OF ALGAE IN AGRICULTURE

***Christos Chatzissavvidis*[1] *and Ioannis Therios*[2]**
[1]Laboratory of Pomology, Department of Agricultural Development, Democritus University of Thrace, Orestiada, Greece
[2]Laboratory of Pomology, School of Agriculture, Aristotle University of Thessaloniki, Thessaloniki, Greece

ABSTRACT

Seaweeds form an integral part of marine coastal ecosystems. They include the multicellular marine algae that commonly inhabit the coastal regions of the world's oceans, where suitable substrata exist. These marine seaweed species are often regarded as an underutilized bioresource, and many of them have been widely used as a source of food, industrial raw materials, and in therapeutic, cosmetic and botanical applications for centuries. Seaweeds likewise have been used for millennia as fodder supplements to improve animal nutrition and productivity. Moreover, some seaweeds have the capacity to purify water by removing heavy metals and can be used in the bioremediation of such pollutants. In agriculture, there is a growing concern that environmental pollution caused by nutrient imbalance and overuse of chemical fertilizers and pesticides is directly or indirectly related to human health problems. Consequently, worldwide, there is a tendency towards a shift from chemical-based conventional farming towards to low input sustainable, organic, or alternative agriculture. With increasing demand, availability of organic fertilizers from limited sources was not adequate. To meet the increasing demand many viable options should be explored and one such option is the use of seaweed extracts as fertilizers. More specifically, plants grown in soils treated with seaweed manures, or extracts applied either to the soil or to foliage, exhibit a wide range of responses that have been well documented in a number of reports. Positive responses include improved seed germination, root development, leaf quality, plant vigor and resistance to pathogens. Also, benefits in flower set, fruit production and marketable qualities of fruit are recorded for a large number of both herbaceous and woody crop species.

Many reports indicate that seaweeds have a high content of macro- and microelements that are necessary for the development and growth of plants, amino acids, antibiotics, and vitamins. It is now well recognized that many of the common plant hormones (abscisic acid, auxins, cytokinins, gibberellic acid) occur in algae. Moreover, seaweed synthesize polymers such as agars, alginates, phlorotannins and others that show a variety of biological activities in both plants and animals and involved in host defense mechanisms. This short summary indicates the significance of algae as sources of various chemicals in agriculture and their potential use in the near future.

1. Introduction

Marine plants comprise of algae, sea grasses, mangroves and sand dune vegetation. The algae are of different shapes and sizes. The microscopic algae are known as phytoplankton and macroscopic ones as seaweeds (Dhargalkar and Pereira, 2005). Seaweeds are of ecological importance because they assist in supplying oxygen to the sea and act as one of the primary producers in the marine food chain (Chan et al., 2006). Like plants, algae use the sunlight for the process of photosynthesis. Photosynthesis is an important biochemical process in which plants, algae, and some bacteria convert the energy of sunlight to chemical energy. Algae capture light energy through photosynthesis and convert inorganic substances into simple sugars using the captured energy.

The interest in the agricultural use of seaweeds in composts increased in the last decades due to an unusual increase in algal biomass, particularly that of green seaweeds, resulting from the progressive eutrophication of some coastal ecosystems (Morand and Briand, 1996). Moreover, since the increasing occurrence of biotic and abiotic stresses is likely due to climate change and will further reduce yields and/or will have an impact on crops, research into developing sustainable methods to alleviate these stresses should be a priority (Khan et al., 2009). Recent studies have shown that seaweed extracts protect plants against a number of biotic and abiotic stresses and offer potential for field application. At present, growing interest in ecological methods of crop production, as well as the decreasing number of synthetic chemical products allowed for use in the fields, is motivating the search for new biological formulations (Colavita et al., 2011). Also, the adverse effect of inorganic fertilizers on soil and environment is leading science to examine alternative biofertilizers (Metting et al., 1990).

1.1. Botanical and Physiological Aspects

About 90% of the species of marine plants are algae and about 50% of the global photosynthesis is algal derived. Thus, every second molecule of oxygen we inhale comes from algae and algae reuse every second molecule of carbon dioxide we exhale (Dhargalkar and Pereira, 2005). Photosynthetic algae must be considered the true survivors amongst the Plantae as they have survived dramatic changes in climatic conditions from a hot, CO_2 and methane-rich nitrogen atmosphere to presently occupy virtually all niches on Earth from the tropics to the cold, dry, frozen deserts of Antarctica (Craigie, 2011). Gradually, they produced the oxygen necessary to support animal life and, perhaps more importantly, participated in the evolution of macroalgae and the eukaryotic plants upon which we depend for our survival (Rousé, 2006).

It has been estimated that there are about 9,000 species of marine macroalgae, better known as seaweeds, classified according to their pigmentation into brown (Phaeophyta), red

(Rhodophyta), and green (Chlorophyta) seaweeds (Chan et al., 2006). The seaweeds most widely used in agriculture due to their good biostimulant activity are red algae: *Corralina mediterranea, Jania rubens, Pterocladia pinnata*, green algae: *Cladophora dalmatica, Enteromorpha intestinalis, Ulva lactuca* and brown algae: *Ascophyllum nodosum, Ecklonia maxima, Saragassum* spp (Tuhy et al., 2013). Brown seaweeds are the second most abundant group comprising about 2,000 species which reach their maximum biomass levels on the rocky shores of the temperate zones (Khan et al., 2009). They are the type most commonly used in agriculture (Blunden and Gordon, 1986) and, among them, *Ascophyllum nodosum* (L.) Le Jolis is the most researched (Ugarte et al., 2006). It dominates the rocky intertidal shores of Atlantic Canada and northern Europe. Interestingly, the above alga develops a lichenous relationship with the ascomycete *Mycophycias ascophylli* (Craigie, 2011). It was reported that infection of the *Ascophyllum* zygote markedly improves sporeling growth, and that all *Ascophyllum* collected in nature is infected with *M. ascophylli* (Garbary and Deckert, 2001). The term proposed for this unusual obligate and mutualistic symbiosis is a *symbiotum* (Deckert and Garbary, 2005).

1.2. History

It was reported recently that seaweeds formed an important resource for humans in pre-historic times (Dillehay et al., 2008). Moreover, there is a long history of coastal people using seaweeds, especially the large brown seaweeds, to fertilize nearby land (McHugh, 2003). In detail, the ancient Greeks used these plants as a fertilizer (Aitken and Senn, 1965), whereas the use of seaweeds as manure was a common practice for ancient Romans (Newton, 1951). According to the latter researcher, the earliest reference to seaweed manure is in the second half of the first century when the Roman writer L.J.M. Columella recommended that cabbage roots be mulched and manured with seaweed. She also references a recommendation of Palladius, in the fourth century AD, to apply March marine algae to the roots of pomegranate and citron trees. Pre-Roman Britons also added seaweed to soil as manure. A common practice was to heap seaweed in fields and to turn it periodically in a process called "weathering" which aerates it and reduces the production of toxic sulfhydryl compounds (Craigie, 2011).

For centuries, agricultural areas close to these coastal zones have utilized seaweeds as a valuable source of organic matter for various soil types and for many different fruit and vegetable crops. Wet seaweed is heavy so it was not usually carried very far inland, although on the west coast of Ireland enthusiasm was such that it was transported several kilometers from the shore (McHugh, 2003). Generally drift seaweed or beach-washed seaweed is collected, although in Scotland farmers sometimes cut *Ascophyllum* exposed at low tide. In Cornwall (United Kingdom), the practice was to mix the seaweed with sand, let it rot and then dig it in. For over a few hundred kilometers of the coast line around Brittany (France), the beach-cast, brown seaweed is regularly collected by farmers and used on fields up to a kilometer inland. Similar practices are reported for Argentina, where large quantities of green seaweeds are cast ashore every summer and interfere with recreational uses of beaches. Also, in a more tropical climate like the Philippines, large quantities of *Sargassum* have been collected, used wet locally, but also sun dried and transported to other areas (McHugh, 2003).

Table 1. Seaweed research timeline

Years	Activity and research highlights
1940s	Discovery of hydrocolloids from seaweeds; seaweed biology, research on seaweed as food
1952	First International Seaweed Symposium in Scotland, UK
1960s-1970s	Taxonomic classification, biogeography and ecophysiology of seaweeds
1980s-early 1990s	Mutant studies; seaweed cultivation and biotechnology; properties of hydrocolloids and their applications in industries; drug discovery and bioactive compounds from seaweeds; classical genetics
Mid 1990s	Genetic transformation and tissue culture; molecular phylogenetics; small-scale gene cloning and characterization
1997	Introduction of an EST approach in a large-scale study on molecular genetics
2000s-	Search for a model plant for a large-scale genomic study; shift to multi-disciplinary research

Chan et al., 2006

Table 2. Commercial seaweed products used in the agriculture and horticulture industries

Product name	Seaweed name	Company	Application
Acadian®	*Ascophyllum nodosum*	Acadian Agritech	Plant growth stimulant (PGS)
Acid Buf	*Lithothamnium calcareum*	Chance & Hunt Limited	Animal feed
Agri-Gro Ultra	*A. nodosum*	Agri Gro Marketing Inc.	PGS
AgroKelp	*Macrocystis pyrifera*	Algas y Biod. Mar. S.A.	PGS
Alg-A-Mic	*A. nodosum*	BioBizz Worldwide N.V.	PGS
Bio-Genesis™	*A. nodosum*	Green Air Products, Inc.	PGS
Biovita	*A. nodosum*	PI Industries Ltd	PGS
Emerald RMA	Red marine algae	Dolphin Sea Vegetable Company	Health product
Espoma	*A. nodosum*	The Espoma Company	PGS
Fartum®	Unspecified	Inversiones Patagonia S.A.	Biofertilizer
Guarantee®	*A. nodosum*	MaineStream Organics	PGS
Kelp Meal	*A. nodosum*	Acadian Seaplants Ltd	PGS
Kelpak	*Ecklonia maxima*	BASF	PGS
Kelpro	*A. nodosum*	Tecniprocesos Biol., S.A. de C.V.	PGS
Kelprosoil	*A. nodosum*	Productos del Pacifico, S.A. deC.V.	PGS
Maxicrop	*A. nodosum*	Maxicrop USA, Inc.	PGS
Nitrozime	*A. nodosum*	Hydrodynamics International Inc.	PGS
Profert®	*Durvillea antarctica*	BASF	Plant biostimulant
Sea Winner	Unspecified	China Ocean Univ. Product Dev. Co., Ltd	Plant biostimulant
Seanure	Unspecified	Farmura Ltd.	PGS
Seasol®	*Durvillea potatorum*	Seasol International Pty Ltd	PGS
Soluble SW Extract	*A. nodosum*	Technaflora Plant Products, LTD	PGS
Stimplex®	*A. nodosum*	Acadian Agritech	PGS
Synergy	*A. nodosum*	Green Air Products, Inc.	PGS
Tasco®	*A. nodosum*	Acadian Agritech	Animal feed

Khan et al., 2009 (modified)

The industrial production of preparations from algae was reported in the 17th century (Basak, 2008). A timeline of the highlights of research on seaweeds since forties is presented in Table 1. It is worth mentioning that a major advance in the utilization of seaweeds in agriculture resulted from research conducted during World War II on new sources of fiber (Craigie, 2011). A significant amount of fiber was required for netting to camouflage assemblages of aircraft, factories, and other potential targets important to the war effort. The British government focused research efforts to develop local sources of fiber, and a biochemist, Dr. R.F. Milton, was tasked with the challenge. Factories were built in Scotland to extract alginate and produce seaweed rayon for the netting. Unfortunately, the nets failed as the calcium and beryllium alginate fibers biodegraded and dissolved rather rapidly in the prevailing wet climate and, therefore, the project was terminated (Craigie, 2011).

Table 3. Some larger and smaller producers of commercial seaweed extracts for agricultural uses

Larger producers	**Country**
Acadian Seaplants Ltd	Canada
Algea	Norway
Arramara Teo	Ireland
Atlantic Labs	USA
Beijing Leili	P R China
Bioatlantis	Ireland
China Ocean University	P R China
Goëmar	France
Kelpak	South Africa
Seasol	Australia
Smaller producers	**Country**
AfriKelp	South Africa
Agrocean	France
Agrosea	New Zealand
Brandon Products	Ireland
Cytozyme	USA
Dash	Egypt
Fairdinkum	Australia
Fartum	Chile
Gofar	P R China
Natrakelp	Australia
Nitrozyme	USA
Plantalg	France
Sammibol	France
Seagold	Australia
Setalg	France
Thorvin	Iceland
West Coast Marine Products	Canada

Craigie, 2011.

The first practical method for liquefying seaweed for agricultural use was developed by 1949 (Craigie, 2011). Today, seaweed meals and soil amendments are available in ready-to-apply dry form for use in crop soils and home gardens alike. Moreover, high quality powder and liquid seaweed extract products can be found in pure form or in recipe formulations with or without ingredients ranging from traditional (e.g. fertilizers, pesticides, etc.) to non-traditional products (e.g., fish products, etc.)(McHugh, 2003). Some commercial seaweed extract products available for use in agriculture and horticulture, their producers and origin country are presented in Tables 2 and 3.

Table 4. Value of the world's seaweed industry in 2004.

Products	Value (US$, millions)
Sea vegetables	5,290
Phycocolloids	650[a]
Phycosupplements	53
(Soil additives)	(30)
(Agrochemicals)	(10)
(Animal feeds)	(10)[b]
(Other)	(3)
Total	5,993

Adapted from Chopin and Sawhney (2009)

[a]Others suggest US$1.02 billion for phycocolloids

[b]Seaweed meal, used principally as a vitamin and mineral supplement, is produced mainly from the kelps *Ascophyllum nodosum*, *Fucus* spp., *Laminaria* spp., and *Macrocystis* spp.

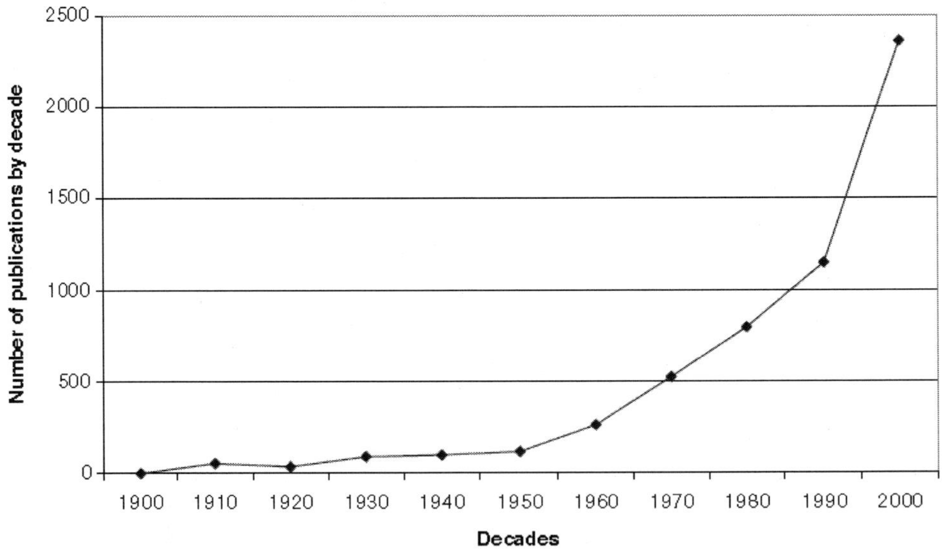

Figure 1. Number of publications related to seaweed use by decade for 1900-2000 (Craigie, 2011).

1.3. Economy

The total seaweed production of the world in the year 2000 was 7.9 million metric tons. Out of which, 6.8 million metric tons came from cultivated area of 200 x 10^3 hectares (Dhargalkar and Pereira, 2005). The global seaweed industry was evaluated at just under US$

6 billion in 2004 (Chopin and Sawhney, 2009; Soto, 2009). Three major subcomponents were identified as sea vegetables, phycocolloids, and phycosupplements (Table 4). Products related to agriculture comprised $50 M and were represented as soil additives, agrochemicals (biostimulants and fertilizers) and animal feeds (supplements and ingredients). The agricultural uses of seaweeds thus comprise less than 1% of the overall value of the current seaweed industry. However, it is noticeable that the interest in agricultural uses of seaweeds is increasing rapidly as judged by the number of related publications appearing since 1950 (Figure 1, Craigie, 2011).

2. Effects of Seaweed Manure and Extracts on Agricultural Plants

Plants grown in soils treated with seaweed manures, or extracts applied either to the soil or foliage, exhibit a wide range of responses that have been well documented in a number of reviews such as those of Khan et al. (2009) and Craigie (2011). In detail, when applied to fruit, vegetable and flower crops, some improvements have included higher yields, improved vigor, increased uptake of soil nutrients, increased resistance to some pests and diseases, longer shelf life of fruit, improved seed germination, and enhanced defense against abiotic stresses. However, no one is really sure about why they are effective, despite many controlled studies having being made. It has been shown that most of the extracts contain several types of plant growth regulators such as cytokinins, auxins and betaines, but even here there is no clear evidence that these alone are responsible for the improvements. Nevertheless, their mode of action may be related to the fact that they contain active organic substances such as cytokinins, auxins, gibberellins, betaines, major and minor nutrients, polysaccharides, alginates, vitamins and amino acids (Colavita et al., 2011). Seaweed extracts have the 1.3-1.6 D-glucane, oligosaccharide known as laminarine, which performs as elicitor for polyamine synthesis, a group of growth regulators that stimulate cell division and defense mechanisms (Patier et al., 1993). The earlier assumption that plant hormones and related low molecular weight organic compounds native to the seaweeds are mainly responsible for the bioactivities observed in extract treated plants now requires revision (Craigie, 2011). It is evident that larger molecules (oligomers and polysaccharide elicitors) in the extracts can be biologically potent. The mechanisms of action of these complex seaweed products is only now being unraveled through the use of techniques developed for molecular biology, metabolomics, and genomics. As genomes of a number of plants are now completely sequenced or nearing completion, it is possible to look at the effects of seaweed extracts and components of the seaweeds on the whole genome/transcriptome of plants to better understand the mechanisms of action of seaweed-induced growth response and stress alleviation (Khan et al., 2009, Figure 2).

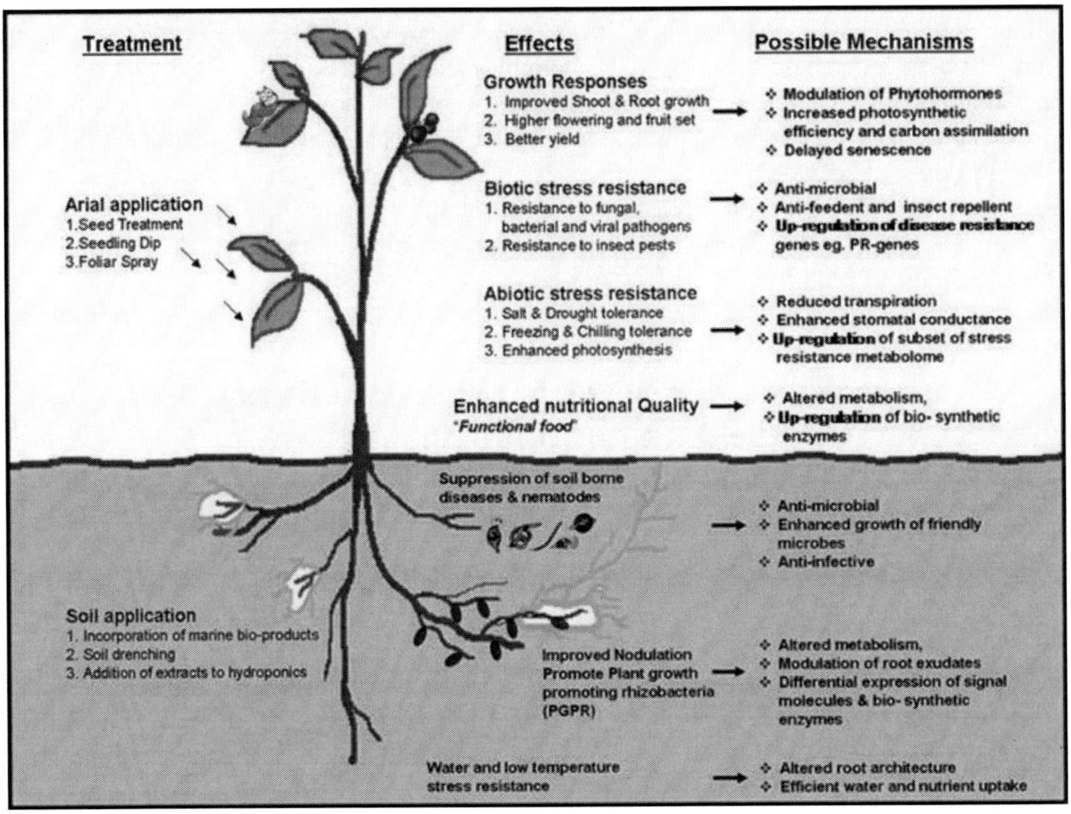

Figure 2. Schematic representation of physiological effects elicited by seaweed extracts and possible mechanism(s) of bioactivity (Khan et al., 2009).

2.1. Plant Growth Parameters

Seaweed components such as macro- and microelement nutrients, amino acids, vitamins, cytokinins, auxins, and abscisic acid (ABA)-like growth substances affect cellular metabolism in treated plants leading to enhanced growth and crop yield (Khan et al., 2009 and references therein). Since seaweed products exhibit growth-stimulating activities, the use of seaweed formulations as biostimulants in crop production is well established. Biostimulants are defined as ''materials, other than fertilizers, that promote plant growth when applied in small quantities'' and are also referred to as ''metabolic enhancers'' (Khan et al., 2009). Seaweed extracts are bioactive at low concentrations, i.e. diluted as 1:1000 or more (Crouch and van Staden, 1993).

There are many reports suggesting the capacity of seaweed extracts to enhance growth in the vegetative parameters. Therefore, promoting effects of seaweed extracts on vegetative growth were reported in watermelon, grape and apple (Mancuso et al., 2006; Basak, 2008; Abdel-Mawgoud et al., 2010). What is more for apple, young trees after replanting presented improved growth when treated with a commercial seaweed product (Kelpak®)(Basak, unpublished data). Other studies on pepper seedlings showed that repeated foliar applications of Maxicrop (*A. nodosum*) increase fruit length, diameter and yields, while affording an earlier first harvest of the crop (Eris et al., 1995).

There are also beneficial effects of algae products on root growth and development. Biostimulants, in general, are capable of affecting root development by both improving lateral root formation (Atzmon and van Staden, 1994; Vernieri et al., 2005) and increasing total volume of the root system (Thompson, 2004; Slavik, 2005; Mancuso et al., 2006). Therefore, Rayorath et al. (2008) reported that the application of extract from *Ascophyllum nodosum* increases in *Arabidopsis thaliana* root length (up to 32%), while seaweed concentrate prepared from *Ecklonia maxima* was found to enhance root growth of tomato plants (Crouch and van Staden, 1992). Also, seaweed concentrate applications reduced transplant shock in seedlings of cabbage and tomato by increasing root size and vigor (Aldworth and van Staden, 1987; Crouch and van Staden, 1992). In line with the above results, fertilizer plus a simple autoclaved extract of the brown seaweed *Rosenvigea intricata*, stimulated lateral root development, increased pigment content and the number of leaves and vegetables over the fertilizer controls when tested with the okra (*Abelmoschus esculentus)* plants (Thirumaran et al., 2009).

Regarding the whole plant response, it is worthwhile that wheat plants treated with seaweed concentrate Kelpak® exhibited an increase in root: shoot dry mass ratio, indicating that the components in the seaweed had a considerable effect on root development (Nelson and van Staden, 1986).

2.2. Plant Nutrient Status

As a biostimulant, seaweed sap may contain chelating compounds (i.e., mannitol) that can increase nutrient availability. A better absorption of the chelated compounds at leaf level has recently been suggested (Shah et al., 2013). In addition, concentrates can increase root size, thus increasing the volume of soil sampled by a plant (Nelson and van Staden, 1984), which indeed helps in uptake of nutrients by plant.

2.2.1. Macronutrients

Kelpak® (*Ecklonia maxima*) increased 2 to 11% N levels in terapy beans under low nutrient levels (Beckett et al., 1994). It was speculated that the increase in N levels may have been a result of N-containing plant growth regulators (auxins and cytokinins) within the seaweed concentrate that were absorbed by the plants (Beckett et al., 1994) but considering the low concentrations and application rates, this is unlikely. An alternative explanation is that organic molecules such as organic acids, methionine and even polyamines in seaweed concentrates can increase nutrient absorption in plants by chelating to the available nutrients, thereby increasing their absorbance (Papenfus et al., 2013).

Protein content in wheat was found to increase by 15.6% and 13.1% when the plants were sprayed with *K. alvarezii* and *G. edulis* sap, respectively. Similarly, Singh and Chandel (2005) reported protein content in wheat grain was greatly influenced by the application of a seaweed extract (Biozyme), which might be because of promotive effects on root proliferation, thus higher uptake of nutrients particularly of those needed as constituents in protein synthesis (N, P, and S) resulting in higher protein synthesis. In a study on potatoes, when seaweed was added to soil, the availability of P increased and this effect was attributed to the alginic acid content of the seaweed used (Eyras et al., 2008). This acid may sequester cations (i.e., Al and Fe) that precipitate phosphates, increasing the availability of P (Lopez-

Mosquera and Pazos, 1997). In line with the above results, Papenfus et al. (2013) experimenting on okra seedlings, found that Kelpak® was effective in relieving P and K deficiency.

Crouch et al. (1990) noted that in nutrient element deficient conditions, seaweed extracts had no effect on ion uptake of lettuce, but at optimum nutrition levels of K, Mg and Ca, uptake increased notably by lettuce. Unlike, application of seaweed extract on grapevine increased N, P, K, Ca and Mg uptake at optimum nutrient element conditions (Turan and Kose, 2004). There has been suggested that seaweed application to a soil had effects similar to those of liming, i.e. increased pH, increased exchangeable Ca, and reduced exchangeable Al (Lopez-Mosquera and Pazos, 1997). Besides, even foliar sprays with seaweed extract (*Ecklonia maxima*) increased Ca uptake by *Brassica oleracea* (Kotze and Joubert, 1980).

2.2.2. Micronutrients

Seaweed meals provide readily available micronutrients compared to most animal manures (Sridhar and Rengasamy, 2010). Moreover, the presence of kahydrin and alginic acid in Actiwave® (*A. nodosum*) contributes to a more efficient mobilisation and assimilation of acid-soluble ions such as Fe and B by acidifying the rhizosphere (Spinelli et al., 2009). In particular, B seems to play a crucial role in determining the alternate bearing tendency. In fact, B-deficiency appeared to be involved in the low yields and poor growth in "off"-year date palm trees (Pillaya et al., 2005). Therefore, sufficient B-availability in treated plants might contribute to reducing their tendency to alternate bearing.

Also, there were observed influences of seaweed extracts on Cu uptake of grapevine that were attributed to increasing membrane permeability of roots, leaves and stoma cells, and hormone-like activities of the seaweed extract through their involvement in cell respiration, photosynthesis, and various enzymatic reactions (Turan and Kose, 2004). Experimenting with grapevine, Turan and Kose (2004) found that seaweed extracts increased Cu, Fe, Mn and Zn uptake at optimum nutrient element conditions, while at limited nutrient levels in plant growth media seaweed extracts improved only Cu uptake. In accordance with the above results, foliar applications of commercial seaweed (*A. nodosum*) extracts to olive trees increased Cu, Fe, Mn, Zn and B concentration in leaves (Tasioula-Margari et al., 2012).

2.3. Yield and Fruit Quality

There are many reports suggesting the beneficial effects of seaweed products on productivity and fruit quality of crops. Seaweed extracts probably encourage flowering by initiating robust plant growth (Craigie, 2011) and, also, by triggering fruit set in a number of crop plants.

Yield increases in seaweed-treated plants are thought to be associated with the hormonal substances present in the extracts (TABLE 5), especially cytokinins (Featonby-Smith and van Staden, 1983a, b, 1984). Cytokinins in vegetative plant organs are associated with nutrient partitioning, whereas in reproductive organs, high levels of cytokinins may be linked with nutrient mobilization.

Table 5. Classic plant growth hormone (PGH) occurrence in brown seaweeds

Classic PGH	Reported in genera
Abscisic acid	*Ascophyllum, Laminaria*
Auxins	*Ascophyllum, Fucus, Laminaria, Macrocystis, Undaria*
Cytokinins	*Ascophyllum, Cystoseira, Ecklonia, Fucus, Macrocystis, Sargassum*
Ethylene	Not reported
Gibberellins	*Cystoseira, Ecklonia, Fucus, Petalonia, Sargassum*

Craigie, 2011

In other works (e.g. Eyras et al., 2008, tomato plants), the increased production of seaweed-treated plants has been attributed to a combination of higher nutrient availability (mainly P) due to slight increases in pH of the soil amended, together with increases in readily available K and an improvement in soil physical conditions. Specifically, a compost aged 20 months at doses of about 50 ton /ha added to soil would be enough to increase productivity of tomato by about 66% above the control (Eyras et al., 2008).

Yield of wheat grain was increased significantly by 19.74% and 13.16% in the plants receiving 7.5% and 5.0% concentrations of *K. alvarezii* and *G. edulis* sap, respectively, over the control (Shah et al., 2013). The increase in the yield was associated with the increase in the number of spike, spike weight, spike length and 100 seed weight. Similarly, Beckett and van Staden (1989, 1990) have reported significant increases in yield of nutrient-stressed wheat with the application (root flush) of 'Kelpak' a seaweed extract derived from *Ecklonia maxima*. Tomato plants sprayed with 0.4% concentration of seaweed concentrate of *Ecklonia* showed 10% increase in fruit number (Crouch and van Staden, 1992).

According to Pise and Sabale (2010, and references therein) seaweed foliar applications increased harvestable bean yields by an average of 25%, staked tomato yields by up to 99%, early yield of one variety of greenhouse cucumber, and, moreover, greenhouse tomato total fruit fresh weights by 17%.

Besides, several experiments with seaweed applications on fruit trees and grapes revealed their positive effect on productivity and fruit quality. Commonly, application of products derived from *A. nodosum* resulted in increased yield and quality of citrus (Fornes et al., 2002), grapes (Norrie and Keathley, 2006), apple (Gény et al., 2007) and pear (Baszczyk, 2008). Beneficial effect of foliar application of extract obtained from *Ascophyllum nodosum* (as a support to N and B fertilization) on olive trees was also investigated. *Ascophyllum* extract applications caused increase of olive size and improved the quality of olive oil (Chouliaras et al., 2009), as well as, increased the production of olive oil per tree (Tasioula-Margari et al., 2012). According to Basak (2008), applying Goëmar (*A. nodosum*) during bloom and in the early stages of apple fruit development it stimulates cell division in young fruit (a key factor for total fruit size), intensifies the uptake of minerals from the soil, and mobilizes transport of assimilates and mineral salts to rapidly growing tissues. As a result, growers may achieve a better distribution of apples in size classes and an increase in the commercial yield of 8% to 11% (Basak, 2008).

Extracts obtained from other algae species were also found to improve productivity and fruit quality, e.g. an extract of a brown alga (*Sargassum wightii* Greville) applied as foliar spray onto Indian jujube (*Zizyphus mauritiana* Lamk.)(Rao, 1991).

2.4. Seed Germination and Seedling Vigor

It is well known that seed emergence and improved seedling vigor have a large impact on seedling establishment, growth, and development. Early emergence assists the plant in better establishment in the field through swift transformation from a heterotroph, which relies upon stored food reserves, to an autotroph with functional photosynthetic machinery (Rayorath et al., 2008).

Earlier studies have suggested that application of seaweed extracts enhances seed germination and seedling vigor (Rayorath et al., 2008 and references therein). Specifically, in barley, the organic components of *A. nodosum* extract induced amylase activity independent of GA_3 and might act in concert with GA-dependent amylase production leading to enhanced germination and seedling vigor (Rayorath et al., 2008). Moreover, Kelpak® also improved root growth and thus the higher seedling vigor of nutrient-stressed okra seedlings treated with Kelpak® increased due to improved absorption capacity of available nutrients (Papenfus et al., 2013). That effect was attributed to the combination of auxins and polyamines which might stimulate rooting, increase nutrient uptake and plant establishment.

2.5. Chlorophyll Content and Photosynthetic Rate

A positive effect of seaweed extract application on chlorophyll content has been suggested by several reports. For example, application of a low concentration of *Ascophyllum nodosum* extract to soil or on foliage of tomatoes produced leaves with higher chlorophyll content than those of untreated controls. This increase in chlorophyll content was a result of reduction in chlorophyll degradation, which might be caused in part by betaines in the seaweed extract (Whapham et al., 1993).

Nutrient elements (mainly Mg and Fe) and betaines contained in seaweed products seem to have a significant role in chlorophyll content and photosynthetic rate of plants. In detail, plants treated with a commercial product of seaweeds (Actiwave®) had a higher availability of carbohydrates and that was attributed to the increased chlorophyll contents and stomatal densities, which resulted in a significant increase in photosynthetic rate. As for the higher chlorophyll contents, Fe is an essential element for chlorophyll biosynthesis and its improved uptake may be a direct effect of the betaines contained in the bio-stimulant (Whapham et al., 1993; Spinelli et al., 2009). In plants, betaines serve as a compatible solute that alleviates osmotic stress induced by salinity and drought stress; however, other roles have also been suggested (Blunden and Gordon, 1986), such as enhancing leaf chlorophyll content of plants following their treatment with seaweed extracts. It is also well known that glycine betaine delays the loss of photosynthetic activity by inhibiting chlorophyll degradation during storage conditions in isolated chloroplasts (Genard et al., 1991). Yield enhancement effects due to improved chlorophyll content in leaves of various crop plants have been attributed to the betaines present in the seaweed (Genard et al., 1991; Whapham et al., 1993). It has been stressed that betaine may work as an N source when provided in low concentration and may serve as an osmolyte at higher concentrations (Naidu et al., 1987).

In another experiment with fenugreek plants, the application of two seaweed liquid fertilizers produced of the species *Sargassum ilicifolium* and *Ulva fasciata* promoted the chlorophyll content. It was suggested that high Mg and Fe content in *Sargassum* might have influenced the synthesis of chlorophyll (Pise and Sabale, 2010). Similarly, Whapham et al. (1993) observed that application of *Ascophyllum* extract increased the chlorophyll content of cucumber cotyledons and tomato plants. This may be the consequence of uptake of Mg, a constituent of chlorophyll, contained in the seaweed liquid fertilizer.

2.6. Tolerance to Abiotic Stresses

It is well known that reactive oxygen species (ROS) are a common factor in many abiotic stresses such as salinity, temperature extremes, and drought (Hodges, 2001). Experiments with application of various seaweed products to crops showed that such stresses must be alleviated efficiently. In detail, treating turf grasses with an *A. nodosum* extract (Tasco®) increased the activity of the antioxidant enzyme superoxide dismutase (SOD), which scavenges superoxide (Fike et al., 2001). Similarly, Ayad (1998) reported an increase in SOD, glutathione reductase (GR), and ascorbate peroxidase (AsPX) activities in tall fescue in response to 3.4 kg Tasco ha^{-1}. Thus, the primary effect of seaweed application to tall fescue in these experiments seemed to be through an increase in antioxidant capacity (Fike et al., 2001). On the other hand, a number of reports suggest that the beneficial antistress effects of seaweed extracts may be related to cytokinin activity. Cytokinins mitigate stress-induced free radicals by direct scavenging and by preventing reactive oxygen species (ROS) formation by inhibiting xanthine oxidation (McKersie and Leshem, 1994; Fike et al., 2001). However, whereas Kelpak® mediated stress tolerance by enhancing K^+ uptake, synthetic benzylaminopurine (BA), at a similar concentration to natural cytokinin present in Kelpak®, showed no effect on yield, irrespective of the K^+ supply. Hence, the beneficial effects of seaweed extract against abiotic stress could also be partly elicited by bioactive chemicals other than cytokinin (Beckett and van Staden, 1989).

Concerning stress by low temperatures, commercial formulations of *Ascophyllum* extracts (Seasol®) improved freezing tolerance in grapes (Mancuso et al., 2006). Specifically, grapevines sprayed with Seasol (0.8%) showed a reduction in leaf osmotic potential, a key indicator of osmotic tolerance. Besides, field studies on winter barley have shown that application of seaweed extract (Maxicrop) improves winter hardiness and increases frost resistance (Burchett et al., 1998).

Finally, the seaweed (*Ascophyllum nodosum*) extract treatment had a significant effect on orange tree water relations (Spann and Little, 2011). However, the maintenance of the plant growth by the seaweed extract under drought stress conditions was unrelated to photosynthesis. Therefore, the authors suggested that seaweed extract may be a useful tool for improving drought stress tolerance of container-grown citrus trees.

3. Seaweeds in Animal Diets

Seaweeds can be used to prepare meals as supplementary to the daily ratio of the livestock, poultry and fish. It has been established that seaweed meal increases fertility and birth rate of animals and also improves yolk colour in eggs (Chapman and Chapman, 1980). European usage of seaweeds in animal husbandry dates at least to Roman times (for horses) and seaweed was fed regularly to domestic animals in Iceland, France, and Norway (Craigie, 2011).

Improvements in gastrointestinal health, immune system responses in animals, and their ability to withstand heat stress are reported benefits in seaweed-fed animal husbandry (Craigie, 2011, TABLE 6). Also, seaweeds are an important source of iodine. Therefore, feeding milk cattle with seaweeds can increase the quantity of iodine in milk (Abdel-Raouf et al., 2012).

A significant merit of algae as animal feed is that low-quality algal protein can be converted by the animals into higher quality protein in the form of meat without the necessity for extensive pre-processing of the algal product (Venkataraman, 1978). Similarly, He et al. (2009) studying the effects of an *Ascophyllum nodosum* extract enriched in phlorotannins on fatty acid accumulation, suggested that the ability of phlorotannins to alter fatty acid metabolism in favor of monounsaturates may be useful for improving meat products.

Table 6. Some bioactivities reported for seaweed extracts in plants and animals

Plants	Animals
Improves plant vigor, root development, chlorophyll synthesis	Improves overall health
Promotes earlier flowering, fruit set and uniformity of fruit	Benefits reproduction
Retards senescence, extends product shelf life	Increases milk production
Improves nutritional quality	Reduces core body temperature under heat stress
Imparts stress tolerance, drought, salinity and frost	Alters fat deposition patterns
Alleviates disease, bacterial and fungal	Improves carcass grading, extends shelf life
Assists in pest control of insects, soil nematodes	Results in favorable changes ingastrointestinal flora
Adjuvant action in pesticide mixtures	Reduces *E. coli* O157:H7 shed by cattle Lowers disease incidence during livestock shipments

Craigie, 2011

Castro et al. (2009) reported about *Macrocystis pyrifera* flour that as no deleterious effects were observed, it could also serve as a partial replacement fodder for goats. Specifically, *M. pyrifera* could be fed to goats or sheep at 30% of their diets with no reduction in overall performance relative to the control animals (Craigie, 2011).

Interestingly, Braden et al. (2007) experimenting on cattle nutrition found that short-term feeding of Tasco (a commercial product of *A. nodosum*) prior to slaughter led to a reduction in enterohemorragic *Escherichia coli* in fecal samples. *Salmonella* spp. also was reduced compared to the post-slaughter control samples. It was suggested that phlorotannin content in seaweed meal may be important in reducing the numbers of *E. coli* shed by cattle after incorporating *Ascophyllum nodosum* meal in their diets (Bach et al., 2008). What is more, Tasco-treated fescue resulted in animals yielding steaks with greater uniformity and improved color over the display period relative to the controls (Montgomery et al., 2001).

4. Recaptures Carbon and Nutrients- Pollution Control

Algae naturally extract materials from the immediate environment. Energy, nitrogen, carbon and phosphorus elements are all consumed by algae. Research of the U.S. Department of Agriculture revealed how algae can recover the nitrogen and phosphorus from manure-treated soil before it is washed away. According to the researchers Mulbry et al. (2008), there was achieved a recover of 60-90% of the N and 70-100% of the P from the manure effluents. Therefore, using algae as fertilizers results in less nitrogen and phosphorous runoff than using livestock manure. This, in turn, increases the quality of water flowing into rivers and oceans.

Interestingly, the macro-algae *Ulva* was proposed to be used in space agriculture to harvest K and other minerals from composted human waste (recycling). *Ulva* was selected for this study, since it tolerates a wide range of salinity levels and due its high intracellular concentration of K, e.g. 20 times higher than seawater (Yamashita et al., 2009).

With concern over global warming, new methods for the thorough and efficient capture of CO_2 are being sought out. The carbon dioxide that a carbon-fuel burning plant produces can feed into open or closed algae systems, fixing the CO_2 and accelerating algae growth. Also, untreated sewage can supply additional nutrients, thus turning two pollutants into valuable commodities (McKenna, 2006). Concerning another pollutant, mercury, the alga *Chlorella*, particularly a transgenic strain which carries an extra mercury reductase gene, has been studied as an agent for environmental remediation due to its ability to reduce Hg^{2+} to the less toxic elemental mercury (Huang et al., 2006).

5. Algae Fertilizers and Biostimulants For Plants

Fifteen million metric tonnes of seaweed products are produced annually (FAO, 2006), a considerable portion of which is used for nutrient supplements and as biostimulants or biofertilizers to increase plant growth and yield. Apart from seaweed extracts or the application of seaweed directly to a soil, either fresh or dried and chopped, seaweeds have been used for preparation of composts (Sultana et al., 2011a). The seaweed products can be placed into three groups: meals for supplementing soil in large volumes or for blending into defend rooting media for glasshouse crops, powered or liquid extracts, and concentrates employed as root dips, soil drenches and foliar sprays (Booth, 1969; Senn, 1987; Metting et al., 1988).

The farmers in developed countries began to shift from chemical-based conventional farming methods towards organic, alternative, or low input sustainable agriculture (Bhatia, 2002). With increasing demand, availability of organic fertilizers from one or two sources was not adequate. To meet the increasing demand many viable options have to be explored (Chhaya, 1997) and one such option is the use of seaweed extracts as fertilizer (Zodape, 2001). Unlike chemical fertilizers, fertilizers derived from seaweeds (*Fucus, Laminaria, Ascophyllum, Sargassum* etc.) are biodegradable, non-toxic, non-polluting and non-hazardous to human, animals and birds (Dhargalkar and Pereira, 2005).

The beneficial uses of seaweeds and their products used in agriculture include enhanced seed germination, plant yield, root growth, tolerance to different plant stresses and increase in plant resistance to infections or insect attack (Eyras et al., 1998). These products contain all major plant nutrients, trace elements, organic compounds (carbohydrates, amino acids, and vitamins) as well as substances of stimulatory and antibiotic nature (Crouch and van Staden, 1993). In addition, the ability of certain forms of blue-green algae to carry out both photosynthesis and N fixation confers on them an ecological and agricultural advantage as a renewable natural resource of biological N (Venkataraman, 1978). Researches have clearly shown that one of the most effective N-fixing biological systems in the rice fields are certain blue-green algae that contribute about 25 - 30 kg N/ha/season.

Brown seaweeds are the type most commonly used as biofertilizers in agriculture (Blunden and Gordon, 1986) and, among them, *Ascophyllum nodosum* (L.) Le Jolis is the most researched (Ugarte et al., 2006). Besides *A. nodosum*, other brown algae such as *Fucus* spp., *Laminaria* spp., *Sargassum* spp., and *Turbinaria* spp. are used for this purpose (Hong et al., 2007).

5.1. Soil Amendments

The use of seaweeds as soil conditioners has a long history in coastal regions with sandy soils of low organic matter content where access to the seaweed is easy (Haslan and Hopkins, 1996). In those places, seaweed residues deposited naturally on the beaches provide a potentially useful soil amendment. Application of seaweeds as organic soil amendment has increased in recent years due to raising awareness about the adverse effect of chemical pesticides (Sultana et al., 2011b).

A potential problem with seaweed is its salt content, though it is unlikely that growers will add sufficient seaweed to seriously upset the balances of salt in soil. The accepted standard for salt content for mature composts is 2 dS/m (Eyras et al., 2008). It is suggested that electrical conductivity (EC) of seaweed composts should be carefully controlled and monitored if this amendment is going to be applied often. To lower salinity content, it can be hosed down before adding to the soil to reduce the salt content, or left to be desalinated by rainwater, or more irrigation would be required during composting. Rinsing seaweed is risky as valuable *alginates* are potentially lost to runoff (Eyras et al., 2008).

Alginic acid acts as a soil-conditioning agent since it combines with metal ions in the soil, forming polymers of greatly increased molecular weight that absorb moisture, swell, retain soil moisture, and improve crumb structure. These cross-linked polymers improve soil aeration and capillary activity of soil pores as well as the water-holding characteristics of the soil (Verkleij, 1992; Lattner et al., 2003) which, consequently, stimulates root growth and boosts soil microbial activity (Eyras et al., 1998; Gandhiyappan and Perumal, 2001; Chen et al., 2003; Moore, 2004). Also, when Ca is added to alginate, it forms strong gels, thus, improving aggregate stability of soil (McHugh, 2003). In practice, spraying steep slopes suffering from soil loss by runoff with composted algae (e.g. *Ascophyllum*), clay, fertilizer, seed, mulch and water has given good results, even on bare rock. Plants quickly grow and topsoil forms after a few years (McHugh, 2003).

Seaweed, particularly *bladderwrack* (*Fucus vesiculosus*), *kelp* or *laminaria* (order Laminariales), can be either applied to the soil as a mulch (although it will tend to break down very quickly) or can be added to the *compost* heap, where it is an excellent *activator* (Khan et al., 2009; Craigie, 2011). In terms of soil structure it does not add a great deal of bulk, but its jelly like *alginate* content helps to bind *soil crumbs* together, and it contains all soil nutrients

(0.3% N, 0.1% P, 1.0% K, plus a full range of trace elements) and amino acids (Khan et al., 2009; Craigie, 2011). In line with the above results, *Solieria robusta*, a seaweed applied as soil amendment was found as effective as chemical fertilizers like urea and potash in improving the soybean plant growth (Sultana et al., 2011b).

Another product, seaweed meal is dried, milled seaweed, and it is usually based on the brown seaweeds because they are the most readily available in large quantities (McHugh, 2003). Species of *Ascophyllum*, *Ecklonia* and *Fucus* are the common ones. They are sold as soil additives and function as both fertilizer and soil conditioner. They have a suitable content of N and K, but are much lower in P than traditional animal manures and the typical N:P:K ratios in chemical fertilizers. The large amounts of insoluble carbohydrates in brown seaweeds act as soil conditioners (improve aeration and soil structure, especially in clay soils) and have good moisture retention properties. Likewise, the addition of seaweed compost may increase pore volume and aggregate stability of soil (Eyras et al., 2008). This may be associated with their high fiber content and/or the increases in the microbial biomass and biological activity in the soil (Haslam and Hopkins, 1996). However, it was not possible to separate the direct effects of kelp addition on soil physical properties and the indirect effects of kelp on soil microorganisms.

Finally, another beneficial effect of algal products when applied into soil is that on root-mycorrhiza symbiosis. In detail, application of red and green algal extracts to roots of papaya (*Carica papaya* Linn.) and passion fruit (*Passiflora edulis* Sims.) improved mycorrhizal development more than the control (Kuwada et al., 2006). Similarly, indigenous arbuscular mycorrhizal fungi demonstrated a 27% improvement in root colonization, while spore number was increased about 21% over the controls when liquid fertilizer containing sea tangle (*Laminaria japonica*) extracts was applied via a sprinkler system in a citrus orchard (Kuwada et al., 2000).

5.2. Seaweed Extracts and Liquid Fertilizers

Seaweed extracts and suspensions have achieved a broader use and market than seaweed and seaweed meal. They are sold in concentrated form, are easy to transport, dilute and apply and act more rapidly (McHugh, 2003). Spraying fertilizer on the plants has increased nutrient absorption efficiency in the plants. The nutrients are not leached down into the soil but are available to the plant through leaf openings such as lenticels, hydathodes and stomata. Leaves absorb nutrients within 10 to 15 minutes of its application (Dhargalkar and Pereira, 2005). What is more, seaweed extracts are environmentally friendly due to biological origin of material. Therefore, algal extracts can constitute an alternative to synthetic plant stimulants, the application of which very often causes environmental pollution, and support to traditional fertilizers (Tuhy et al., 2013). On the other side, seaweed extract, as being organic and biodegradable, is important in sustainable agriculture (Cassan et al., 1992), and their use in organic farming has widely increased (Russo and Beryln, 1990).

Seaweeds extracts are marketed as liquid fertilizers and biostimulants since they contain many growth regulators such as cytokinins (Durand et al., 2003; Stirk et al., 2003), auxins (Stirk et al., 2004), gibberellins (Wildgoose et al., 1978), betaines (Blunden et al., 1991; Wu et al., 1997), macronutrients such as Ca, K, P, and micronutrients like Fe, Cu, Zn, B, Mn, Co, and Mo (Khan et al., 2009), necessary for the development and growth of plants. Earlier reports indicated that seaweeds, especially Phaeophyta, possess a very high content of macroelements (Ca, K, P), microelements that are necessary for the development and growth of trees (Fe, Ca, Zn, Mo, B, Mn, Co), amino acids, antibiotics, and vitamins (Hong et al., 2007 and references therein).

Today there are several products and brands available, such as Maxicrop (United Kingdom), Goëmill (France), Algifert (Norway), Kelpak 66 (South Africa) and Seasol (Australia), Seagrow (New Zealand), Shaktizyme (India), etc. (McHugh, 2003; Dhargalkar and Pereira, 2005). The current commercial extracts are manufactured mainly from the brown seaweeds *A. nodosum*, *Laminaria* spp., *Ecklonia maxima*, *Sargassum* spp., and *Durvillaea* spp., although other species such as *Fucus serratus*, *Enteromorpha intestinalis*, *Ulva lactuca*, and *Kappaphycus alvarezii* have been used (Craigie, 2011 and references therein). In agriculture, the most often used are preparations containing extracts from the algae species *Ecklonia maxima* and *Ascophyllum nodosum* (Verkeij, 1992; Zodape, 2001). Some are made by alkaline extraction of the seaweed and anything that does not dissolve is removed by filtration (e.g. Maxicrop and Seasol). Others are suspensions of very fine particles of seaweed (McHugh, 2003).

Considering the results of a great number of works with application of seaweed extracts, one may suggest that these preparations, when used in combination with inorganic fertilizers, can improve the effectiveness of the fertilizers, with a lowering of costs (Sunarpi et al., 2010; Tuhy et al., 2013).

6. Growth Promoting Substances

Bio-stimulants are environmentally-friendly, natural substances that can promote vegetative growth, mineral nutrient uptake, and tolerance to biotic and abiotic stresses (Khan et al., 2009). Seaweeds are a known source of plant growth regulators, organic osmolites (e.g., betaines), amino acids, mineral nutrients, vitamins, and vitamin precursors (Spinelli et al., 2009 and references therein). Main phytohormones identified in seaweed extracts are: auxins, cytokinins, gibberelins, abscisic acid and ethylene (Tuhy et al., 2013).

6.1. Auxins

Auxins are responsible for elongational growth of plant tissues and apical dominance, cell division, plant movements and plant aging (Tuhy et al., 2013). Marine algae are reportedly rich in auxins and auxin-like compounds (Crouch and van Staden, 1993). Below, a number of algae species and their auxin content are mentioned.

An *A. nodosum* extract had as high as 50 mg IAA (indole 3 acetic acid) per gram of dry extract (Kingman and Moore, 1982). The auxin content of *Kappaphycus alvarezii*, *Gracilaria asiatica*, *Turbinaria* sp. and *Sargassum mcclurei* cultivated in south of Vietnam was 5500, 5900, 4567 and 5780 $\mu g.g^{-1}$, respectively (Hong et al., 2007), while the quantity of IAA found in the harvested seaweed *Undaria pinnatifida* was small (approx. 10 $\mu g\ kg^{-1}$ fresh weight)(Craigie, 2011). Finally, Kelpak® (*Eclonia maxima*) contains 11 mg/L of natural auxins (e.g., indole-3-acetic acid, indole-3-carboxylic acid, indole-3-aldehyde, N,N-dimethyltryptryptamine, N-hydroxyethylphalimide)(Basak, 2008).

6.2. Cytokinins

Cytokinins are involved in cell division regulation affecting plant growth and rest period. Moreover, they inhibit aging of plant tissues and play crucial role in transport of nutrients (Tuhy et al., 2013). Cytokinin-like bioactivity was reported in the early 1970s in commercial seaweed extracts (Williams et al., 1981). Cytokinins have been implicated in nutrient mobilization in vegetative plant organs (Gersani and Kende, 1982) as well as reproductive organs (Davey and van Staden, 1978). Such a response indicates that seaweed extracts are involved either in enhancing the mobilization of cytokinins from the roots to the developing fruit, or, more likely, by improving the amount or synthesis of endogenous fruit cytokinins (Hahn et al., 1974). In any case, fruit treated with seaweed extracts had higher concentrations of cytokinins compared to untreated fruit in tomato (Featonby-Smith and van Staden, 1984).

Evidence that cytokinins are widely distributed in seaweeds was also presented by Stirk et al. (2003) who used immunoaffinity chromatography, HPLC-MS, and single ion measurements to investigate 31 species of seaweeds. It is interesting to note that trans-ZOG and meta-topolin are dominant cytokinins in cell burst seaweed concentrates (Stirk et al., 2004). Moreover, liquid chromatography/mass spectroscopy (LC/MS) analysis of 31 seaweed species representing various groups revealed that zeatin (Z) and isopentenyl (IP) conjugates of cytokinins are the predominant cytokinins (Stirk et al., 2003). Seaweed concentrates also contained aromatic cytokinins BAP (benzyl amino purine) and topolin (6-[3-hydroxybenzyl-amino] purine) derivatives) (Stirk et al., 2004).

Some examples of algae species with their cytokinins content are as follows: *Acanthophora* and *Sargassum mcclurei* were found to contain 15.300 $\mu g.g^{-1}$ and 8998 $\mu g.g^{-1}$ of cytokinins, respectively (Hong et al., 2007). A commercial seaweed extract from *Eclonia maxima* (Kelpak®) contains 0.031 mg/L of natural cytokinins (e.g., trans-zeatin, cis-zeatin, transribosyizeatin, dihydrozeatin, isopentenyladenosine, isopentenyladenine)(Basak, 2008). Also, it has been reported that seaweed extract derivated from *Ascophyllum nodosum* had hormonal activity equivalent to 50 mg/L kinetin (Khan et al., 2009).

6.3. Gibberellins

One of the basic functions of gibberellins is initiation of seed germination, growth regulation, breaking bud dormancy, flowering and fruit development (Tuhy et al., 2013). Direct chemical evidence for gibberellins does not appear to exist for brown algae, although traces of GA_4 were identified by mass spectroscopy in a soluble extract produced by Acadian Seaplants Ltd from *A. nodosum* (Khan et al., 2009). Indicatively, gibberellin content was 157 $\mu g\ g^{-1}$ in *Acanthophora*, 279.1 $\mu g\ g^{-1}$ in *Kappaphycus alvarezii*, 259.1 $\mu g\ g^{-1}$ in *Gracilaria*, and 99.5 $\mu g\ g^{-1}$ in *Turbinaria*.

6.4. Abscisic Acid and Ethylene

Abscisic acid and ethylene are responsible for response to stress factors, inhibition of cell growth, and acceleration of plant aging (Tuhy et al., 2013). Presence of ABA in seaweeds has been reported by Tietz et al. (1989) (green algae) and Kingman and Moore (1982) (*A. nodosum*). However, the function of ABA in the algae remains to be discovered although it induces genes for protein synthesis required for drought tolerance of seeds in terrestrial plants (Verslues et al., 2006; Rensing et al., 2008). Furthermore, abscisic acid participates in regulation of seed germination.

On the other hand, evidence for ethylene production in brown seaweeds is tenuous although 1-aminocyclopropane-1-carboxylic acid, a potential ethylene precursor, was reported in Kelpak 66, a commercial extract of *E. maxima* (Nelson and van Staden, 1985).

6.5. Betaines

Betaines, while not traditionally included among the classical plant hormones, they have a cytokinin-like activity (Spinelli et al., 2009). Betaines are compatible solutes that act as osmo-protectants to enhance plant resistance to drought and salinity stress (Huang et al., 2000). They are widely distributed in plants including seaweeds and are found in seaweed extracts (Craigie, 2011 and references therein). Glycine betaine, γ-amino butyric acid betaine, δ-aminovaleric acid betaine and laminine have been reported in *A. nodosum* and *F. serratus* (Blunden et al., 1984). Interestingly, the betaine yields from various samples of *Laminaria digitata*, *L. hyperborean*, and *Fucus serratus* showed little variation, regardless of either the place or date of collection (Blunden et al., 2010).

An effect suggested to be dependent on the betaines presence was the enhanced chlorophyll content in the leaves of tomato, bean, barley and maize plants after the application of an alkaline extract of *Ascophyllum nodosum* (Blunden et al., 1996).

6.6. Polymers

Seaweeds synthesize polymers such as agars, alginates, carrageenans, fucans, phlorotanins and others not found in terrestrial plants (Connan et al., 2006; Khan et al., 2009). The principal polymers of brown seaweeds are mainly polysaccharides, phlorotannins, proteins and nucleic acids (Craigie, 2011). The richest sources are found in the Fucaceae, especially *A. nodosum* and *F. vesiculosus* where phlorotannins can account for up to 15% of the algal dry mass (Ragan and Glombitza, 1986).

Polymers show a variety of biological activities in both plant and animal systems and are involved in host defense mechanisms (Craigie, 2011 and references therein). Of particular interest in agriculture are those that elicit defensive responses resulting in protection against pathogens or insect damage. For example, Linear β-1, 3-glucan elicited four families of pathogenesis-related proteins in tobacco leaves providing resistance to soft rot caused by *Erwinia carotovora*; the minimum effective glucan polymer length was laminaripentaose (Klarzynski et al., 2000). The heavily sulfated λ-carrageenan also can trigger signaling pathways in tobacco leaves mediated by ethylene, jasmonic and salicylic acids as well as inducing several defense related genes (Mercier et al., 2001).

6.7 .Other Chemical Compounds

Kahydrin is a derivative of vitamin K1 and, by influencing the efficiency of proton pumps, enhances nutrient uptake by the roots. Exogenous application of vitamin K induced the secretion of H^+ ions into the apoplast and consequent acidification of the rhizosphere (Spinelli et al., 2009), which helped in the reduction of Fe(III) to soluble Fe(II), which thus became available to the plant.

The polyamines putrescine (Put) and spermine (Spm) have been quantified in Kelpak®, with concentrations ranging between 0.98-5.51 μg/mL Put and 1.27-4.95 μg/mL Spm., respectively (Papenfus et al., 2013). Polyamines are involved in relieving nutrient deficiency and they accumulate in plants in response to stress (Sudha and Ravishankar, 2002) and therefore may lessen the stress on plants brought on by nutrient deficiency. This increase in polyamines (e.g., Put) concentration might be a physiological adaptation to ionic stress for plants experiencing nutrient deficiency (Young and Galston, 1984).

7. Agar in Plant Propagation and Other Uses

Agar (or agar-agar) is a gelatinous substance derived by boiling a polysaccharide in red algae, where it accumulates in the cell walls of *agarophyte* and serves as the primary structural support for the algae's cell walls. Agar is a mixture of two components: the linear polysaccharide *agarose*, and a heterogeneous mixture of smaller molecules called agaropectin.

Agar, as well as carrageenan and alginate are products difficult to synthesize chemically because of the formidable chemical barriers and hence for these commercially important products we have to depend on seaweed resources (Dhargalkar and Pereira, 2005). About 90 percent of the agar produced is for food applications, the remaining 10 percent being for bacteriological and other biotechnology uses (McHugh, 2003). There are only a small number of processors who produce purified, high quality agarose for a small but growing market, mainly in biotechnology applications, with a total worldwide consumption of about 50 tonnes/year (McHugh, 2003). The production process of agar is presented schematically in Figure 3.

The uses of agar centre on its ability to form gels, and the unique properties of these gels. Agar dissolves in boiling water and when cooled it forms a gel between 32 and 43°C, depending on the seaweed source of the agar. In contrast to gelatin gels, that melt around 37°C, agar gels do not melt until heated to 85°C or higher. This large difference between the temperature at which a gel is formed and the temperature at which it melts is unusual, and unique to agar. Many of its applications take advantage of this difference (McHugh, 2003).

In plant nurseries, agar gels containing appropriate nutrients and hormones are used as the growth substrate to obtain clones of particular plants (McHugh, 2003). This type of propagation is called micropropagation or *in vitro* propagation. Technically, meristems - the part of the plant with actively dividing cells, usually the stem tips - are grown in the gel that had been placed into glass tubes or vessels until there has been sufficient root development and growth for them to be transplanted. An advantage of this system is that the plants can be cultured in a sterile and totally controlled environment.

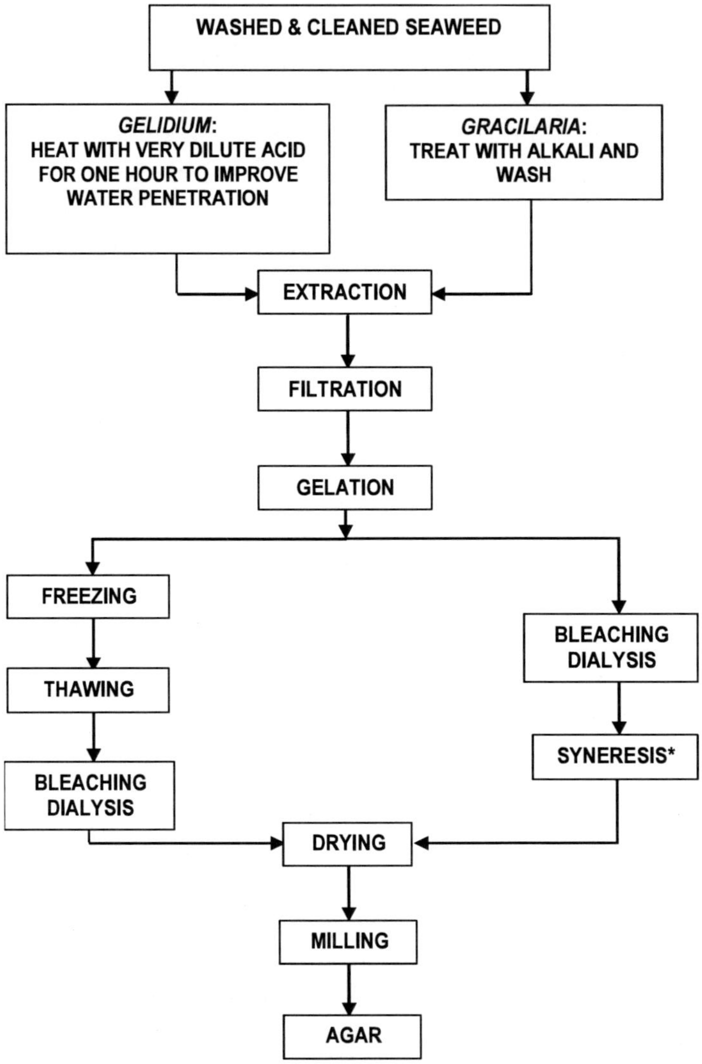

*This term is used to describe the separation of liquid from a gel.

Figure 3. The production processes for agar (McHugh, 2003).

Another use of agar is that in the pharmaceutical industry where it has been used for many years as a smooth laxative. Moreover, bacteriological agar is used in testing for the presence of bacteria (McHugh, 2003). It is specially purified to ensure that it does not contain anything that might modify bacterial growth. This can only be made from species of *Gelidium* because the resulting agar has a low gelling temperature (34-36°C) that allows the addition of other materials to the agar with a minimum risk of heat damage. 'Bacto' agars must not contain anything that might inhibit the growth of bacteria, such as trace metals, soluble carbohydrates or proteins, nor should they contain any bacterial spores. They must not interact with any materials that must be added as nutrients for the bacteria under study. The gels must be strong and have good clarity. Manufacturers of bacteriological agar keep all processing details confidential.

In McHugh's (2003) technical paper, it was predicted that the market for agarose would expand in the near future as its uses in biotechnology increase and probably diversify as new techniques are developed.

8. Effects on Plant Diseases and Insect Pests

Plants possess many adaptive defense mechanisms to counteract pathogen or insect attack. Following appropriate induction, plants are capable of mounting an enhanced defense capability, commonly referred to as induced resistance (Ton et al., 2006). Systemic acquired resistance can be induced by salicylic acid (SA) and related compounds, including benzo (1,2,3) thiadiazole-7-car-bothioic acid S-methyl ester (BTH), 2,6 –dichloro-isonicotinic acid and Dl-3-amino-n-butyric acid (BABA), oligosaccharides such as glucomannans and oligogalacturanides, and natural products containing these compounds (Walters et al., 2005). There are numerous reports of enhanced resistance in crops to various pathogens following application of commercial and preparations containing carbohydrate elicitor molecules. These compounds are an environmentally friendly means of disease control (Vidhyasekaran, 1997) and could be utilized in organic farming and for vegetable cropping systems where application of synthetic fungicides or chemicals needs to be avoided.

Seaweed extracts have been reported to increase plant resistance to pests and diseases (Pardee et al., 2004). Therefore, Mercier et al. (2001) reported that two algal polysaccharides, laminarin and carrageenans induce efficient signaling and defense gene expression in tobacco leaves. Besides influencing the physiology and metabolism of plants, seaweed products promote plant health by affecting the rhizosphere microbial community (Khan et al., 2009).

8.1. Fungal Diseases

Nowadays, in light of concerns over the use of chemical fungicides on horticultural crops, alternative strategies that utilize non-fungicidal products, such as seaweed preparations, need to be evaluated. Plants protect themselves against pathogen invasion by the perception of signal molecules called elicitors (or plant systemic inducers) which include a wide variety of molecules such as oligo and polysaccharides, peptides, proteins, and lipids, often found in the cell wall of attacking pathogens (Khan et al., 2009 and references therein). Treating a plant with those elicitors such as salicylic acid, jasmonic acid, arachidonic acid, a seaweed extract, or humic acid and a reactive oxygen donor provides an increase in pathogenesis-related proteins (PR1, phenylalanine ammonium lyase PAL, or hydroxyproline rich glycoproteins) thereby affording protection from diseases (Moon and Anderson, 2003, 2006).

It is well known that a variety of polysaccharides present in algal extracts include effective elicitors of plant defense against plant diseases (Kloareg and Quatrano, 1988).

Seaweeds, particularly the red and brown algae, are a source of unusual and complex polysaccharides not present in land plants (Khan et al., 2009 and references therein) (TABLE 2). For example, the brown seaweeds *Ascophyllum nodosum*, *Fucus vesiculosus*, and *Saccharina longicruris* contain the polysaccharides laminarin, fucoidan, and alginate. Laminarin has been shown to stimulate natural defense responses in plants and is involved in the induction of genes encoding various pathogenesis–related (PR) proteins with antimicrobial properties (Fritig et al., 1998; van Loon and van Strien 1999). Specifically, laminarin has been shown to upregulate the production of PAL, caffeic acid O-methyl transferase, lipoxygenase and salicylic acid as defense responses in tobacco, and of antifungal compounds in alfalfa (Craigie, 2011 and references therein).

Carrot plants sprayed with seaweed extract (*A. nodosum*) showed less disease due to *Alternaria* and *Botrytis* compared to the control (Jayaraj et al., 2008). Molecular analysis of seaweed treated plants showed the accumulation of defense gene transcripts at higher levels, with enhanced defense gene transcripts at higher levels, with enhanced defense enzyme activities and accumulation of phenolics and phytoalexins. Similarly, cDNA array in seaweed (*Ulva* spp.)-treated alfalfa revealed an upregulation of 152 genes, mostly plant defense genes, such as those involved in phytoalexin, PR proteins, cell wall proteins, and oxylipin pathways, resulting in increased resistance to *Colletotrichum* (Cluzet et al., 2004).

Phytoalexin (capsidiol) accumulation and induced peroxidase activity was also observed after treatment of pepper plants with an *A. nodosum* extract and as a result, their foliar resistance to *Phytophthora capsici* was enhanced (LIzzi et al., 1998). Capsidiol, a phytoalexin in pepper, is known that it is synthesized by the host plant in response to infection and has a fungistatic action on the development of *P. capsici* (Turelli et al., 1984).

In another experiment, significantly less dollar spot (*Sclerotinia homeocarpa* F.T. Bennett) disease incidence was recorded in seaweed extract plus humic acids-treated creeping bentgrass (*Agrostis stolonifera*)(Zhang et al., 2003). The results indicated that increased SOD activity due to seaweed extract improved overall physiological health of bentgrass, irrespective of fertilization regime.

On the contrary, Masny et al. (2004) evaluating the liquefied seaweed products Goëmar BM 86® and Kelpak SL for their effects on two strawberry cultivars, found that neither seaweed preparation reduced the incidence of the gray mold (*Botrytis cinerea*) on fruits.

Application of seaweed to plants can also result in significantly decreased levels of soil-borne pathogen fungi. Management of soil-borne plant pathogens is one of the single greatest challenges facing modern agriculture worldwide (Sultana et al., 2011a). The importance of soil-borne pathogens in modern agriculture systems is made especially clear by the current rush worldwide to find alternatives to methyl bromide for pre-plant treatment of agricultural soils. Also, most soilborne pathogens are difficult to control by conventional strategies such as the use of resistant cultivars and synthetic fungicides (Weller et al., 2002). The proposed mode of action of seaweed products is related to the fact that the presence of easily degradable organic matter of seaweed provides the food base for the multiplication of the antagonistic bacteria, which increase their population in the rhizosphere, or that the alginate of seaweed directly suppresses the pathogens (Sultana et al., 2011b). Indeed, alginate pellets developed as carrier material for biocontrol agents have been reported to reduce multiplication of *Rhizoctonia* disease in potato (Jacobs et al., 2003), while in incorporation of *A. nodosum* extract into the planting medium caused delayed and reduced incidence of *Verticillium* wilt of pepper plants (Jayaraj et al., 2008).

In conclusion, seaweed extract may be beneficial supplement for reducing standard fungicide and fertilizer inputs, while maintaining adequate plant health.

8.2. Pests

Several reports suggest that seaweeds contain elaborated secondary metabolites that play a significant role in the defense of the host against predators and parasites (Paracer et al., 1987; Ara et al., 2005). There are also other, earlier, studies that revealed that sap-feeding insects generally avoid plants treated with seaweed extracts (e.g. apple trees, Stephenson, 1966). Moreover, red spider mite populations on strawberries grown in polyethylene tunnels were significantly reduced following a twice weekly spraying with a commercial product of *A. nodosum* (Hankins and Hockey, 1990). The mechanism of action remains unknown although seaweed extracts might contain chelated metals that can reduce fecundity of red spider mites (Khan et al., 2009). In addition, increased levels of anthocyanins and phenolic constituents in leaves may alter the palatability or acceptability of leaves to insect predators.

8.3. Nematodes

Losses caused by plant parasitic nematodes are estimated about US $100 billion annually (Saifullah et al., 2007) and there is a current rush worldwide to find alternatives to methyl bromide for pre-plant treatment of soils.

Kahn et al. (2009) in their review, taking into consideration some inconsistent results, suggest that seaweed extract imparts nematode resistance possibly by altering the auxin: cytokinin ratio in the plant. Otherwise, Wu et al. (1997) suggested that betains of *Ascophyllum nodosum* caused a reduction of *M. javanica* and *M. incognita* infection on tomato. In that study, seaweeds showed more or less similar suppressive effect on root rotting fungi and root knot nematode like the chemical nematicide carbofuran. However, mixed application of *Spathoglossum variabile* with carbofuran caused maximum reduction in nematode's penetration in roots and produced greater fresh shoot weight, root length and maximum yield of tomato under field condition (Sultana et al., 2011a).

In other experiments, soil amendment with *Solieria robusta* alone or with fertilizers or pesticides significantly ($P<0.05$) reduced nematode infection by reducing the gall formation on soybean roots (Sultana et al., 2011b). Moreover, a seaweed extract, Kelpak 66 (*Eclonia maxima*), is reported to reduce root damage from nematode (*Meloidogyne incognita*) predation in tomatoes whether applied to the foliage or as a soil drench in a 1:500 dilution (Featonby-Smith and van Staden, 1983a). Finally, reduction in fecundity of the root-knot nematode *Meloidogyne javanica* occurred when *Arabidopsis* seedlings were inoculated *in vitro* with juvenile nematodes if the seedlings were treated with Maxicrop (*Ascophyllum nodosum*) or an equivalent concentration of a betaine mixture (Wu et al., 1998).

8.4. Bacterial Diseases

Algae are a rich source of antioxidant polyphenols with bactericidal properties (Zhang et al., 2006). The virulence of pathogenic bacteria is under the control of a communication mechanism (known as Quorum Sensing System –QSS) used by bacterial populations which triggers and controls gene expression that regulates various physiologic functions and responses (Khan et al., 2009 and references therein). Agents that affect the QSS can potentially alter pathogenicity. The marine red alga *Delisea pulchra* synthesizes halogenated furanones and enones that are homologous to acyl-HSL. They bind to the LuxR in the acyl-HSL binding site and prevent the binding of acyl-HSL autoinducers, thereby inhibiting the process of QSS (Khan et al., 2009 and references therein).

9. Algae Production Systems- Algaculture

To date, several of the more than 12,000 varieties of seaweed in the ocean have been shown to be valuable additions to the organic garden and can be abundantly available free for those living near the coast. However, caution should be observed when collecting seaweed, particularly from areas that are liable to pollution, such as downriver of industrial activities as seaweed is susceptible to contamination. There are also legal implications relating to gathering seaweed, and concerns about sustainability.

Therefore, all the uses of seaweeds that have been mentioned in this chapter will demand continuous supply of good quality seaweeds raw material (Dhargalkar and Pereira, 2005). To meet this challenge, it is required to develop an appropriate cultivation technology, suited to contemporary conditions that must have a blend of traditional knowledge and modern biotechnological tools. Commercial and industrial algae cultivation has numerous uses, including production of food ingredients such as omega-3 fatty acids or natural food colorants and dyes, food, fertilizer, chemical feedstock, bioplastics, pharmaceuticals, and algal fuel, and can also be used as a means of pollution control.

Algaculture is a form of *aquaculture* involving the farming of species of *algae*. The majority of algae that are intentionally cultivated fall into the category of *microalgae* (also referred to as *phytoplankton*, *microphytes*, or *planktonic algae*). *Macroalgae*, commonly known as *seaweed*, also have many commercial and industrial uses, but due to their size and the specific requirements of the environment in which they need to grow, they do not lend themselves as readily to cultivation. However, this may change with the advent of new seaweed cultivators using upflowing air bubbles in small containers (Wikipedia, 2013). By now, several cultivation methods are applied for large-scale cultivation along such as the floating raft, fixed off-bottom monolines, broadcasting seedlings directly onto the bottom and fixed off-bottom nets (Hong et al., 2007). A key prerequisite for creating an industry continues to be the availability of a steady source of raw material, in our case seaweeds (Craigie, 2011). The resource must be renewable and sustainable and so requires enlightened management of the seaweed harvest.

According to Venkataraman (1978), there are a number of relatively large production systems for micro-algae, using completely synthetic nutrients. These systems are expensive, not only because of the high cost of land and the technology required but also in terms of water use and the price of inorganic nutrients. For reasons of economy alone, it is therefore necessary to use organic wastes for industrialized production of algal protein feed. For that reason, in 1976, the Indian Agricultural Research Institute developed an integrated waste recycling system, in which algal production forms an integral part (Venkataraman, 1978). This is significant from both the biological and the environmental point of view. This system can provide, simultaneously, fuel (methane), feed algae (for fish, livestock), fish, manure for crops, and water for irrigation. The two main objectives of that project were: a. to develop rural-based recycling systems involving agriculture-aquaculture-pisciculture, and b. to develop viable, economical, rural-oriented systems for algal biofertilizers for rice that would help to relieve the pressure on chemical N supplies without impairing crop productivity.

It has to be noticed that, a drawback of the common open algae culture systems is the major risk of contamination by undesirable photosynthetic micro-organisms which can be introduced via air or rain (Oilgae, 2013). On the contrary, in photobioreactors (suitable only for micro-algae cultivation), the process conditions of algae cultivation can be accurately controlled, and no infection carrying algae species will occur. A major drawback of the closed photobioreactors resides in the high investment costs which lead to high production costs.

In general, the most important factors that determine the growth rate of algae are the following (Oilgae, 2013; Wikipedia, 2013):

- Light: It is needed for the photosynthesis process
- Temperature: There is an ideal temperature range that is required for algae to grow
- Medium/Nutrients: *Composition* of the water is an important consideration (including salinity). Nitrogen (N), phosphorus (P), and potassium (K) serve as fertilizer for algae, and are generally necessary for growth. Silica and iron, as well as several trace elements, may also be considered important marine nutrients. These elements must be dissolved into the water, in bio-available forms, for algae to grow
- pH: *Algae* typically need a pH between 7 and 9 to have an optimum growth rate
- Algae type: Different types of algae have different growth rates
- Aeration: The algae need to have contact with air, for its CO_2 requirements
- Mixing: It prevents sedimentation of algae and makes sure all cells are equally exposed to light
- Photoperiod: Light and dark cycles

Besides conventional methods, recent biotechnological methods such as tissue culture, protoplast fusion, genetic engineering etc. are being employed to produce genetically altered and improved strains with faster growth rate, altered phycocolloid composition and high yield (Dhargalkar and Pereira, 2005). These advances will be a precursor to sustainable development of ocean resources, as mankind will have to depend on them for their livelihood in near future.

10. Seaweed Extraction Methods

In general, extracts are made by processes using water under high pressure, alkali or acids, alcohols, microwaves, CO_2, or by physically disrupting the seaweed by low temperature milling to give a "micronized" suspension of fine particles (Hervé and Rouillier, 1977). Alternatively, the algal cells are ruptured using a high-pressure apparatus and the soluble cytosolic components recovered in the filtered liquid (Hervé and Percehais, 1983; Stirk and van Staden, 2006). The most widely used process involves heating the seaweed with alkaline sodium or potassium solutions. However, the use of alkali to liquefy seaweed components can generate a variety of compounds not present in the parent seaweeds (Craigie, 2011). These novel and biologically interesting compounds will appear in commercial extracts. Some are normal metabolites, while others may be expected to be biologically active either positively or negatively. The nature and quantities of these reaction by-products will depend on the composition and chemical structure of the polymers originally in the seaweed as well as on the processing conditions used to manufacture the soluble extract (Craigie, 2011).

In any case, conditions of the process depend on the active substances of interest (Booth, 1969; Crampon et al., 2011). Extracts rich in auxins can be produced by alkaline extraction. The process is carried out under low pressure. Previously dried probes are extracted with the use of sodium hydroxide (Booth, 1969). By microwave assisted extraction (MAE) combined with water extraction under high pressure, fucoidan (polysaccharide) can be extracted. For the highest efficiency, adequate temperature and pressure should be applied. Cytokinins can be extracted using chilled 70% ethanol, while extraction in 85% methanol leads to obtainment of algae extracts rich in gibberellins.

Depending on the application, extracts frequently are fortified with common plant fertilizers and micronutrients to take advantage of the natural chelating properties of the seaweed extracts which prevent trace metal ions from precipitating (Milton, 1962).

REFERENCES

Abdel-Mawgoud A.M.R., Tantaway A.S., Hafez M. and Habib H.A.M. 2010. Seaweed extract improves growth, yield and quality of different watermelon hybrids. *Research Journal of Agriculture and Biological Sciences* 6: 161-168.

Abdel-Raouf N., Al-Homaidan A.A. and Ibraheem I.B.M. 2012. Agricultural importance of algae. *African Journal of Biotechnology* 11(54): 11648-11658.

Aitken J.B. and Senn T.L. 1965. Seaweed products as a fertilizer and soil conditioner for horticultural crops. *Botanica Marina* 8: 144–148.

Aldworth S.J. and van Staden J. 1987. The effect of seaweed concentrate on seedling transplants. *S. Afr. J. Bot.* 53: 187–189.

Ara J., V. Sultana, R. Qasim, S. Ehteshamul-Haque and Ahmad V.U. 2005. Biological activity of *Spatoglossum asperum*: a brown alga. *Phytother. Res.* 19: 618-623.

Atzmon N. and van Staden J. 1994. The effect of seaweed concentrate on the growth of *Pinus pinea* seedlings. *New For.* 8: 279–288.

Ayad J.Y. 1998. *The effect of seaweed extract (Ascophyllum nodosum) extract on antioxidant activities and drought tolerance of tall fescue* (*Festuca arundinacea* Schreb.). Ph.D. dissertation, Texas Tech University, Lubbock.

Bach S.J., Wang Y. and McAllister T.A. 2008. Effect of feeding sun-dried seaweed (*Ascophyllum nodosum*) on fecal shedding of *Escherichia coli* O157:H7 by feedlot cattle and on growth performance of lambs. *Anim. Feed Sci. Technol.* 142(1–2): 17–32.

Basak A. 2008. Effect of preharvest treatment with seaweed products, Kelpak® and Goemar BM 86®, on fruit quality in apple. *International Journal of Fruit Science* 8(1-2): 1-14.

Baszczyk J. 2008. Quality of 'Conference' pears as affected by Goëmar BM 86 and Fruton. pp. 18-24. In: A. Sadowski (Ed.), Biostimulators in modern agriculture: *Fruit Crops.* Editorial House Wie. Jutra, Limited, Warsaw, Poland.

Beckett R.P. and van Staden J. 1989. The effect of seaweed concentrate on the growth and yield of potassium stressed wheat. *Plant Soil* 116: 29–36.

Beckett R. P. and van Staden J. 1990. The effect of seaweed concentrate on the yield of nutrient stressed wheat. *Botanica Marina* 33: 147–152.

Beckett R.P., Mathegka A.D.M. and van Staden J. 1994. Effect of seaweed concentrate on yield nutrient stressed terapy bean (*Phaseolus acutifolius* Gray). *J. Applied Phycology* 6: 429-430.

Bhatia P. C. 2002. Revitalizing Indian agriculture for higher productivity. *Indian Farming* 52: 3.

Blunden G., Rogers D.J. and Barwell C.J. 1984. Biologically-active compounds from British marine algae. In: Natural Products and Drug Development, Krogsgaard-Larsen P., Christensen S.B. and Kofod H. (Eds.). *Alfred Benzon Symposium* 20: 179-190.

Blunden G. and Gordon S.M. 1986. Betaines and their sulphono analogues in marine algae. In: Round F.E. and Chapman D.J. (Eds.), *Progress in phycological research*, vol. 4. Biopress Ltd, Bristol, pp. 39–80.

Blunden G., Smith B.E., Mason T.G., Ying Xu, Yvin J.C. and Chabot R. 1991. Betaines in seaweed extracts. *J Eur. Cosmet.* 13: 113–118.

Blunden G., Jenkins T. and Liu Y. 1996. Enhanced leaf chlorophyll levels in plants treated with seaweed extract. *J. Appl Phycol.* 8: 535–543.

Blunden G., Morse P.F., Mathe I., Hohmann J., Critchleye A.T. and Morrell S. 2010. Betaine yields from marine algal species utilized in the preparation of seaweed extracts used in agriculture. *Natural Product Comm.* 5(4):581-585.

Booth B. 1969. The manufacture and properties of liquid seaweed extracts, *Proceedings International Seaweed Symposium*, 6: 655–662.

Braden K.W., Blanton J.R. Jr., Montgomery J.L., van Santen E., Allen V.G. and Miller M.F. 2007. Tasco supplementation: effects on carcass characteristics, sensory attributes, and retail display shelf-life. *J. Anim. Sci.* 85: 754–768.

Burchett S., Fuller M.P. and Jellings A.J. 1998. Application of seaweed extract improves winter hardiness of winter barley cv. Igri. The Society for Experimental Biology, Annual Meeting, The York University, March 22–27, 1998. *Experimental Biology Online.* Springer ISSN 1430-34-8.

Cassan L., Jeannin I., Lamaze T. and Morot-Gaudry J.F. 1992. The effect of the *Ascophyllum nodosum* extract Goemar GA 14 on the growth of spinach. *Botanica Marina* 35: 437-439.

Castro N.M., Valdez M.C., Álvarez A.M., Ramírez R.N.Á., Rodríguez I.S., Contreras H.H. and García L.S. 2009. The kelp *Macrocystis pyrifera* as nutritional supplement for goats. *Revista Cientifica* 19(1): 63–70.

Chan C., Ho C. and Phang S. 2006. Trends in seaweed research. *Trends in Plant Science* 11(4): 165-166.

Chapman V.J. and Chapman D.J. 1980. *Seaweed and their uses. 3rd edition*, Chapman & Hall, New York, 63-85.

Chen S. K., Edwards C.A. and Subler S. 2003.The influence of two agricultural bio-stimulants on nitrogen transformations, microbial activity, and plant growth in soil microcosms. *Soil Biology and Biochemistry* 35: 9–19.

Chhaya N.D. 1997. *Minding our Marine Wealth, an Appraisal of Gujarat Coastal Resources.* Ahmedabad, India: Centre for Environment Education.

Chopin T. and Sawhney M. 2009. Seaweeds and their mariculture. In: Steele J.H., Thorpe S.A. and Turekian K.K. (Eds.) *Encyclopedia of ocean sciences.* Elsevier, Oxford, pp. 4477–4487.

Chouliaras V., Tasioula-Margari M., Chatzissavvidis C., Therios I. and Tsabolatidou E. 2009. The effects of a seaweed extract in addition to nitrogen and boron fertilization on productivity, fruit maturation, leaf nutritional status and oil quality of the olive (*Olea europaea* L.) cultivar Koroneiki. *Journal of the Science of Food and Agriculture* 89: 984-988.

Cluzet S., Torregrosa C., Jacquet C., Lafitte C., Fournier J., Mercier L., Salamagne S., Briand X., Esquerre´-Tugaye´ M.T. and Dumas B. 2004. Gene expression profiling and protection of *Medicago truncatula* against a fungal infection in response to an elicitor from the green alga *Ulva* spp. *Plant Cell Environ.* 27: 917–928.

Colavita G.M., Spera N., Blackhall V. and Sepulveda G.M. 2011. Effect of seaweed extract on pear fruit quality and yield. *Acta Horticulturae* 909: 601-607.

Connan S., Delisle F., Deslandes E., Ar Gall E. 2006. Intra-thallus phlorotannin content and antioxidant activity in Phaeophyceae of temperate waters. *Botanica Marina* 49(1): 34–46.

Craigie J.S. 2011. Seaweed extract stimuli in plant science and agriculture. *Journal of Applied Phycology* 23: 371-393.

Crampon Ch., Boutin O. and Badens E. 2011. Supercritical carbon dioxide extraction of molecules of interest from microalgae and seaweeds. *Industrial & Engineering Chemistry Research* 50(15): 8941–8953.

Crouch I.J., Beckett R.P. and van Staden J. 1990. Effect of seaweed concentrate on the growth and mineral nutrition of nutrient stressed lettuce. *J. Appl. Phycol.* 2: 269–272.

Crouch I.J. and van Staden J. 1992. Effect of seaweed concentrate on the establishment and yield of greenhouse tomato plants. *J. Appl. Phycol.* 4: 291–296.

Crouch I.J. and van Staden J. 1993. Evidence for the presence of plant growth regulators in commercial seaweed products. *Plant Growth Regul.* 13: 21–29.

Davey J.E. and van Staden J. 1978. Cytokinin activity in *Lupinus albus*. III. Distribution in fruits. *Physiol. Plant.* 43: 87–93.

Deckert R.J. and Garbary D.J. 2005. *Ascophyllum* and its symbionts. VI. Microscopic characterization of the *Ascophyllum nodosum* (Phaeophyceae), *Mycophycias ascophylli* (Ascomycetes) symbiotum. *Algae* 20(3):225–232.

Dhargalkar V. K. and Pereira N. 2005. Seaweed : Promising plant of Millennium. *Science and Culture* 71(3-4): 60-66.

Dillehay T.D., Ramirez C., Pino M., Collins M.B., Rossen J. and Pino-Navarro J.D. 2008. Monte Verde: seaweed, food, medicine, and the peopling of South America. *Science* 320: 784–789.

Durand N., Briand X. and Meyer C. 2003. The effect of marine bioactive substances (NPRO) and exogenous cytokinins on nitrate reductase activity in *Arabidopsis thaliana*. Physiol. Plant. 119: 489–493.

Eris A., Sivritepe H.Ö. and Stvritepe N. 1995. The effects of seaweed (*Ascophyllum nodosum*) extract on yield and quality in peppers. *Acta Horticulturae* 412: 185–192.

Eyras M.C., Rostagno C.M. and Defosse G.E. 1998. Biological evaluation of seaweed composting. *Compost Science and Utilization* 6: 74–81.

Eyras M.C., Defosse G.E. and Dellatorre F. 2008. Seaweed compost as an amendment for horticultural soils in Patagonia, Argentina. *Compost Science and Utilization* 16: 119-124.

FAO (Food and Agricultural Organisation of the United Nations) 2006. *Yearbook of fishery statistics,* Rome, 98: 1–2.

Featonby-Smith B.C. and van Staden J. 1983a. The effect of seaweed concentrate on the growth of tomato plants in nematode-infested soil. *Sci Hortic.* 20: 137–146.

Featonby-Smith B.C,. and van Staden J. 1983b. The effect of seaweed concentrate and fertilizer on the growth of *Beta vulgaris. Z. Pflanzenphysiol.* 112: 155–162.

Featonby-Smith B.C. and van Staden J. 1984. The effect of seaweed concentrate and fertilizer on growth and the endogenous cytokinin content of *Phaseolus vulgaris. S. Afr. J. Bot.* 3: 375–379.

Fike J.H., Allen V.G., Schmidt R.E., Zhang X., Fontenot J.P., Bagley C.P., Ivy R.L., Evans R.R. Coelho R.W. and Wester D.B. 2001. Tasco-Forage: I. Influence of a seaweed extract on antioxidant activity in tall fescue and in ruminants. *J. Anim. Sci.* 79: 1011–1021.

Fornes F., Sanchez-Perales M. and Guardiola J.L. 2002. Effect of a seaweed extract on the productivity of 'de Nules' Clementine mandarin and Navelina orange. *Botanica Marina* 45: 486-489.

Fritig B., Heitz T. and Legrand M. 1998. Antimicrobial proteins in induced plant defense. *Curr. Opin. Immunol.* 10: 16–22.

Gandhiyappan K. and Perumal P. 2001. Growth promoting effect of seaweed liquid fertilizer (*Enteromorpha intestinalis*) on the sesame crop plant. *Seaweed Resource Util.* 23: 23–25.

Garbary D.J. and Deckert R.J. 2001. Three part harmony – *Ascophyllum* and its symbionts. In: Seckbach J. (Ed.) Symbiosis: mechanisms and model systems. Kluwer, Dordrecht, The Netherlands, pp. 309–321.

Genard H., Le Saos J., Billard J.-P., Tremolieres A. and Boucaud J. 1991. Effect of salinity on lipid composition, glycine betaine content and photosynthetic activity in chloroplasts of *Suaeda maritima. Plant Physiol. Biochem.* 29: 421–427.

Gény L., Bernardon Méry A. and Larrive G. 2007. Un filtrat d'algues agit sur la vigne et le pommier. *Phytoma, La Défense des Végétaux* 609: 37-40.

Gersani M. and Kende H. 1982. Studies on cytokinin-stimulated translocation in isolated bean leaves. *J. Plant Growth Regul.* 1: 161–171.

Hahn H., de Zacks R. and Kende H. 1974. Cytokinin formation in pea seeds. *Naturwissenschaften* 61: 170–171.

Hankins S.D. and Hockey H.P. 1990. The effect of a liquid seaweed extract from *Ascophyllum nodosum* (Fucales, Phaeophyta) on the two-spotted red spider mite *Tetranychus urticae. Hydrobiologia* 204/205: 555–559.

Haslam S.F.I. and Hopkins D.W. 1996. Physical and biological effects of kelp (seaweed) added to soil. *Applied Soil Ecology* 3: 257-261.

He M.L., Wang Y., You J.S., Mir P.S. and McAllister T.A. 2009. Effect of a seaweed extract on fatty acid accumulation and glycerol-3-phosphate dehydrogenase activity in 3T3-L1 adipocytes. *Lipids* 44: 125-132.

Hervé R.A. and Rouillier D.L. 1977. *Method and apparatus for communiting (sic) marine algae and the resulting product.* United States Patent 4,023,734.

Hervé R.A. and Percehais S. 1983. *Noveau produit physiologique extrait d'algues et de plantes, procédé de préparation, appareillage d'extraction et applications.* French Patent 2 555 451.

Hodges D.M. 2001. Chilling effects on active oxygen species and their scavenging systems in plants. In: Basra A. (Ed.) *Crop responses and adaptations to temperature stress.* Food Products Press, Binghamton, NY, pp. 53–78.

Hong D.D., Hien H.M. and Son P.N. 2007. Seaweeds from Vietnam used for functional food, medicine and biofertilizer. *J. Appl. Phycol.* 19: 817–826.

Huang C., Chen M.W., Hsieh J.L., Lin W.H. Chen P.C. and Chien L.F. 2006. Expression of mercuric reductase from *Bacillus megaterium* MB1 in eukaryotic microalga *Chronella* sp. DT: an approach for mercury phytoremedation. *Appl. Microbiol. Biotechnol.* 72(1): 197-205.

Huang J., Hirji R., Adam L., Rozwadowski K.L., Hammerindl J.K., Keller W.A. and Selvaraj G. 2000. Genetic engineering of glycinebetaine production toward enhancing stress tolerance in plants: metabolic limitations. *Plant Physiology* 122: 747–756.

Jacobs H., Gray S.N. and Crump D.H. 2003. Interactions between nematophagous fungi and consequences for their potential as biological agents for the control of potato cyst nematodes. *Mycol. Res.* 107: 47-56.

Jayaraj J., Wan A., Rahman M. and Punja Z.K. 2008. Seaweed extract reduces foliar fungal diseases on carrot. *Crop Protection* 27: 1360– 1366.

Khan W., Rayirath U.P., Subramanian S., Jithesh M.N., Rayorath P., Hodges D.M., Critchley A.T., Craigie J.S., Norrie J. and Prithiviraj B. 2009. Seaweed extracts as biostimulants of plant growth and development. *Plant Growth Regul.* 28: 386–399.

Kingman A.R. and Moore J. 1982. Isolation, purification and quantification of several growth regulating substances in *Ascophyllum nodosum* (Phaeophyta*). Botanica Marina* 25: 149–153.

Klarzynski O., Plesse B., Joubert J-M., Yvin J-C, Kopp M. and Kloareg B. and Fritig B. 2000. Linear β-1, 3 glucans are elicitors of defence responses in tobacco. *Plant Physiol.* 124: 1027–1037.

Kloareg B. and Quatrano R.S. 1988. Structure of the cell walls of marine algae and ecophysiological functions of the matrix olysaccharides. *Oceanogr. Mar. Biol. Annu. Rev.* 26: 259–315.

Kotze W.A.G. and Joubert M. 1980. Influence of foliar spraying with seaweed foliar spraying with seaweed products on the growth and mineral nutrition of rye and cabbage. *Elsenburg Joernoal* 4: 17–20.

Kuwada K., Utamura M., Matsushita I. and Ishii T. 2000. Effect of tangle stock ground extracts on *in vitro* hyphal growth of vesicular arbuscular mycorrhizal fungi and their *in vivo* infections of citrus roots. In: *Proceedings of the* 9th *International Society of Citriculture Congress*, Orlando, FL, 3–7 December 2000, pp. 1034–1037.

Kuwada K., Wamocho L.S., Utamura M., Matsushita I. and Ishii T. 2006. Effect of red and green algal extracts on hyphal growth of arbuscular fungi, and on mycorrhizal development and growth of papaya and passionfruit. *Agron. J.* 98: 1340–1344.

Lattner D., Flemming H. and Mayer C. 2003. 13C-NMR study of the interaction of bacterial alginate with bivalent cations. *International Journal of Biological Macromolecules* 33: 81–88.

Lizzi Y., Coulomb C., Polian C., Coulomb P.J. and Coulomb P.O. 1998. Seaweed and Mildew: what does the future hold? Encouraging laboratory results. *Phytoma, The defense of plants* 508: 29–30.

Lopez-Mosquera M.E. and Pazos P. 1997. Effects of seaweed on potato yield and soil chemistry. *Biol. Agric. Hortic.* 14: 199-206.

Mancuso S., Azzarello E., Mugnai S. and Briand X. 2006. Marine bioactive substances (IPA extract) improve ion fluxes and water stress tolerance in potted *Vitis vinifera* plants. *Adv. Hortic. Sci.* 20: 156–161.

Masny A., Basak A. and Zurawicz E. 2004. Effects of foliar applications of Kelpak SL and Göemar BM 86® preparations on yield and fruit quality in two strawberry cultivars. *J. Fruit Ornamental Plant Res.* 12: 23–27.

McHugh D.J. 2003. A guide to the seaweed industry. *FAO fisheries technical paper* 441, Rome.

McKenna P. 2006. From smokestack to gas tank. *New Scientist* 192: 28-29.

McKersie B.D. and Leshem Y.Y. 1994. *Stress and stress coping in cultivated plants.* Kluwer Academic Publishers, Dordrecht.

Mercier L., Lafitte C., Borderies G., Briand X., Esquerré-Tugayé M-T. and Fournier J. 2001. The algal polysaccharide carrageenans can act as an elicitor of plant defence. *New Phytol.* 149: 43–51.

Metting B., Rayburn W.R. and Reynaud P.A. 1988. Algae and agriculture. In: Lembi C.A. and Waaland, R.A. (Eds.) *Algae and human affairs*. Cambridge Univ. Press, Cambridge, pp. 335-370.

Metting B., Zimmerman W.J., Crouch I.J. and van Staden J. 1990. Agronomic uses of seaweed and microalgae. In: Akatuska I. (Ed.) *Introduction to applied phycology. SPB, The Hague*, pp. 267–307.

Milton R.F. 1962. *The production of compounds of heavy metals with organic residues.* British Patent no. 902,563. 3 pp.

Montgomery J.L., Allen V.G., Pond K.R., Miller M.F., Wester D.B., Brown C.P., Evans R., Bagley C.P., Ivy R.L. and Fontenot J.P. 2001. Tasco-Forage: IV. Influence of a seaweed extract applied to tall fescue pastures on sensory characteristics, shelf-life, and vitamin E status in feedlot-finished steers. *J. Anim. Sci.* 79: 884–894.

Moon D.J. and Anderson A.J. 2003. *Novel methods of protecting plants from pathogens.* United States Patent no. 6,582,961. Assignee: Redox Chemicals Inc.

Moon D.J. and Anderson A.J. 2006. *Method of increasing plant yield.* United States Patent 7,052,911.

Moore K.K. 2004. Using seaweed compost to grow bedding plants. *BioCycle* 45: 43–44.

Morand P. and Briand X. 1996. Excessive growth of macroalgae: a symptom of environmental disturbance. *Botanica Marina* 39: 491-516.

Mulbry W., Kondrad S., Pizarro C. and Kebede-Westhead E. 2008. Treatment of dairy manure effluent using freshwater algae: Algal productivity and recovery of manure nutrients using pilot-scale algal turf scrubbers. *Bioresource Technology* 99: 8137–8142.

Naidu B.P., Jones G.P., Paleg L.G. and Poljakoff-Mayber A. 1987. Proline analogues in Melaleuca species: response of *Melaleuca anceolata* and *M. uncinata* to water stress and salinity. *Aust. J. Plant Physiol.* 14: 669–677.

Nelson W.R. and van Staden J. 1984. The effect of seaweed concentrate on growth of nutrient- stressed greenhouse cucumbers. *HortScience* 19: 81–82.

Nelson W.R. and van Staden J. 1985. 1-Aminocyclopropane-1-carboxylic acid in seaweed concentrate. *Botanica Marina* 28:415–417.

Nelson W.R. and van Staden J. 1986. Effect of seaweed concentrate on the growth of wheat. S. Afr. J. Sci. 82: 199–200.

Newton G.W. 1951. *Seaweed manure for perfect soil and smiling fields*. Sampson Low, London, 188 pp.

Norrie J. and Keathley J.P. 2006. Benefits of *Ascophyllum nodosum* marine-plant extract applications to 'Thompson seedless' grape production. Acta Horticulturae 727: 243–247.

Oilgae 2013. Cultivation of algae. http:*//www.oilgae.com/algae/oil/biod/cult/cult.html.*

Papenfus H.B., Kulkarni M.G., Stirk W.A., Finnie J.F. and van Staden J. 2013. Effect of a commercial seaweed extract (Kelpak®) and polyamines on nutrient-deprived (N, P and K) okra seedlings. *Scientia Horticulturae* 151: 142–146.

Paracer S., Armin C., Tarjan A.C. and Hodgson L.M. 1987. Effective use of marine algal products in the management of plant parasitic nematodes. *J. Nematol*. 19: 194-200.

Pardee K.I., Ellis P., Bouthillier M., Towers G.H.N. and French C.J. 2004. Plant virus inhibitors from marine algae. *Can. J. Bot.* 82: 304-309.

Patier P., Yvin J.C., Kloareg B., Liénart Y. and Rochas C. 1993. Seaweeds liquid fertilizer from *Ascophyllum nodosum* contains elicitors of plant B-glycanases *J. Appl. Phycology* 5: 343-349.

Pillaya A.E.,Williams A.J.R., El-Mardib M.O., Hassana S.M. and Al-Hamdi A. 2005. Boron and the alternate-bearing phenomenon in the date palm (*Phoenix dactylifera*). *Journal of Arid Environments* 62: 199–207.

Pise N.M. and Sabale A.B. 2010. Effect of seaweed concentrates on the growth and biochemical constituents of *Trigonella foenum-graecum* L. *Journal of Phycology* 2(4): 50-56.

Ragan M.A. and Glombitza K-W. 1986. Phlorotannins, brown algal polyphenols. In: Round F.E. and Chapman D.J. (Eds.) *Progress in Phycological Research* 4: 129-241.

Rao K.R. 1991. Effect of seaweed extract on *Zizyphus mauratiana* Lamk. *Journal of the Indian Botanical Society* 71: 19–21.

Rayorath P., Jithesh M.N., Farid A., Khan W., Palanisamy R., Hankins S.D., Critchley A.T. and Prithiviraj B. 2008. Rapid bioassay to valuate the plant growth promoting activity of *Ascophyllum nodosum* (L.) Le Jol. using a model plant, *Arabidopsis thaliana* (L.) Heynh. *J Appl. Phycol*. 20: 423–429.

Rensing S.A. and 69 others. 2008. The *Physcomitrella* genome reveals evolutionary insight into the conquest of land by plants. *Science* 319:64–69.

Rousé P. 2006. *Ostreococcus tauri*, at the root of plant ancestry. http://www.international.inra.fr/layout/set/print/research/some_examples/ostreococcus_tauri

Russo R.O. and Beryln G.P. 1990. The use of organic biostimulants to help low input. *Sustainable Agriculture* 1: 9-42.

Saifullah M., Stephen M.M. and Khattak B. 2007. Isolation of *Trichoderma harzianum* and *in vitro* screening for its effectiveness against root-knot nematodes (*Meloidogyne* sp.) from Swat, Pakistan. *Pak. J. Nematol.* 25: 313-322.

Senn T. L. 1987. *Seaweed and plant growth*. Faith printing co. Taylor, South Carolina, 166 pp.

Shah M.T., Zodape S.T., Chaudhary D.R., Eswaran K. and Chikara J. 2013. Seaweed as an alternative liquid fertilizer for yield and quality improvement of wheat. *J. Plant Nutrition* 36: 192-200.

Singh P.K. and Chandel A.S. 2005. Effect of Biozyme on yield and quality of wheat (*Triticum aestivum*). *Indian Journal of Agronomy* 50: 58–60.

Slavik M. 2005. Production of Norway spruce (*Picea abies*) seedlings on substrate mixes using growth stimulants. *J. For. Sci.* 51: 15–23.

Soto D. 2009. Integrated Mariculture. A Global Review. *FAO Fisheries and Aquaculture Technical Paper 529.* FAO of the United Nations, Rome.

Spann T.M and Little H.A. 2011. Applications of a commercial extract of the brown seaweed *Ascophyllum nodosum* increases drought tolerance in container-grown 'Hamlin' sweet orange nursery trees. *HortScience* 46(4): 577-582.

Spinelli F., Fiori G., Noferini M., Sprocatti M. and Costa G. 2009. Perspectives of the use of a seaweed extract to moderate the negative effects of alternate bearing in apple trees. *Journal Horticultural Science and Biotechnology ISAFRUIT special issue*: 131-137.

Spoehr H.A. and Milner H.A. 1949. The chemical composition of Chlorella: Effect of environmental conditions. *Plant Physiology* 24: 120.

Sridhar S. and Rengasamy R. 2010. Significance of seaweed liquid fertilizers for minimizing chemical fertilizers and improving yield of *Arachis hypogeal* under field trial. *Recent Research in Science and Technology* 2(5): 73-80.

Stephenson W.M. 1966. The effect of hydrolyzed seaweed on certain plant pests and diseases. *Proceedings International Seaweed Symposium* 5: 405–415.

Stirk W.A., Novak M.S. and van Staden J. 2003. Cytokinins in macroalgae. *Plant Growth Regul.* 41: 13–24.

Stirk W.A., Arthur G.D., Lourens A.F., Novak O., Strnad M. and van Staden J. 2004. Changes in cytokinin and auxin concentrations in seaweed concentrates when stored at an elevated temperature. *J. Appl. Phycol.* 16: 31–39.

Stirk W.A. and van Staden J. 2006. Seaweed products as biostimulants in agriculture. In: Critchley A.T., Ohno M. and Largo D.B. (Eds.) *World Seaweed Resources* [DVD-ROM]: ETI Information Services Ltd, Univ. Amsterdam. ISBN: 90 75000 80-4.

Sudha G. and Ravishankar G.A. 2002. Involvement and interaction of various signaling compounds on the plant metabolic events during defense response, resistance to stress factors, formation of secondary metabolites and their molecular aspects. *Plant Cell, Tissue and Organ Culture* 71: 181-212.

Sultana V., Baloch G.N., Ara J., Ehteshamul-Haque S., Tariq R.M. and Athar M. 2011a. Seaweeds as an alternative to chemical pesticides for the management of root diseases of sunflower and tomato. *Journal of Applied Botany and Food Quality* 84: 162 – 168.

Sultana V., Baloch G.N., Ambreen, Ara J., Tariq R.M. and Ehteshamul-Haque S. 2011b. Comparative efficacy of a red alga *Solieria robusta*, chemical fertilizers and pesticides in managing the root diseases and growth of soybean. *Pak. J. Bot.* 43(1): 1-6.

Sunarpi, Jupri A., Kurnianingsih R., Julisaniah N.I. and Nikmatullah A. 2010. Effect of seaweed extracts on growth and yield of rice plants. *Nusantara Bioscience* 2: 73–77.

Tasioula-Margari M., Stamatakos G., Chatzissavvidis C., Mantzoutsos I., Chytiri A. and Chouliaras V. 2012. The effect of commercial seaweed extracts and commercial organic nitrogen foliar sprays on productivity, oil quality and nutritional status of the olive cultivar 'Mastoidis'. In: Chartzoulakis K. (Ed.) Proceedings of the 4th International Conference on *'Olive Culture and Biotechnology of Olive Tree Products'*, OLIVEBIOTEQ 2011, October 31st-November 4th 2011, Chania, Greece, vol. 2, pp. 475-479.

Thirumaran G., Arumugam M., Arumugam R. and Anantharaman P. 2009. Effect of seaweed liquid fertilizer on growth and pigment concentration of *Abelmoschus esculentus (l) medikus*. *American-Eurasian J. Agron.* 2(2): 57–66.

Thompson B. 2004. Five years of Irish trials on biostimulants: the conversion of a skeptic. *USDA For Serv. Proc.* 33: 72–79.

Tietz A., Ruttkowski U., Kohler R. and Kasprik W. 1989. Further investigations on the occurrence and the effects of abscisic acid in algae. *Biochem. Physiol.* Pflanzen. 184: 259–266.

Ton J., Corne M.J., Pieterse D. and van Loon L.C. 2006. The relationship between basal and induced resistance in Arabidopsis. In: Tuzun T. and Bent E. (Eds.) *Multigenic and Induced Systemic Resistance in Plants*. Springer, NY, USA, pp. 197–225.

Tuhy L., Chowanska J. and Chojnacka K. 2013. Seaweed extracts as biostimulants of plant growth: review. *Chemik* 67(7): 636-641.

Turan M. and Kose C. 2004. Seaweed extracts improve copper uptake of grapevine. *Acta Agriculturae Scandinavica, Section B-Soil & Plant Science* 54(4): 213-220.

Turelli M., Coulomb C. and Coulomb P.J. 1984. Effects of capsodiol on the lipid and protein content of isolated membranes of *Phytophthora capsici*. *Physiol. Plant. Pathol.* 24: 211–221.

Ugarte R.A., Sharp G. and Moore B. 2006. Changes in the brown seaweed *Ascophyllum nodosum* (L.) Le Jol. Plant morphology and biomass produced by cutter rake harvests in southern New Brunswick, Canada. *J Appl. Phycol.* 18: 351–359.

van Loon L.C. and van Strien E.A. 1999. The families of pathogenesis related proteins, their activities, and comparative analysis of PR-1 type proteins. *Physiol. Mol. Plant Pathol.* 55: 85–97.

Venkataraman G.S. 1978. Indian experience with algal ponds. *Conference on the State of the Art of Bioconversion of Organic Residues for Rural Communities.* Guatemala City (Guatemala). 13 November 1978.

Verkleij F.N. 1992. Seaweed extracts in agriculture and horticulture – a review. *Biological Agriculture and Horticulture* 8: 309–324.

Vernieri P., Borghesi E., Ferrante A. and Magnani G. 2005. Application of biostimulants in floating system for improving rocket quality. *J. Food Agric. Environ.* 3: 86–88.

Verslues P.E., Agarwal M., Katiyar-Agarwal S., Zhu J. and Zhu J-K. 2006. Methods and concepts in quantifying resistance to drought, salt and freezing, abiotic stresses that affect water plant status. *Plant J.* 45: 523–539.

Vidhyasekaran P. 1997. *Fungal pathogenesis in Plants and Crops: Molecular Biology and Host Defense Mechanisms*. Marcel Decker, NY, USA.

Walters D., Walsh D., Newton A. and Lyon G. 2005. Induced resistance for plant disease control: maximizing the efficacy of resistance elicitors. *Phytopathology* 95: 1368–1373.

Weller D.M., Raaijimakers J.M., Gardner B.B.M. and Thomashow L.S. 2002. Microbial population responsible for specific soil suppressiveness to plant pathogens. *Annu. Rev. Phytopathol.* 40: 309-348.

Whapham C.A., Blunden G., Jenkins T. and Hankins S.D. 1993. Significance of betaines in the increased chlorophyll content of plants treated with seaweed extract. *J. Appl. Phycol.* 5: 231–234

Wikipedia 2013. Algaculture. *http://en.wikipedia.org/wiki/Algaculture*

Wildgoose P.B., Blunden G. and Jewers K. 1978. Seasonal variation in gibberellin activity of some species of Fucaceae and Laminariaceae. *Botanica Marina* 21: 63–65.

Williams D.C., Brain K.R., Blunden G., Wildgoose P.B. and Jewers K. 1981. Plant growth regulatory substances in commercial seaweed extracts. *Procedings International Seaweed Symposium* 8: 760–763.

Wu Y., Jenkins T., Blunden G., von Mende N. and Hankins S.D. 1998. Suppression of fecundity of the root-knot nematode, *Meloidogyne javanica*, in monoxenic cultures of *Arabidopsis thaliana* treated with an alkaline extract of *Ascophyllum nodosum*. J. Appl. Phycol. 10: 91–94.

Wu Y., Jenkins T., Blunden G., Whaphan C. and Hankin S.D. 1997. The role of betaines in alkaline extracts of *Ascophyllum nodosum* in the reduction of *Meloidogyne javanica* and *M. incognita* infestations of tomato plants. *Fundam. Appl. Nematol.* 20: 99–102.

Yamashita M., Tamita-Yokotani K., Hashimoto H., Sawaki N. and Notoya M. 2009. Sodium and potassium uptake of *Ulva* – Application of marine macro-algae for space agriculture. *Advances in Space Research* 43(8): 1220-1223.

Young N.D. and Galston A.W. 1984. Physiological control of arginine decarboxylase activity in K-deficient oat shoots. *Plant Physiology* 76: 331-335.

Zhang Q., Zhang J., Shen J., Silva A., Dennis D.A. and Barrow C.J. 2006. A simple 96-well microplate method for estimation of total polyphenol content in seaweeds. *J. Appl. Phycol.* 18:445–450.

Zhang X., Ervin E.H. and Schmidt R.E. 2003. Physiological effects of liquid applications of a seaweed extract and a humic acid on creeping bentgrass. *J. Amer. Soc. Hort. Sci.* 128(4): 492-496.

Zodape, S. T. 2001. Seaweeds as a biofertilizer. *Journal of Scientific and Industrial Research* 60: 378–382.

In: Seaweeds
Editor: Vitor H. Pomin
ISBN: 978-1-63117-571-8

Chapter 2

An Illustrated Review on Cultivation and Life History of Agronomically Important Seaplants

Felix Bast*
Centre for Biosciences, Central University of Punjab, Bathinda, Punjab, India

Abstract

Seaplants (a better alternative to the misnomer "Seaweeds"), by all means, are "future plants"; they have been projected as the future viand for ever-increasing human populations, viable and sustainable source for biofuel without disturbing global food scenario, as potential candidates for carbon capture and sequestration that is considered as a practical remedy for global warming, and they have a number of pharmaceutical, industrial and biotechnological applications. However, information on its cultivation methods or life history remains obscure to a majority of marine botanists. While life histories of seaweeds have traditionally been an exotic topic for specialists-language of which is ciphered with scientific jargons incomprehensible to general scientific audience, its agronomy had been a trade secret for coastal communities in East Asian countries, especially Japan, the Philippines and Indonesia. In this up-to-date illustrated review, current scientific understanding on the life-histories of agronomically pertinent seaweeds are presented in a fashion akin to popular science journalism with an overview of major coastal and offshore seaweed mariculture techniques, presented with the aid of clear-to-understand illustrations. Also discussed in this report are recent advances in the algal natural products; including uses in hydrocolloid and pharmaceutical industries, Integrated Multi Trophic Aquaculture, energy production, environmental impacts of the seafarming and its counter measures, before concluding with an overview of future research avenues.

Keywords: Seaweed cultivation, seaplant, mariculture, integrated muti-trophic aquaculture, edible alga, marine agronomy, algal natural product

---

* Telephone: +91 98721 52694; Fax: +91 164 2430586. Email: felix.bast@cup.ac.in.

INTRODUCTION

Seaweeds are heterogeneous group of marine plants; comprising many of the primitive non-vascular lineages, known as algae. Of the nine algal lineages in the six-kingdom model of the tree of life (Cavalier-Smith 2004, 1998), three have macroscopic taxa. These are conventionally grouped into green (Chlorophyta), red (Rhodophyta) and brown algae (Heterokontophyta) depending on the type of pigment being expressed. However, the external color is not a synapomorphic character and most of these groups are polyphyletic in nature. Despite the overwhelming importance of seaweed, the very name is a misnomer. Weed is a plant-or even a person- that is unsightly, useless and causes injury to others; while seaweeds are crucial primary producers for which the entire ocean biota depend on and indeed very useful for the humans-as we will see in this paper. To avoid wrong connotations, it is suggested to refer seaweeds as "seaplant"-a more accurate lexicological portrayal of this heterogeneous group of macroscopic marine non-vascular plants. Conventionally, microscopic aquatic plants be referred as algae (marine or limnetic based upon the habitat). This term "seaplant" is therefore adopted in the subsequent discussion to refer plants popularly referred as seaweeds.

Seaplant farming, or macroalgal mariculture, refers to the cultivation of marine macroalgae for the support of various industries including food, hydrocolloid, pharmaceutical and nutraceutical, biotechnology, cosmetic, textile, paper, energy and so on. In the wake of competing demands during the world food crisis, seaplants have attracted much attention lately from the researchers and environmentalists alike as a future food source. Multitude of environmental crises such as human population explosion –that lead to the shortage and overexploitation of cultivable land, chemical and genetic pollutions due to yield-oriented agricultural practices, and ever increasing shortage of freshwater have been encouraging farming of edible seaplants as a sustainable alternative to the conventional agriculture. Farming of seaplants for food has a number of definitive advantages. Seaplants can be readily cultivated nearshore or offshore without fertilizers and drugs and its cultivation do not require land or freshwater resources. However, issues such as non-acceptance of seaplants in the existing culinary cultures of many countries and nonpalatability of a number of seaplant taxa remains to be effectively addressed before its universal acceptance as a future food can be achieved. People from countries that have geographic proximity to the sea and long cost line, such as Japan, North and South Koreas, the Philippines, Scandinavia, Ireland and Chile, are more accustomed to the taste of seaplants to have them in their routine diet. The history of seaplant utilization as food has been extensively reviewed (Chapman 1950; Boylan 1971) and it is apparent that edible seaplant cultivation in East Asia, especially Japan, is at least two thousand years old. A critical analysis of medieval Japanese *Waka*-a form of short poetry suggests that the word seaplant stood as an imagery to express emotions such as love, compassion and sensuality in ancient Japan (Bast 2013). Seaplant occupies a prime position in traditional Japanese culinary (Nisizawa et al. 1987). With increasing number of reports validating the projection of seaplants as a healthier alternative to the conventional terrestrial plants for dieting, prevention of diseases and longevity, its wider acceptance is likely in the near future (Ozaki et al. 2007).

Marine Agronomy

It was only a few decades ago that the concepts of modern agronomy have started to find applications in the field of seaplant farming (Doty 1979; Santelices 1999), however improvement of the existing farming methods have been very minimal for the last four decades or so. Manipulating factors contributing to the site fertility is often impractical due to the continuous nature of ocean. Therefore, we are limited to choose right, fertile site before the farming and continue until nutrients get exhausted, rather than attempting to alter site to make it more fertile. Current understanding of the farm fertility suggests far more viable factors than the four originally hypothesized; viz., temperature, water quality, water motion and light. These include additional abiotic factors such as availability of suitable substrata, pollution and eutrophication, and biotic factors such as biodiversity, presence or absence of grazing vectors, pathogens or epiphytes, and presence of invasive species. Recent findings suggest that role of biodiversity in providing stability for the established seaplant beds is much more than previously thought (Stachowicz et al. 2008; Boyer et al. 2009). Seaplant polyculture is found to be significantly more productive than monoculture, perhaps due to factors such as facilitation and differential use of macrohabitats in heterogeneous environment. A small-scale experiment in which herbivore density have been manipulated in tide pool suggest that seaplant species evenness and primary productivity were increased by the presence of snails, as they preferentially consumed otherwise dominant less-productive seaplants (Altieri et al. 2009). Another factor is UV-B irradiation that affects spatial and functional structure of seaplant communities in coastal environments (Bischof et al. 2007).

One significant difference from the land plants is that there is no alternative to the seeds in seaplants, making its long term storage and propagation significantly complicated. Seaplants have complicated life cycles and maintaining unialgal cultures in laboratories, although laborious, is often the method of choice. Some progresses have been made recently in these lines, such as development of vegetative propagation protocols from cells and protoplast and thereby using these as seed stock (Polne-Fuller and Gibor 1986; Hernández-González et al. 2007) and selection of improved and/or disease resistant strains by selective breeding or hybridization by protoplast fusion (Cheney 1990). Recent advances in seaplant micropropagation techniques have been reviewed by (Reddy et al. 2008). Selective breeding and genetic improvement of the cultivars for increased yield, flavor etc. have been successfully applied in several high-value seaplant genera, including *Saccharina* (Yan et al. 2006), *Undaria* and *Porphyra* (Chaoyuan and Guangheng 1987; Dai et al. 1993). Attempts by genetic engineering have also been done, whereby transferring biolistic gene to spores of several commercially important seaweed genera, taking advantage of the life cycle (Qin et al. 2010).

In an agronomic point of view, there are two types of seaplants; viz., clonal and non-clonal (Santelices 1999). Clonal seaplants have ability to propagate by fragmentation and therefore its cultivation is less laborious. Typically, these are farmed in one-step, i.e., tying fragments to the ropes and nets and its installation in the farm site. During each successive harvesting, a small piece of fragment is allowed to remain attached with the nets so that the thalli get regenerated in the next growth season. This farming method do not involve nursery-rearing of the seedlings and do not demand expensive infrastructural support or expertise and this might be the reason for its immense popularity in developing countries including China, the Philippines and Chile. Examples include *Eucheuma*, *Kappaphycus*, *Sargassum* and *Caulerpa*. In contrary, non-clonal (unitary) seaplants do not have fragmentation ability and the only way to cultivate them is by completing the life cycle during each growth season. Farming is multistep and it typically involves *in-vitro* fertilization and nursery rearing of young seedlings on the nets, before its installation in farm sites. Examples include several high-value seaplants such as *Porphyra*, *Gracillaria*, *Saccharina*, *Undaria*, *Monostroma* and *Ulva*. Cultivation techniques for clonal and non-clonal seaplants are illustrated in Figure 1.

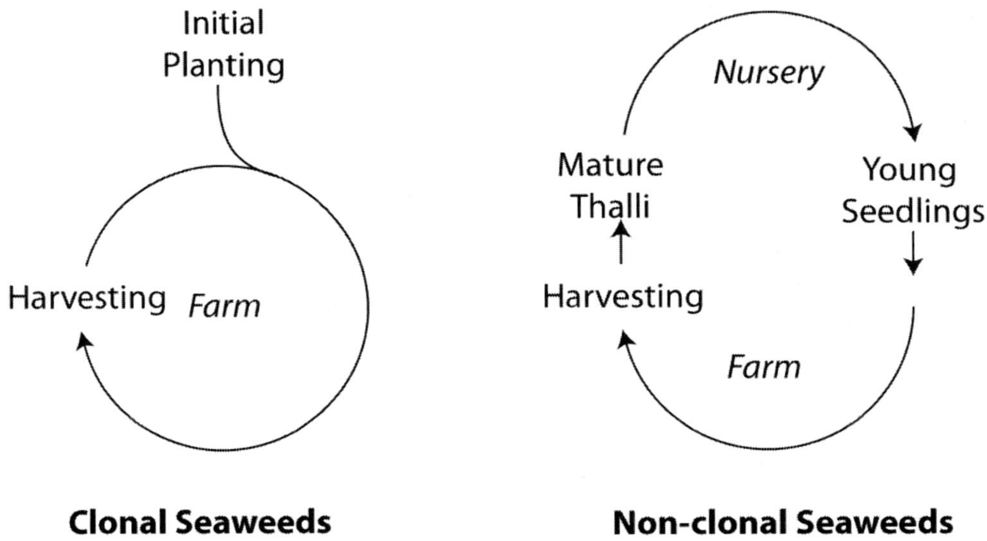

Figure 1. Summary of farming method employed for clonal and non-clonal seaplants.

There are four basic types of agronomic methods widely employed in commercial seaplant cultivation; Floating raft method, Semi-floating raft method, Off-bottom (fixed bottom) method and Bottom planting method. In floating raft method (Figure 2A), rafts are floated by means of buoys (styroform or even inflated PET bottles or coconut husks) installed in nearshore or offshore sites, such that they are floating all the time irrespective of the tides. The frame of the raft can be either synthetic material or wood, such as bamboo poles. Seaplant is cultivated in nets made of nylon or other materials (such as polyethylene or coir), which is interwoven in the raft frame. In this method, rafts are held in place via deadweight mooring, to prevent its drifting. A variant of this method is known as "longline method", in which seaplants are grown in the main rope that is floated via buoys installed at every 4-5 meters and ends fixed via deadweight mooring. Floating raft or longine is the method of choice for kelps (*Saccharina*, *Undaria*) and employed for Eucheumoid seaplants (*Eucheuma* and *Kappaphycus*) and *Sargassum*.

In semi-floating raft method (Figure 2B), rafts with seaplant cultivation nets are attached with top ends of the poles. Bottom ends of the poles have a tripod-like structure and are free, not anchored into the shoal. As in the floating raft method, rafts are attached to an array of buoys such that the system gets floated during high tide. During low tides, tripod-like structure of the poles firmly touches the shoal and this makes raft with cultivation nets get exposed. This method is therefore a combination of floating raft and fixed net methods and guarantees good sunlight irradiance at all the times. This method is used extensively for the commercial farming of seaplants that require periodic exposure to the air, such as *Porphyra*, *Monostroma* and *Ulva*.

In off-bottom method (Figure 2C), nets are hung between the poles that are fixed to the shoal. Poles of suitable heights are chosen so that at high tides, the nets are immersed, while at low tides, nets get exposed. As the nets are immersed in body of water at high tides, light irradiance is a limiting factor. This method therefore demands an appropriate site with sandy bottom and sufficient sunlight. As the farm being easily accessible at low tides, one potential problem is attack from intertidal epiphytes and grazers. Sites with minimal natural flora of these pests need to be chosen for the successful implementation. Seaplants that require periodic drying such as *Porphyra*, *Monostroma* and *Ulva* are extensively cultivated by this low-cost method. In a variant of this method, nets are installed such a way to have subtidal environment; i.e., nets are lower than water level during low tides. This method has been widely used in developing countries to substitute more expensive floating-raft for the cultivation of Eucheumoid seaplants.

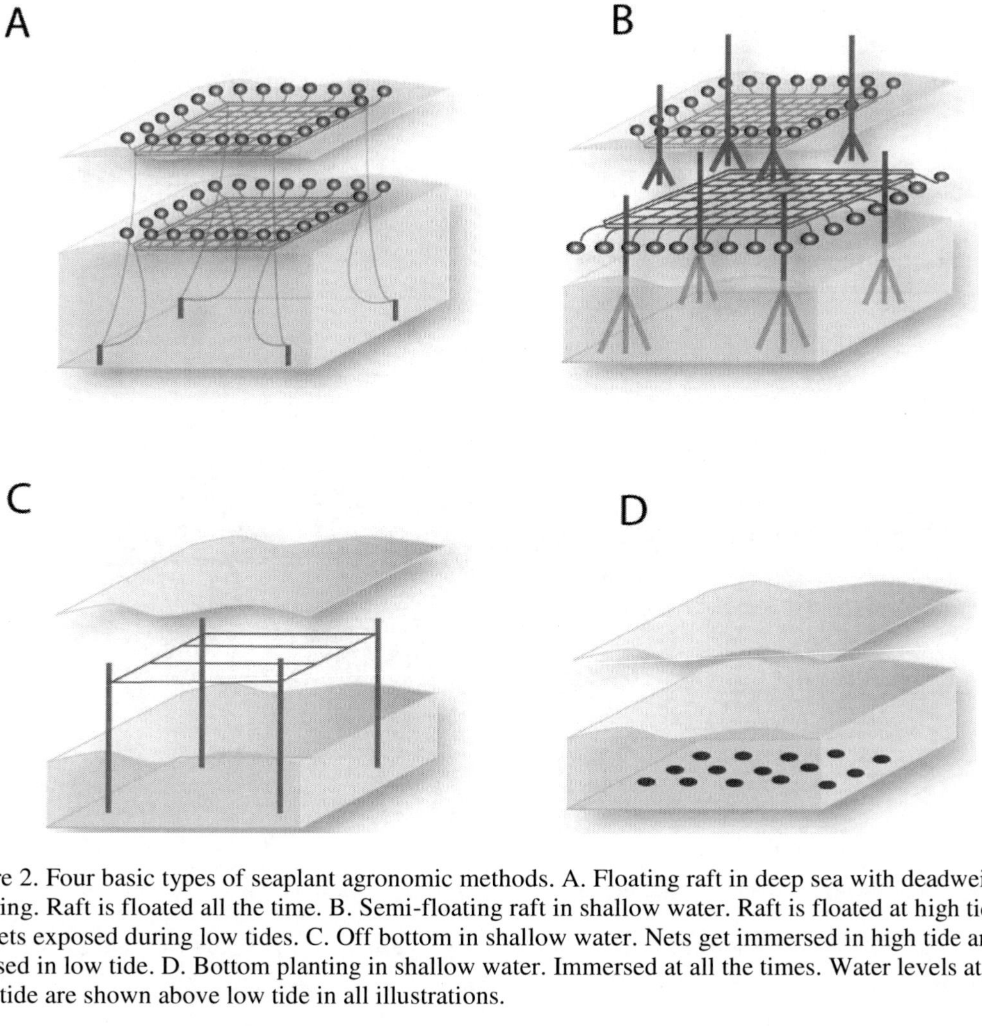

Figure 2. Four basic types of seaplant agronomic methods. A. Floating raft in deep sea with deadweight mooring. Raft is floated all the time. B. Semi-floating raft in shallow water. Raft is floated at high tides but gets exposed during low tides. C. Off bottom in shallow water. Nets get immersed in high tide and exposed in low tide. D. Bottom planting in shallow water. Immersed at all the times. Water levels at high-tide are shown above low tide in all illustrations.

In bottom planting method (Fig 2D), seaplants are cultivated on substratum placed directly on the shoal. This method is typically employed in areas where low level of water remains at low tides such that the planting can be performed without diving. Bottom planting assures immerse of seaplant at all the times and is performed for such benthic genera with thin corticated cylindrical thalli (*Gracillaria* and *Sarcodiotheca*) as well as those with creeping stolon (*Caulerpa*).

EXTENT OF SEAPLANT FARMING

While seaplants can grow in rocky coastlines in all the seven continents throughout the year and many of the human civilizations have been utilized them for hundreds of centuries, commercial farming has not yet been spread in many countries. For a long time, seaplant farming had been limited to East and Southeast Asia, while Northern Europeans harvested natural stocks. It was only less than a century ago that the commercial farming of seaplants for hydrocolloid industry have spread from Asia to South America and Africa, while Europeans and North Americans still have not started wide-scale farming.

In 2008, around 15.8 million tons wet weight of seaplants with estimated market value of USD 7.4 million were farmed worldwide (FAO 2010). Production figures for intensively cultivated taxa are presented in Figure 3. Comparing with world production of fish and other aquacultured animals- which is roughly 52.5 million tons in the same year, seaplant farming still remains a minor aquaculture industry. The FAO data also indicate consistent increase in the seaplant farming sector since 1970 with an annual growth rate of 7.7 per cent.

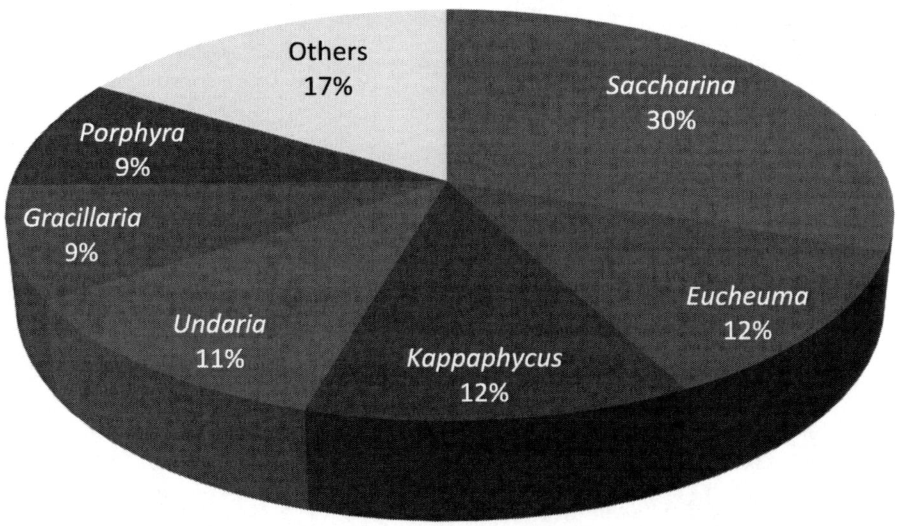

Figure 3. Worldwide production of seaplants in 2008, with contribution of major genera (FAO 2010)

China is the world leader in seaplant farming with almost 63 per cent of the total seaplant production coming from that country. Southeast Asian countries (Indonesia and the Philippines) farmed mostly Eucheumoid seaplants for carrageenan industry. In contrast, almost all of the East Asian countries (China, North and South Koreas and Japan) farmed seaplants that are used as food. East and Southeast Asia- taken together, accounts for almost 99.6 per cent of the world seaplant production. This indicates that contribution of the seaplant farming industry outside Asia to the world seaplant farming is virtually meager. Outside Asia, Chile was the major seaplant farming country with 21,700 tones wet weight of seaplants farmed in 2008. Africa produced 14 700 tones with South Africa, Tanzania and Madagascar as the leading countries. In South Africa, seaplants are farmed mainly for feed supply to the abalone industry, while in Tanzania and Madagascar, main seaplant farmed are Eucheumoid genera (*Eucheuma* and *Kappaphycus*) to support carrageenan industry.

MAJOR SEAPLANT TAXA FARMED

There are about 200 species of seaplants commercially utilized worldwide, of which about 10 genera are farmed intensively (Zemke-White and Ohno 1999). As described already, cultivation methods largely depend upon the life cycle of the taxa being cultivated and therefore understanding the life cycle is an important aspect of the seaplant farming. Detailed life cycle and cultivation information with suitable illustrations for the major farmed seaplant taxa is summarized here for the first time. Because a comprehensive treatment of all of the cultivated taxa is beyond the scope of this review, only top ten species are included:

Red seaplants:	*Porphyra*, *Kappaphycus*, *Eucheuma* and *Gracilaria*
Brown seaplants:	*Saccharina*, *Undaria* and *Sargassum*
Green seaplants:	*Monostroma*, *Ulva* and *Caulerpa*

Porphyra (Bangiales)

Porphyra is the most expensive seaplant taxa worldwide owing to the high commercial demand of its product *Nori* (laver). This seaplant is found on intertidal splash zones of subtropical and temperate habitats. Life cycle is haplodiplontic (Figure 4); with macroscopic monostromatic gametophyte stage consisting of haploid foliose (blade-like) thalli. In many species, the gametophytic thalli are monoecious (i.e., having both male and female reproductive structures in one plant). Fertilization occurs when spermatia, the male gametes, lands on the blades and produce a tube that penetrate cell wall of the female gametes. After fertilization, repeated mitotic divisions of zygotes result in packets of diploid zygotospores (carpospores) within the blade phase. Female gametes in some species also undergo process of parthenogenesis (Nelson et al. 1999).

Originally it was thought that this genera has only one life cycle stage, i.e., the blade-like gametophytes, however, microscopic sporophyte conchocelis stage-that was previously thought to be a separate taxa- was discovered in 1949 which dramatized farming techniques (Drew 1949). This diploid filamentous stage develops on calcareous substrata such as shells of barnacles or mussels when zygotospores are released and deposit upon them. Meiosis within conchosporangium results in haploid conchospores (tetraspores) that develops into foliose gametophyte to complete the life cycle. A number of molecular studies indicate that the *Porphyra* might contain polyphyletic assemblage of taxa that have converged to monostromatic foliose morphology (Stiller and Waaland 1993). This genus is reportedly containing the most primitive plastid genome ever found (Reith and Munholland 1993).

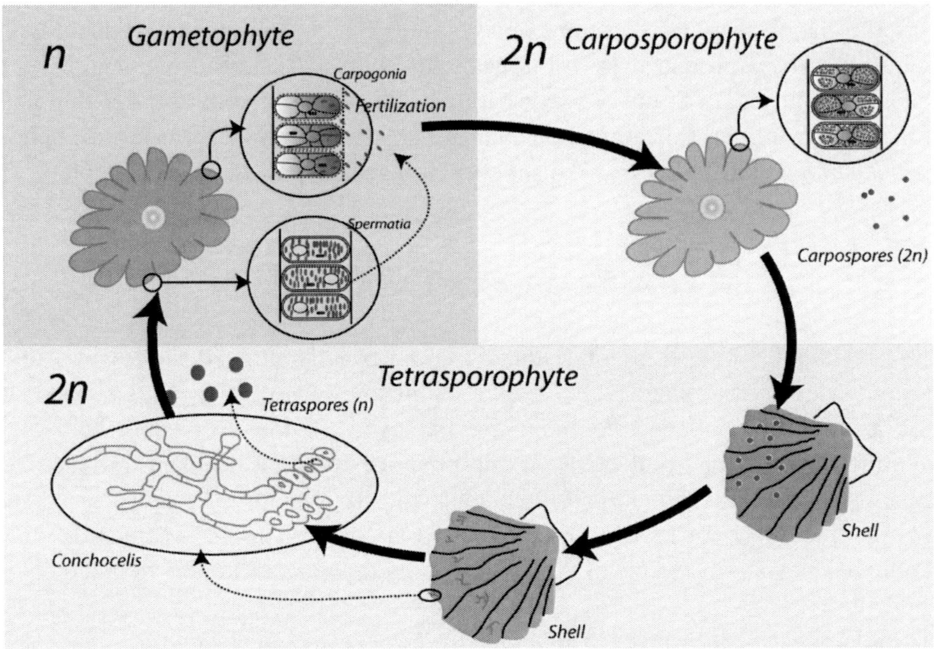

Figure 4. Summary of *Porphyra* life cycle.

Porphyra is a non-clonal type of seaplant and therefore successive regeneration of the thallus from fragments is not possible, necessitating appropriate seed culture prior to each farming season. Conchocelis-filaments are reared in lab on the shells and other calcareous substratum and conchospores are released using controlled temperature and light intensity. Released conchospores are seeded into the cultivation nylon nets that are brought to the farm in the beginning of each season. There are two types of cultivation methods for this seaplant; viz., off-bottom and semi-floating raft. In the off bottom system, cultivation nets are hung between poles fixed in shallow water such that nets experience intertidal habitat (expose to air during ebb tides). One limitation of this method is that it is restricted to the inner regions of bays, with shallow, sandy bottoms. In semi-floating method, nets are tied on floating frames with long poles so that during ebb tides, poles keep the net exposed to air. The thalli loose approximately 90% of the moisture during daytime exposure-during which they are also exposed to intense sunlight, UV and temperature stress, making it one of the highly stress-tolerant plants ever known (Blouin et al. 2010). Experimental cultivation of this seaplant in tanks (Israel et al. 2007) as well as its vegetative propagation (Polne-Fuller et al. 1984) have been reported, although these have never been commercially implemented.

Kappaphycus (Gigartinales)

Kappaphycus is cultivated in many countries for the production of kappa-carrageenan. This seaplant is often found in reef flats and edges from 1m to 15m depth loosely attached with coral reefs. *Kappaphycus* requires high light levels, warm seawater and a high degree of water motion for the successful cultivation (Glenn and Doty 1990). The lifecycle is triphasic, consists of dioecious gametophyte (N), carposporophyte (2N) and tetrasporophyte (2N) phases, similar to that of *Eucheuma* (Refer Figure 5 for a schematic illustration of sexual life cycle). Gametophytic thalli is erect filamentous with cylindrical branches that are heavily cartilaginous. Spermatia and carpogonium are produced by male and female thallus, respectively. Gametes fertilize within terminal parts of the female thalli (trichogyne) to form diploid carposporophyte, which subsequently produces diploid carpospores. Repeated mitotic divisions in the carpospores results in the formation of tetrasporophyte, and subsequent meiotic divisions within tetrasporangia results in the haploid tetraspores. Released tetraspores develop to respective gametophytes to complete the life cycle.

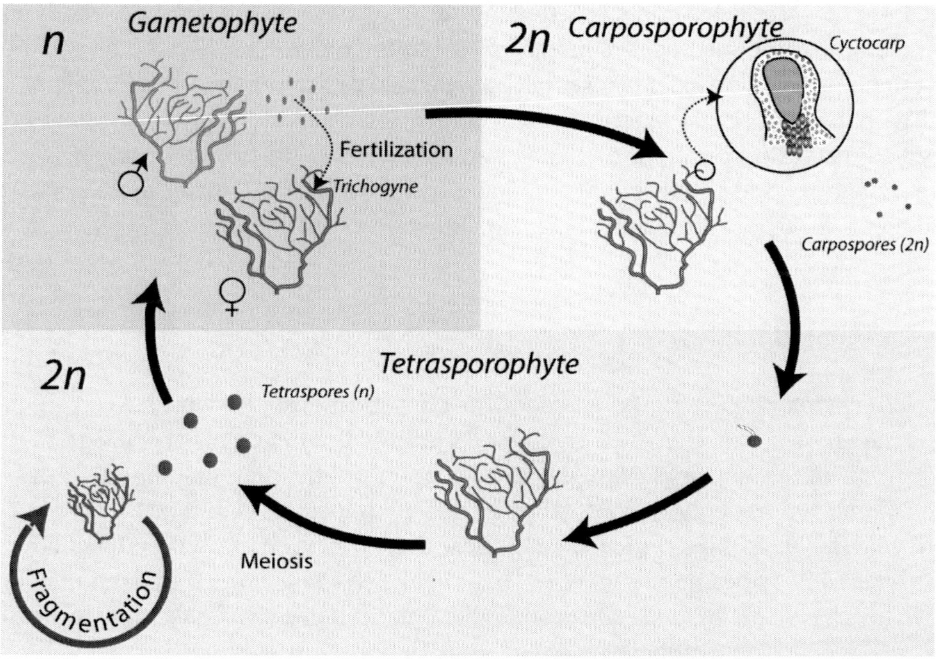

Figure 5. Summary of *Kappaphycus* and *Eucheuma* life cycles.

One of the significant breakthroughs in the commercial farming of this genus is the understanding that the thalli need not undergo the sexual lifecycle or spore production for crop propagation. Fragments as small as 0.5 cm attached to the ropes are capable of developing to complete thalli. Farming techniques of clonal propagation are therefore successfully employed in this seaplant. A fixed off-bottom monoline method is used for the cultivation of this and other Eucheumoid seaplants. Bamboo/wooden sticks are anchored in rows 10 meter apart and 1 meter wide and ends of adjacent sticks in two rows are joined by nylon lines. Seedlings, each about 50g wet weight, are tied to these lines about 30 cm apart. When seedlings reach 1kg (in roughly two months), the plants are harvested (Hurtado-Ponce et al. 1996; Ohno et al. 1994; Trono 1999).

Eucheuma (Gigartinales)

Together with *Kappaphycus* and *Chondrus*, *Eucheuma* is cultivated in many countries for the production of carrageenan. Morphology varies significantly with the habitat, but in general thalli consists of multiaxial filaments with rigid corticated cylindrical branches. The habitat is typically between low tide mark and upper subtidal zone of reef system. The life cycle is already described triphasic polysiphonia-scheme as illustrated in Figure 5 (Kylin 1956), common for many red algae of Gigartinales including *Kappaphycus*. Adequate, moderate to slow, water movement seems a major factor for the growth of this seaplant (Doty et al. 1986). Sufficient sunlight of about irradiance levels of 500-900 $\mu Em^{-2}S^{-1}$ is also a factor to be considered while choosing the farming site. Most popular cultivation method is fixed off-bottom monoline, as described for the *Kappaphycus*. When this method is prohibited due to space constraints etc, floating methods have also been successfully used in many farms. In floating method, seedlings are tied to the nylon lines and the lines are tied to floating rafts of bamboo/wooden sticks that are attached to Styrofoam buoys or empty plastic bottles. The buoys are firmly placed in site via mooring ropes. *Eucheuma* is clonal seaplant and repeated harvesting from the same plant is therefore possible, bypassing need for complicated nursery seed-culturing steps. Algal epiphytism and grazing are reported to be major problems for the commercial *Eucheuma* cultivation (Hurtado-Ponce 1990).

Gracilaria (Gracilariales)

The red agaophyte *Gracilaria* accounts for almost 50% of the world Agar production. Habitat is flat and wide subtidal areas with hard or sandy bottoms. This species grows best in areas where nitrogen content is between 50-100 mg/m^3 and certain amount of freshwater is letting in. Frond morphology widely varies in different species, from erect to prostrate and flattened; generally it is thin corticated cylindrical. Like other red algae described, *Gracilaria* have a triphasic Polysiphonia-scheme life cycle (Figure 6). Tetrasporophytic and gametophytic phases of this alga are isomorphic free-living and carposorophyte phase is a parasite of female gametophyte. After fertilization, distinct cystocarps appear as hemispherical lumps throughout the thallus of female gametophytes. Cystocarps release diploid carpospores (2n) that develops into tetrasporophyte (2n) plants, isomorphic to the gametophytes. Tetrasporophytes undergo meiosis and release tetraspores (n) that develops into respective haploid dioecious gametophyte plants.

As a non-clonal type of seaplant, seedling culture is required prior to each farming season. Plants with numerous cystocarps/tetrasporangium are selected and carpospores/tetraspores are allowed to be released by sun-drying the fronds, cutting to small pieces and placing them in seawater. Released spores will be allowed to sink on the suitable substratum like pebbles or shells (scattering sporeling culture method), farming nets (net culture method) or polypropylene ropes (raft culture method) where they get germinated. Upon germination, substrata are taken and installed in the farm. Alternatively, spore-filled water at optimum density can be collected and sprayed on the substrata at the shoal during ebb tides. Generally few glass slides are also placed along with the substrata before spraying, and microscopically observed next day to ensure adequate spore density. It is normally farmed in 50 cm^2 quadrants with 30 cm space in between and allowed to reach 50-100cm in length within 3-4 months before harvesting can be made. Some preliminary studies have conducted in farming of this alga by vegetative fragment cultivation in floating cages, but the yield was very low (Hurtado-Ponce 1990). Various open pond/ outdoor tank culture methods have also been developed to increase the yield (Dawes 1995), however, algal epiphytism remains a serious issue (Fletcher 1995).

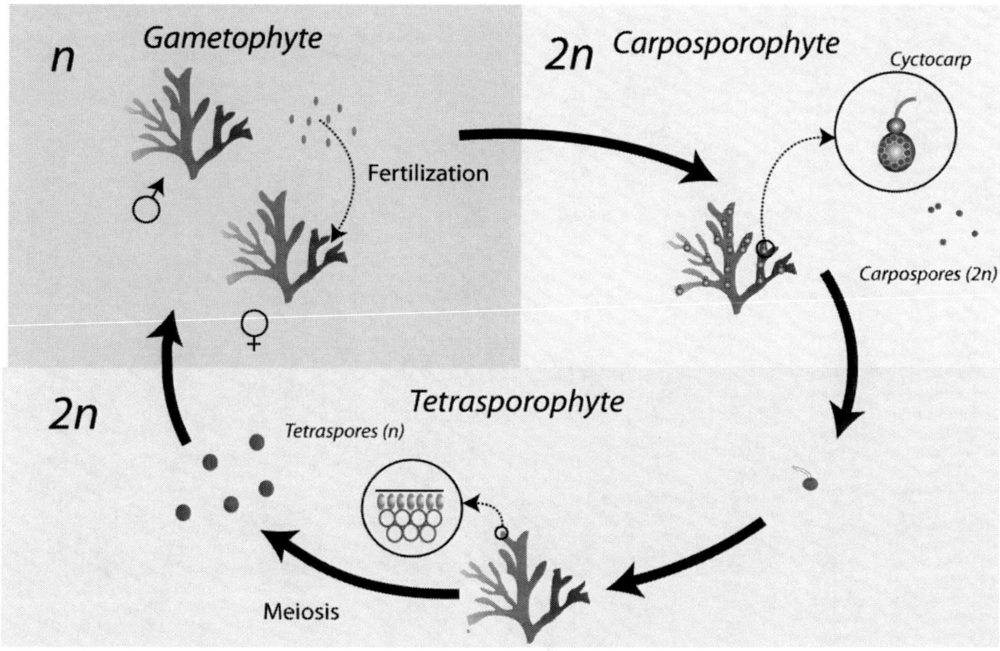

Figure 6. Summary of *Gracilaria* life cycle.

Saccharina (Laminariales)

The brown seaplant genus *Saccharina* (Kelp/Kombu) consists of some 40 species commercially farmed in various countries for human consumption, extraction of hydrocolloid alginate and renal vasodilator, mannitol. This is the most intensively farmed seaplant genera around the world in terms of wet weight harvested. Most of the commercially cultivated species of this genus was earlier grouped under *Laminaria* until a recent molecular phylogenetics study conducted in the family Laminariales that resurrected genus *Saccharina* Stackhouse to include kelps other than the type species *Laminaria digitata* (Lane et al. 2006). According to various sources *Saccharina* is the most intensely cultivated seaplant genera in terms of dry-weight. Along with *Porphyra*, *Undaria* and *Gracilaria*, this seaplant constitutes almost 93% of the worldwide seaplant cultivation (Zemke-White and Ohno 1999; Lüning and Pang 2003). One unique use of this seaplant is the centuries-old practice of cervix dilation ("ripening") by inserting dried *Saccharina* tents and allowing it to expand, to assist in the labor and abortions (Jonasson 1984). Habitat of this seaplant is shallow subtidal cold waters of temperate regions, with southern limit of 36° N. This genus comprises dioecious species having diplohaploid lifecycle (Figure 7), with dominant macroscopic phase being diploid sporophyte. Upon fertility, distal-medial portions of the blades develop numerous "sori"-

aggregation of unilocular sporangia. Diflagellate zoospores settle and germinate to haploid microscopic, isomorphic gametophytes. Gametophytes release diflagellate gamets that fuses to form zygotes and germination of which results in the sporophyte generation, completing the life cycle.

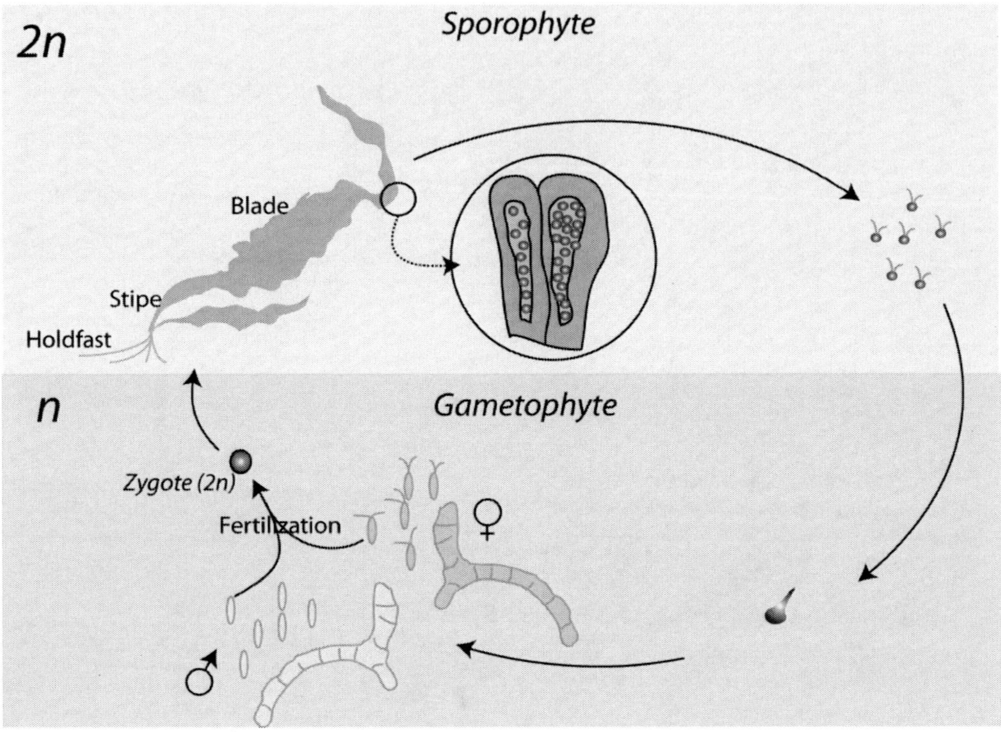

Figure 7. Summary of *Saccharina* life cycle.

Saccharina is non-clonal seaplant and seedlings need to be reared before transplantation in each farming season. Rearing of seedlings is typically done in indoor facility, before it can be transplanted to the farm. Plants with numerous sori are collected and dried to facilitate zoospore release and released zoospores are allowed to deposit on suitable substrata, such as ropes or bamboo sticks. Germination of zoospore and its further development result in mature microscopic gametophyte plants in 2-3 months. Upon gamete release and fertilization, zygotes are deposited on the same substratum and are allowed to germinate and develop for another couple of months till young sporophyte plants reach 15-20 cm. These are then transplanted to thick kelp-culture ropes or rafts when seawater temperature falls below 18° C. These ropes or rafts are installed in shallow coastal areas via buoys tied with mooring ropes. Sporophyte plants are grown hung on the ropes throughout winter and spring, lasting 6-8 months- before a harvest can be made. New methods like gametophyte clone generation (Li et al. 1999) and photolithotrophic cultivation (Qi and Rorrer 1995) have been developed to hasten sporeling culture, although utilization of these in commercial farms have not yet been reported. There are several reports on strain improvement through cross-breeding attempts (Zhang et al. 2007; Li et al. 2008) perhaps owing to its high commercial value.

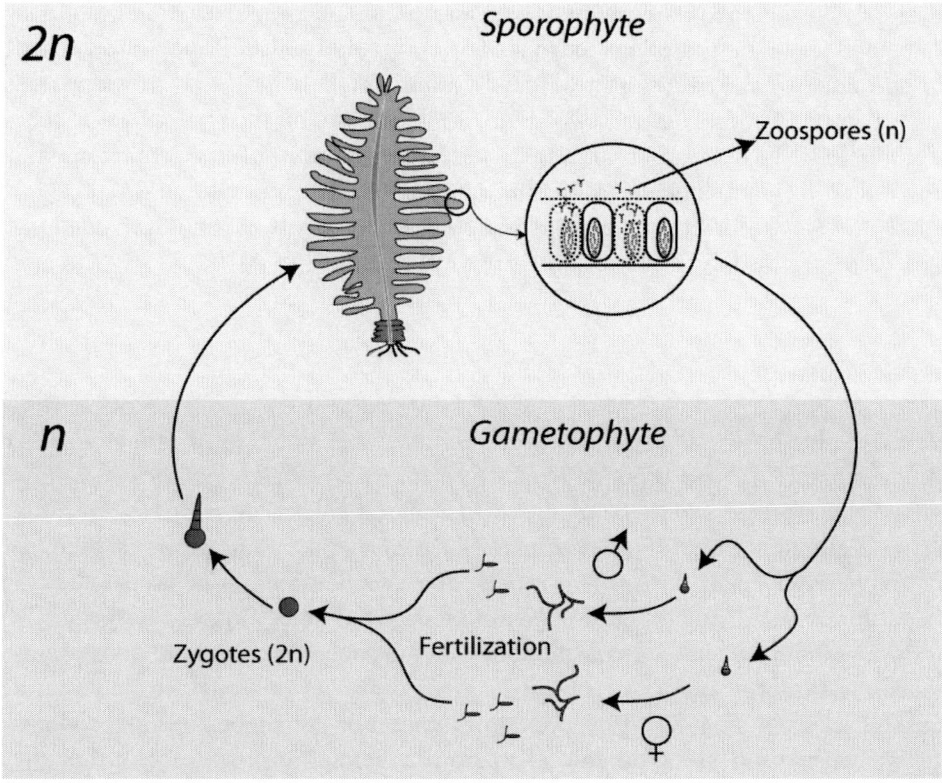

Figure 8. Summary of *Undaria* life cycle.

Undaria (Laminariales)

The brown seaplant *Undaria* is farmed principally in East Asia for the food produce *wakame*. Habitat is shallow subtidal temperate coast with moderate to high wave exposure and there are some progress in understanding its population ecology (Thornber et al. 2004). *Undaria* is native to East Asia (China, Korea and Japan) and got introduced elsewhere predominantly by ship hull fouling (Thornber et al. 2004). This seaplant shares similar life cycle and cultivation strategies to closely related genera *Saccharina*. Life cycle is diplohaplontic alternation (Figure 8) with dominant phase being macroscopic sporophyte that is being harvested. Zoospores are produced from the sporophylls on the basal part of the stipe. Upon the release of zoospores, sporophytes wither and disintegrate. Released zoospores get settled on appropriate substrata like shells, bamboo sticks etc. and germinate to microscopic male and female gametophytes. Male and female gametes released by respective gametophytes fuses to form zygotes that germinate and develop into macroscopic sporophyte generation to complete the life cycle.

Farming of this non-clonal seaplant involves seedling culture and rearing in indoor ponds. Zoospores are collected from matured plants by briefly drying the plants, keeping at dark for 3-4 hours and re-introduction in seawater with intense light supply. Released zoospores are collected by wooden spore collectors with ropes and gametophytes are reared by hanging them in tank filled with seawater. Growth of gametophytes, release and fertilization of gametes, and its germination to microscopic sporophytes are all done in these spore collectors. Once the sporelings reach size of 1mm, the spore collectors are introduced to the farm site and wound over ropes of the floating raft structure and grown in a similar fashion as described for *Saccharina*. There are some reports on the sporophyte regeneration from protoplasts (Matsumura et al. 2001), cryopreservation of the gametophytic cells (Kuwano et al. 2004) and gametophyte culture in photobioreactors (Xu et al. 2002) however none had been widely utilized in the industrial cultivation. Diseases, pathogens and parasites affecting *Undaria* farms have been recently reviewed (Neill et al. 2010).

Sargassum (Fucales)

Members of the brown algal genus *Sargassum* have been cultivated in many parts of the world, predominantly as food produce *Hijiki* (*S. fusiforme* = *Hizikia fusiformis*) in East Asia (Korea and Japan). *Sargassum* is cultivated or its natural stocks are harvested in India as an alginophyte (Kalimuthu et al. 1991). *Sargassum fusiforme* is a rich source of fucoxanthin-a major antioxidant (Yan et al. 1999). This seaplant has slender, long fronds separated from one another and this is one distinct morphological feature from other *Sargassum* species. Habitat is eulittoral to sublittoral shallow coastal zones in subtropics to temperate ecosystems. This alga is endemic to north-west coast of the Pacific ocean and is thought to have introduced elsewhere (Nisizawa et al. 1987). Life cycle information of this dioecious alga is almost non-existent, but considering its taxonomic affiliation in genus *Sargassum*, it is thought to be monobiontic and diplontic (Fig 9). The dominant phase which is harvested is diploid gametophyte. Male and female gametophytes develop respective receptacles in which haploid gametes are produced by meiosis. Female gametes are not broadcasted into the surrounding seawater unlike male gametes. The fertilization is thought to be isogamous within female receptacles, followed by germination of the diploid zygotes within the receptacle itself. Germlings with the mucilaginous coat get released from the receptacle and settle on suitable substrata near the parent plant where it grows back to male and female gametophytes, completing the life cycle. Reproductive maturation in this seaplant is found to be associated with the latitudinal temperature gradient, with populations growing in subtropic regions mature earlier than that in temperate (Nisizawa et al. 1987). The holdfasts of gametophytes can regenerate new seedlings (therefore gametophyte generation is often referred as sporophyte) and this clonal fragmentation ability has been widely utilized for its commercial cultivation. Most of the farms are in South Korea where floating raft culture technique, similar to that in kelp farming, is employed. There are some attempts for the cultivation of zygote-derived seedlings in raceway tanks, however these have not yet been implemented for the commercial farming (Pang et al. 2008).

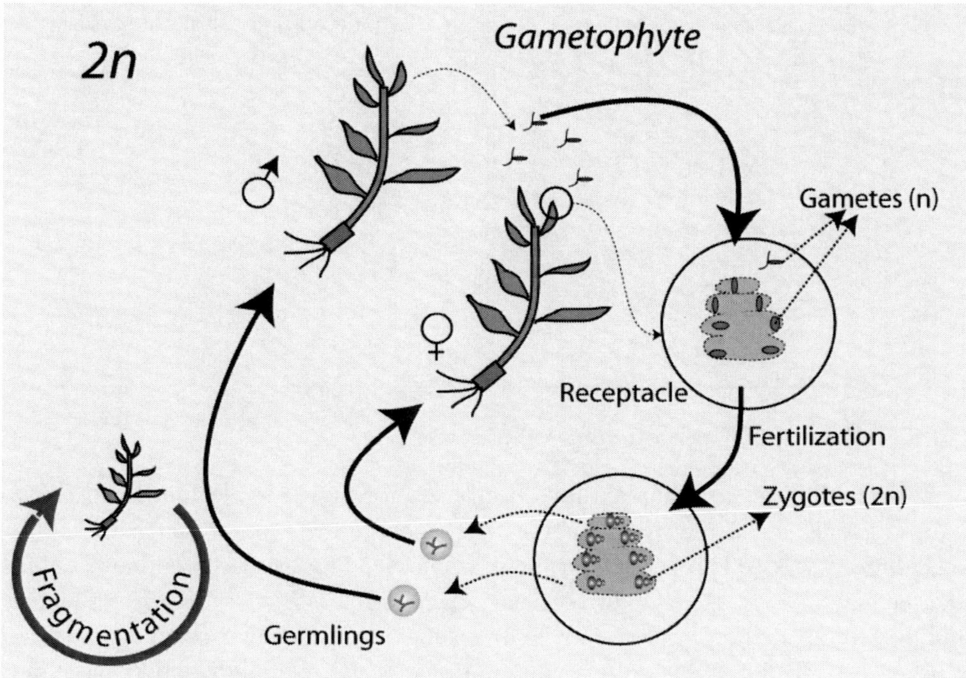

Figure 9. Summary of *Sargassum fusiforme* life cycles.

Monostroma (Ulotrichales)

Among green seaplants, *Monostroma kuroshiensis* Bast is the most intensely cultivated genus, constituting about 90% of the total green algal cultivation (Nisizawa et al. 1987), almost exclusively for the Japanese food produce *hitoegusa*. Habitat is typically intertidal rocks of semi-exposed coasts in subtropical to temperate ecosystems, with most of the occurrences are reported either from the Scandinavia or from West Pacific. The frond is blade-like with eponymous one-cell thickness and therefore it is also known as "Slender sea-lettuce". This is a well-studied seaplant genus with information such as ecophysiology (Bast 2010), habitat-dependent seasonality of its growth pattern and sex-ratio (Bast et al. 2009a; Bast et al. 2009c), gametangial ontogeny (Bast and Okuda 2010) and phylogenetics (Bast 2011) have been reported. Lifecycle is haplodiploid alternation (Figure 10), with dominant, macroscopic phase being haploid dioecious gametophyte. Upon maturity, apical parts of the fronds get matured and phototactic biflagellate gametes get released. Fertilization is anisogamous and settled zygotes mature into microscopic, spherical diploid *Codiolum*-sporophytes. After 3-4 months of growth, sporophytes get matured and quadriflagellate zoospores are produced. Settled zoospores germinate and mature into respective gametophytes, thus completing life cycle. A monomorphic asexual lifecycle in this genus have also been reported from the population in Japan (Bast et al. 2009b).

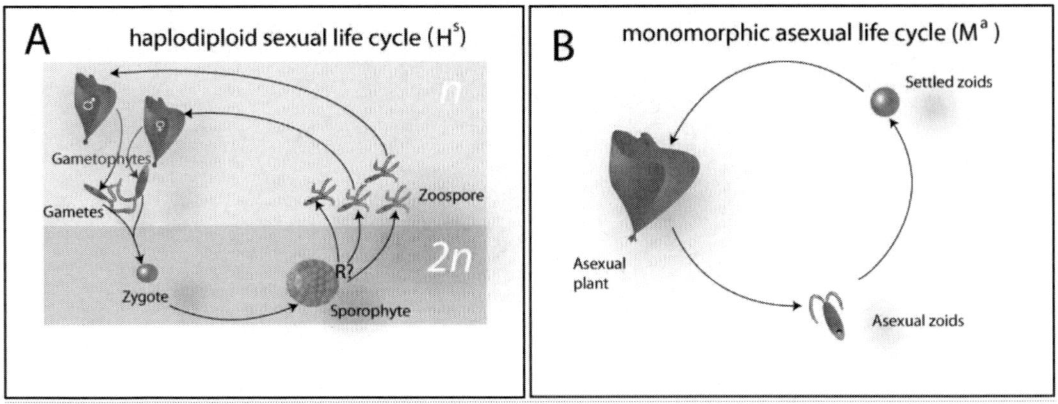

Figure 10. Summary of *Monostroma kuroshiensis* life cycles.

Algae belonging to this complex have been used for human consumption in East Asia- especially in Japan- since time immemorial. As a non-clonal type of seaplant, appropriate seedling culture is needed before each farming season. Seeding method used in the cultivation can either be natural (*e.g.*, Ise Bay) or artificial (*e.g.*, Shimanto Estuary, Kochi Prefecture). In the natural seeding method, plants derived from the naturally deposited zoospores on the culture nets are harvested. On the other hand, in the artificial seeding method a large quantity of zygotes –as obtained from *in vitro* fertilization of isolated gametes by the end of growth period- are grown throughout the summer. The resulting *Codiolum*-sporophytes are then treated with high intensity of light to induce zoospore release. Culture nets are immersed in the concentrated zoospore solution under dark conditions to facilitate successful attachment of released zoospores on the nets. These "seeded" nets are subsequently installed in the attached fabrication of wooden sticks in the coastal waters and the height of nets are adjusted such as to provide adequate immerse-in and drying-out effects with each tidal range. Upon reaching the highest size, thalli are harvested and processed.

Ulva (Ulvales)

Green distromatic seaplants of the genus *Ulva* (Sea lettuce) are commercially cultivated in East Asia for the food produce *Aonori*. Seaplant having resemblance to lettuce are in fact *U. fasciata*, *U. lactuca* etc.- that are distromatic blades, while commercially cultivated species include *U. prolifera* (=*Enteromorpha prolifera*) and *U. intestinalis* that are tubular with walls of the tube one cell thick. Habitat is upper eulittoral zones of subtropics, especially on river mouths where diurnal fluctuation in salinity is high. This species is the main causative agent of green tides, including the one that happened along the shores of Qingdao in 2008 during Beijing Olympics (Liu et al. 2010). This is a dioecious species having diplohaplontic lifecycle (Figure 11) with isomorphic gametophytic and sporophytic stages. Thalli are coarsely filamentous that can be branched or unbranched and have disc-like holdfasts developed from basal cells that attach firmly on suitable substratum like pebbles or rocks. Upon maturity, apical region of the haploid gametophytes changes color to orange-yellow (males) or yellow-green (females) and release biflagellate gametes. Fertilization can be isogamic or anisogamic. Zygotes attaches to suitable substrata where it germinates and develop into diploid sporophyte generation. Apical parts of the sporophytes mature in a similar fashion to that of gametophytes and produces quadriflagellate zoospores by meiosis. Released zoospores settle on substratum and germinate to respective gametophytes, completing the life cycle.

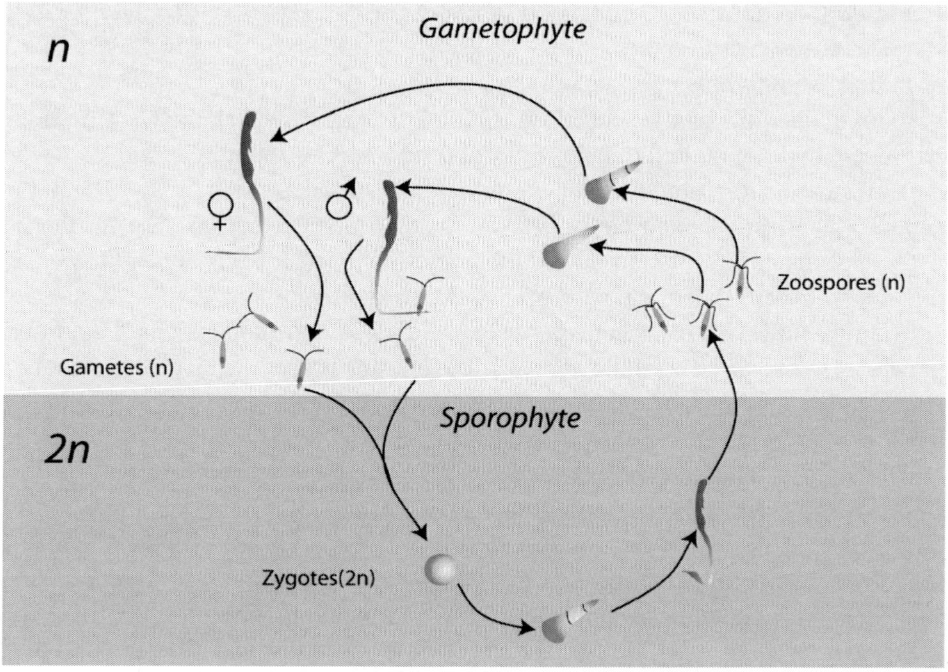

Figure 11. Summary of *Ulva* life cycle.

Cultivation method is similar to that of *Monostroma*. In natural seeding method, rope nets are submerged in places where this alga grows naturally. Once the germination is detected, the nets are taken and brought to the farm. In artificial seeding method, matured thalli are collected and gamete release is induced with a combination of brief drying, followed by floating in seawater with intense illumination. Fragmenting the thalli is found to be an effective way to induce gametogenesis and is used for its artificial seeding (Dan et al. 1997). Released gametes are allowed to get settled on culture nets by providing dark conditions. Once germinated, these nets are taken to the farm. Nets with germinated algae are cultivated on poles fixed in shallow, calm areas of ocean or in estuaries, such that the nets periodically get exposed to air with each low-tide. Tank cultivation using deep sea-water, the so called "germling cluster method" have also been developed, although not widely implemented for the commercial cultivation (Hiraoka and Oka 2008). Information on its phylogenetics (Hayden and Waaland 2004) and phylogeography (Shimada et al. 2009) are known making it one of the well-studied green algal lineages. While reproduction by fragmentation is not known, settled vegetative fragments of this seaplant was found to be a source for the succession of green tides (Zhang et al. 2011).

Caulerpa (Bryopsidales)

While better known to be invasive seaplant wreaking havoc to the Mediterranean marine biodiversity, *Caulerpa* is also the third most cultivated green seaplant. At least two species of edible *Caulerpa* (Sea grapes) are cultivated in many parts of East Asia, *C. lentillifera* and *C. racemosa*. There are many reports on the invasiveness of this genus, especially that by *C. taxifolia* in the Mediterranean (Jousson et al. 1998). Habitat for *C. lentillifera* and *C. racemosa* are muddy or rocky seabed of shallow tropical coastal areas. Morphology is siphonous (coenocytic) with horizontally branched stolon. Almost entire length of stolon axes is covered by many short ramuli. Each ramulus consists of a globular tip with 1-3mm in diameter and a short connecting stalk. Stalk joins to a distinct constriction on the base of globular tip. Not much information on the life cycle is known for this algae, but considering its association in bryopsidales, the safe bet would be monomorphic sexual, as illustrated in Figure 12 (Goldstein Ph 1970). Gametophytes are diploid and monoecious. Upon maturity, gametophytes give rise to biflagellate gametes. During this process, entire protoplast migrates to periphery and integrates into gametes, thus making the thalli appear briefly white before fully disintegrating. Gamete release is synchronized, with simultaneous release of vast numbers of male and female gametes, released as green clouds typically at early morning, a process known as "mass spawning" (Clifton 1997). Fertilization is anisogamous (Miyamura 2005). The zygote settles down on appropriate substrata and germinates back to gametophyte generation, thus completing the life cycle (Ohba et al. 1992). Asexual reproduction by fragmentation is well-known, especially in invasive populations (Wright 2005). Genetically, most of the natural *Caulerpa* populations show little variation and therefore predominant mode of reproduction in nature might be fragmentation.

Caulerpa is cultivated in ponds and lagoons in the Philippines and Okinawa where the fronds are eaten fresh as an ingredient in salads. Bottom planting method is employed in which rhizoidal thalli are planted by hand in upper 2 cm of the muddy shoal (Horstmann 1983). Ponds and lagoons are usually shallow, around 0.5 meters depth, with muddy bottom. This method ensures an all the time immersed environment suitable for this benthic seaplant. While harvesting, about a quarter thalli are left as a seed for the next harvest. Some attempts have been made to initiate land culture of this seaplant using deep sea water (Sudo and Araki 2004), although high cost associated with this method seems to be a major limitation.

UTILIZATION OF SEAPLANTS

As a Food

Historically and in terms of market value, seaplants are primarily utilized as a food. Edible seaplant industries are mostly cottage type with small scale harvesting and primitive modes of processing like rack-drying akin to papermaking (Naylor 1976). A larger demand for the seaplant products branded "traditionally cultivated and processed" might be one of the reasons behind lesser industrial expansion of this sector in Japan. The industry is often controlled by local fishermen's unions and seaplant brands are known after the respective places of production. For example, the famous Shimanto nori (*Monostroma kuroshiensis*) is cultivated by fishermen's union at Shimanto city, Kochi Prefecture, Japan. While seaplants are rarely-if any- consumed raw, the edible seaplants are sometimes referred as "sea vegetables" as an analogy to the land vegetables. Known by the latter name, there is a renewed interest in its culinary usage in the West (Lewallen and Lewallen 1996; Novaczek and Athy 2001).

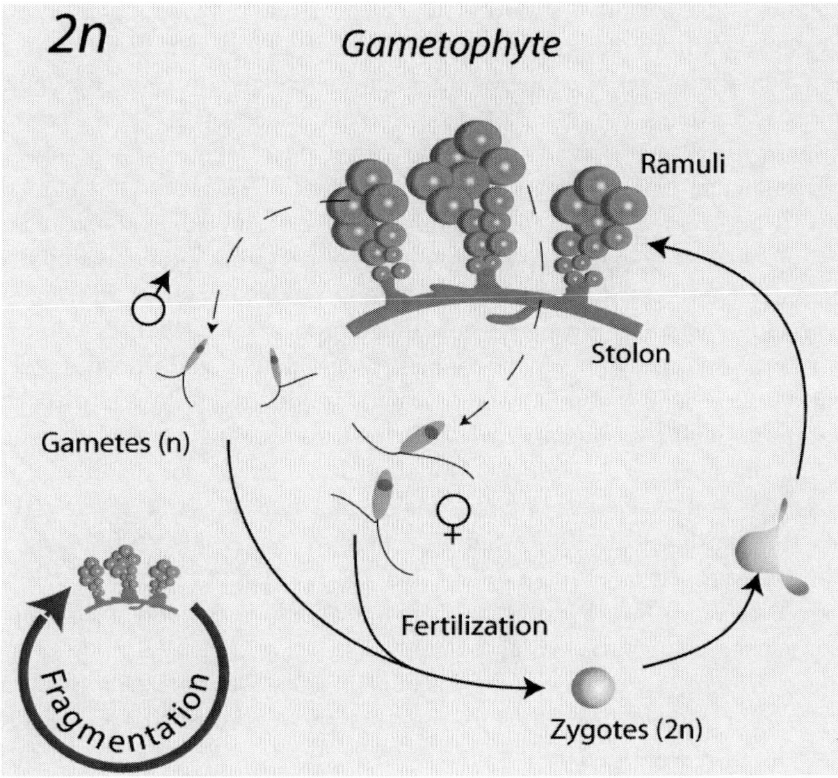

Figure 12. Summary of *Caulerpa* life cycles.

One of the most renowned seaplant food products is *Nori* (Laver), the dried monostromatic blades of the red seaplant *Porphyra.* Nori is the primary constituent in Sushi, the famous Japanese dish which is getting popular throughout the world. Due to high demand, most of the modern research and development for breed selection, improvement of the yield, texture etc. are concentrated on this product. Characteristic flowery aroma of Nori- which is the major factor for its wide appeal- has been attributed to compounds derived from carotenoids (Winterhalter and Rouseff 2002). Recent research suggest that the reason why only people from Japanese descent can digest sulphated polysaccharides of Nori is the horizontal transfer of genes coding for porphyranases- which is crucial in its digestion- from the marine bacteria *Zobellia galactanivorans* to the gut bacteria *Bacteroides plebeius* (Hehemann et al. 2010). While Nori contains high Vitamin B12, the iodine content is quite low compared with other edible seaplants (Watanabe et al. 1999) and therefore, this is the seaplant of choice for thyroid patients. There are several other red seaplants as well commonly used as food produces. Ogonori, made from *Gracilaria* spp (Gracilariales) is a common Japanese produce used as an ingredient in cold summer noodles (Sōmen). Dulse or creathnach, a popular Irish snack, is made from dried *Palmaria palmata* (Palmariales). Chilean produce Quechua (*Durvillaea antarctica*) is an ingredient in traditional stews. The Irish Moss (*Chondrus crispus*) is a very popular seaplant produce in America and Caribbean Islands, where this is sold as snack or as an ingredient in various beverages. Jellified form of the Irish Moss used to be one of the popular confectionary in Ireland where it is known as Cairgean, before its usage have gone down significantly in recent times (Kenicer et al. 2000).

A number of edible seaplant products are made from brown seaplants. *Kombu* is a processed food product from the brown seaplant *Saccharina japonica* (Sugarware or "Kelp", Laminariales). The dried *Kombu* is one of the main ingredients of *dashi*, the Japanese soup stock. *Kombu* is also extensively pickled (*su-kombu*) and used as jam (*kombu tsukudani*). Yet another well-known product is *kombu-cha*, the green "tea" made with *kombu*. The food produce *Arame* is made from *Ecklonia bicyclis* (Laminariales). Having a distinct taste, the dried and pulverized *Arame* is added into a number of dishes, including soups and appetizers. There are several other kelp species commonly used to make food products in the West. One such example is Dabberlocks, a popular western food produce made from *Alaria esculenta* (Laminariales). Fingerware is a popular Irish seaplant produce made from *Laminaria digitata*. Dried *Fucus vesiculosus*, known as bladderwrack, is an additive and flavoring agent used in various European food items. Popular South American produce Carola is made from dried and processed *Callophyllis variegate*. *Wakame* is yet another major seaplant product made from brown seaplant *Undaria Pinnatifida*. Boiled *wakame* is a major ingredient of Japanese miso soup served as a breakfast staple along with rice. Several reports demonstrate anticancerous (Funahashi et al. 1999), antihypertensive (Sato et al. 2002) and antiobesity (Maeda et al. 2005) properties of this product, making it a popular nutraceutical choice. The seaplant produce *Hijiki* is made from the brown seaplant *Sargassum fusiforme*. *Hijiki* is mainly used in Japanese cuisine as an ingredient in salads and soups. As a part of macrobiotic food movement, *Hijiki* got immense popularity in the west during 1990s (McCarty 1996). Due to high arsenic content, the *Hijiki* is often extensively washed and socked prior to cooking (Hanaoka et al. 2001). The popular Hawaian produce Limu-Kala, an ingredient in soups, is made from dried *Sargassum echinocarpum* (Hanaoka et al. 2001). Mozuku is an Okinawan food produce made from *Cladosiphon okamuranus* and is used in salads. Mozuku contains more Fucoidan-a sulphated polysaccharide with many biological properties- than any other seaplant and therefore it is used as a nutritional supplement (Kawamoto et al. 2006).

A plenty of food products are made from green seaplants as well. Perhaps *Hitoegusa,* the monostromatic green alga (*Monostroma latissimum*), is the most important of all edible green seaplants in Japan in terms of economy and quantity of production. Market value of this seaplant is the highest among all the cultivated seaplants with 1 kg costing about USD 30 (Zemke-White and Ohno 1999). Harvested *Hitoegusa* is boiled down in soy sauce to make a jam-like product (tsukudani-nori) while dried sheets (hoshi-nori) are used as sushi wraps. Other popular products are roasted (yaki-nori) and seasoned (ajitsuke-nori) *Hitoegusa*. The Japanese seaplant produce, *aonori* or *aosanori* is made from dried *Ulva prolifera*. Aonori is used as a seasoned produce and sprinkled over boiled rice to serve. Other edible green seaplants include Sea Grapes (*Caulerpa lentillifera*), Sea Lettuce (*Ulva lactuca*) and Gutweed (*Ulva intestinalis*).

Hydrocolloid Industry

Hydrocolloids are polysaccharide colloids dispersed in water and therefore having a gel-like constituency. Several of the seaplant-derived hydrocolloids are used as an ingredient in various food items like jellies, deserts and confectionary. Examples of hydrocolloids include Agar, Alginate and Carrageenan. Agar is a polysaccharide of galactose sugar extracted from several genera of red seaplants-collectively referred as agarophytes- that include *Gelidium* and *Gracilaria*. Exploiting its gelling and emulsifying abilities, agar is used in various ways such as constituent of biological culture media (Agar plate), as a vegetarian gelatin in various food items like jellies, puddings and deserts, toy manufacturing, as an impression material in dentistry and as electrochemical salt bridges. Further chemical refinement of less-ionic fractions of agars produces a highly refined seaplant product Agarose. Agarose is extensively utilized in biotechnology industry as a solid phase in Gel Electrophoresis and therefore played a crucial role in the biotechnological revolution of last century (Renn 1990).

Carrageenan, a sulphated polysaccharide made up of galactose-related monomers, is another seaplant-derived hydrocolloid extensively used in food industry. Carrageenan is extracted from hot water suspensions of red seaplants *Eucheuma*, *Chondrus* and *Kappaphycus* (together, these are known as carrageenophytes). Along with agar, carrageenan is also used as a gelling and emulsifying agent in food and is often used as a substitute for gelatin by the vegans. Carrageenans have a plethora of uses; as an additive in various food items, toothpaste, shampoo, personal lubricants, marbling, shoe polish and as an excipient in tablets to name a few. Low molecular weight degraded carrageenans have been reported to be causing colon ulcers in experimental animals (Watt and Marcus 1971). Although natural carrageenan generally regarded as safe for human consumption, degraded carrageenan has been classified as a possible human carcinogen by the International Agency for Research on Cancer (IARC). It has been found that carrageenan induces cell arrest in human epithelial cells *in-vitro* suggesting it might have role in normal human intestinal pathology (Bhattacharyya et al. 2008). Recently, carrageenan has been demonstrated to inhibit genital transmission of Human Papillomavirus (HPV) infections in mouse, warranting further research to utilize it for human STD prevention and cure (Roberts et al. 2007). Although *in-vitro* studies showed effective anti-HIV properties, largest human trials so far conducted in South Africa showed that carrageenan-based vaginal gel, Carraguard®, fails to protect women against HIV (Cohen 2008).

Alginate, salts of alginic acid-an anionic polysaccharide, is yet another seaplant-derived hydrocolloid widely used in many different ways. This is commercially extracted from brown seaplant genera *Laminaria*, *Macrocystis* and *Ascophyllum*. The process of extraction include precipitation of salts with hot sodium bicarbonate and further filtration of sodium salts (McHugh 1987). Alginates are also used in various industries, such as impression-making material in dentistry, prosthetics, as an appetite suppressant in weight loss industry and waterproofing and fireproofing agent in textiles. Calcium alginate (CA) microbeads have been developed recently which can deliver plasmid DNAs and yeast artificial chromosomes to animal and plant cells (Higashi et al. 2004) and Plant Growth Promoting Bacteria (PGPBs) to wet and dry seeds (Bashan et al. 2002), further accelerating its demand in biotechnology industry.

Other Natural Products

There are a number of seaplant natural products other than what are mentioned above. This includes use of seaplant biomass as a fertilizer and soil conditioner for agriculture and horticulture industry (Aitken and Senn 1965). Recent studies indicate issues with its widespread uses, including arsenic biotransformation and accumulation and therefore care must be must be taken while choosing taxa and/or habitat to use as a fertilizer (Zhao et al. 2010). Uses of seaplant extracts as a plant growth enhancer in agriculture have been reviewed in (Craigie 2010). Addition of low concentrations of commercial kelp extract (Kelpack®) has been proven to be beneficial in agriculture and aquaculture (Robertson-Andersson et al. 2006). Nutritional and pharmaceutical potentials of a number of seaplant natural products have been comprehensively reviewed (Fitton 2003; Abdussalam 1990; Bocanegra et al. 2009; Smit 2004). Seaplant-derived bioactive substances that have multitude of biomedical applications include fucoidans, lectins, sulphated polysaccharides and aplysiatoxins. Seaplant derived substances that are used as vermifuges and insecticides include kainoids, terpenes, diterpenes (Crenulacetal C used in pearl cultivation), sesquiterpenes and polyhalogenated monoterpenes. Another high-value seaplant natural product is phycoerythrins extracted from red seaplants. Phycoerythrins, especially R-Phycoerythrin from *Porphyra*, is extensively used as fluorophore in applications such as immunofluorescence and Fluorescence-Activated Cell Sorting (Kronick and Grossman 1983; Glazer 1994). Seaplant carbon-based carbon nanotubes have recently been developed as a nanotexturing agent for high power supercapacitors (Raymundo Piñero et al. 2011). Ever-growing technology sector is expected to facilitate finding more unique uses for seaplants and development of seaplant-based natural products.

Integrated Multi-Trophic Aquaculture (IMTA)

Fed aquaculture, especially that for shrimps and finfish, generate enormous amounts of nutrient waste that causes multitude of environmental problems in aquaculture fields. Altered nutrient loading in coastal areas lead to problems associated with eutrophication, such as enhanced sediment metabolism, sulphide accumulation and sulphate reduction, high nitrogen and phosphorous influx, turbidity, anoxia, acidification and so on (Troell and Berg 1997). Fed aquaculture has long been regarded as the least sustainable and energy inefficient form of farming practices. In order to mitigate the nutrient wastes generated during fed aquaculture, field of Integrated Multi-Trophic Aquaculture (IMTA) have developed which is centred on the idea of co-cultivating organic extractors (e.g. shellfish/herbivorous fish) and inorganic extractors (seaplants) in the fed aquaculture. While simple polyculture implies culture of different organisms belonging to same trophic level, IMTA involves co-farming of organisms belonging to various trophic levels and considers each contributor's ecological systems and services.

Seaplants in IMTA act as biofilters by assimilating excess inorganic waste and thereby contributing in the bioremediation. Seaplants have been widely integrated in various IMTA systems around the world and its utility have been extensively reviewed (Chopin et al. 2001; Neori et al. 2004). Many of the studies conducted in close circulation systems conclude that seaplants take-up large quantities of nutrients and convert them to seaplant biomass, in addition to improving other water quality parameters. Biofiltration of nitrogenous wastes from recirculating abalone mariculture system using seaplants have been recently demonstrated to be outperforming bacterial biofilms (Cahill et al. 2010). An increasing number of studies suggests "green-tide" seaplants such as *Cladophora*, *Chaetomorpha* and *Ulva*, can be effectively integrated to the tropical pond-based aquaculture for bioremediation (de Paula Silva et al. 2008) as well as for seaplant farming (Yokoyama and Ishihi 2010; Bolton et al. 2009; Msuya and Neori 2010), thus making them attractive candidates for IMTA systems. Shrimps grown with *Ulva* in outdoor tanks have demonstrated lower artificial feed requirements and efficient assimilation of *Ulva* carotenoids, thus making IMTA a sustainable and healthier alternative to traditional aquaculture (Cruz-Suárez et al. 2010).

Energy Production and Carbon Capture and Sequestration

The world is turning to the seaplant farming to combat various environmental issues as well, energy crisis being one of them. With energy prices fluctuating widely amid diminishing fossil energy resources, there had been a recent surge in the research on the potential usage of algae as a renewable fuel source. Although algae fuel won't reduce atmospheric CO_2 level- as CO_2 fixed by the algae is returned to the atmosphere when the fuel is burned, it won't contribute in the introduction of new CO_2 as in the case of burning fossil fuels; i.e., they are "carbon neutral". Various extraction methods have been developed to harness lipids and its esterification into biodiesel, biogasoline etc., from algal biomass. biofuel production has traditionally been concentrated to few species of microalgae as the photosynthetic rate and oil content of microalgae appears to be significantly higher than that of seaplants (Bird and Benson 1987). However, high production cost for the complex microalgal cultivation systems like photobioreactors have incited a renewed interest in seaplants as a potential source. One method is to ferment seaplant biomass to bioethanol and then anaerobic digestion to biogas (Aizawa et al. 2007). A recent study describes the saccharification of Floating Residue-a surplus by-product of alginate extraction process- into to yeast-fermantable sugars for bioethanol production (Ge et al. 2011). A vast majority of seaplants that are currently utilized for energy production come from either harvesting the natural stock or by collecting the drifting biomass. Due to high demand, natural stocks of the seaplants have decreased significantly necessitating increased seaplant farming attempts (Bruton et al. 2009), especially with metabolically engineered seaplant strains that can produce high quantity of lipids (Rosenberg et al. 2008). Seaplant farming has also been proposed as the most viable way to facilitate Carbon Capture and Sequestration (CCS) for combating global warming (Smith 1981). Greenhouse gases mitigated during seaplant biomass production can be safely disposed, the so called "seaplant carbon sink". However recent *in-situ* experiments suggest that the process of using seaplant as global carbon sink may not be as effective as originally thought and might even have undesirable side effects (Dalton 2002).

Environmental Impacts of Seaplant Farming Practices

A number of studies have documented potential environmental issues associated with seaplant farming, although relatively far fewer comparing to that of fed-aquaculture. These can broadly be categorized as the following: 1) Habitat alterations 2) Shading effects and 3) Proliferation.

Introduced seaplants can change local community structure and thereby affecting ecological niche including food chain patterns. In one study, off-bottom farming of *Eucheuma* in Tanzania was found to be affecting growth of seagrass beds underneath and thereby disturbing ecosystem functioning and flow of ecological goods and services (Eklöf et al. 2006). As seagrass meadows are directly linked to productive coral reef ecosystem, changes in its cover might also have indirect consequences on coral reef ecological niche. Seaplant farming was documented to be causing changes in sedimentation patterns, with seabeds underneath the farms having finer sediments and lower organic carbon content (Eklöf et al. 2005). Seaplant farming is also associated with epiphytic abundances which might affect natural algal communities (Buschmann et al. 1996).

Rich canopy of seaplant farms definitely causes shading effects on the flora growing underneath. Altering local habitat due to this shading might have tremendous consequences, such as altered growth patterns of algal communities, differential migration patterns of meiobenthos and whereby structuring epipelagic ecological niche. Another potential effect would be on the sensitive reef ecosystem, alteration of the reef community structure due to algal encroachment will have far fetching impacts, as these harbor almost a quarter of all marine biota (Mulhall 2008). Excessive shading might lead to reduced growth of zooxanthellae- photosynthetic dinoflagellate symbionts of the corals that might lead to catastrophic bleaching of the entire coral reef system. Characterizing these effects on macroscale is warranted for effective policy making to implement sustainable seafarming practices.

Algal proliferation is fundamentally different from algal invasion. Proliferation implies spread of algae to the regions where it compete with existing flora, while invasion results in introduction to waters where it has no predator. Intentional introduction of seaplant, which might lead to proliferation, is an outcome of commercial seaplant farming. Before the advent of seaplant farming, almost all the introduced of alien seaplant communities can be attributed to maritime transport. While the introduction of *Undaria* in many parts of the world is largely due to shipping, introduction of *Kappaphycus* has been reported from many parts of the world where the seaplant is cultivated for carrageenan. Although documented risks from intentional introduction of alien seaplant are fewer than the unintentional introduction, seacrops as a group possesses some degree of threat to the local biota (Pickering et al. 2007). Environmental implications of proliferating seaplants have been recently reviewed in (Morand and Merceron 2009).

Conclusion

Although seaplants have been utilized in many civilizations around the world for hundreds of centuries, its cultivation has been very minimal outside Asia. Reason for disparities in seaplant farming can be attributed to differential treatment of seaplant as a food around the world, or simply lack of necessary impetus from the industry. Effective information dissemination about seaplant farming technologies and critical evaluation of unexplored geographical locations for the feasibility of seaplant farming are two commonsense strategies for the expansion of this sector. Seaplant agronomy is in infancy and more research is needed in such areas as sustainable and economically viable farming methods. Efforts should be taken to harness natural life cycle mechanisms like controlling gametangial maturation/gamete release and increasing growth rate of microscopic stages. Indoor tank cultivation methods developed for seaplants are impractical to be commercially implemented due to prohibitive costs. Large scale utilization of tissue culture and other micropropagation methods in seaplant farming might not be feasible at all. Strain improvements by hybridization or genetic engineering, although successfully implemented for many high-value seaplants, need to be expanded to other farmed taxa as well; improving carrageenan production by Eucheumoid seaplants being one of them. Native seaplant species which are fast growing, cultivable year round and having good market value need to be surveyed for the feasibility of utilizing them in both land-based and open water IMTA systems. Attention should be given to the location of farms and choice of farming methods for minimizing environmental impacts. Some straight forward measures would be avoiding locations with lush natural seagrass flora or coral reefs. Results from large scale seaplant-based energy harnessing projects, like Ocean Sunrise Project in Japan (Aizawa et al. 2007), are awaited to further clarify feasibility for integrating the seaplant farming in energy sector.

ACKNOWLEDGEMENTS

I thank Department of Science and Technology, government of India INSPIRE Faculty Award for materializing this review, including computational, graphical and bibliographical requirements. I am also thankful to the Vice Chancellor, Central university of Punjab for his support with respect to the execution of this work.

REFERENCES

Abdussalam S (1990) Drugs from seaweeds. *Med Hypotheses* 32 (1):33-35. doi:DOI:0306-9877(90)90064-L [pii].

Aitken J, Senn T (1965) Seaweed products as a fertilizer and soil conditioner for horticultural crops. *Botanica Marina* 8 (1):144-147.

Aizawa M, Asaoka K, Atsumi M, Sakou T Seaweed Bioethanol Production in Japan-The Ocean Sunrise Project. In: *Oceans 2007*, 2007. IEEE, pp 1-5. doi:10.1109/OCEANS.2007.4449162.

Altieri AH, Trussell GC, Ewanchuk PJ, Bernatchez G, Bracken MES, Sandin SA (2009) Consumers control diversity and functioning of a natural marine ecosystem. Plos One 4 (4):e5291.

Bashan Y, Hernandez JP, Leyva LA, Bacilio M (2002) Alginate microbeads as inoculant carriers for plant growth-promoting bacteria. *Biology and Fertility of Soils* 35 (5):359-368.

Bast F (2010) *Comparative Ecophysiology and Phylogeography of Monostroma in Southern Japan.*, Kochi University, Kochi, Japan.

Bast F (2011) *Monostroma: the Jeweled Seaweed for Future.* LAP LAMBERT Academic Publishing GmbH & Co. KG, 66121, Saarbrücken, Germany.

Bast F (2013) *Seaweeds in Japanese Culture: An analysis of medieval Waka poetry.* The Phycologist 85:(accepted).

Bast F, Hiraoka M, Okuda K (2009a) Spatiotemporal Sex Ratios of a Dioecious Marine Green Alga: *Monostroma latissimum* (Kützing) Wittrock. *International Journal on Algae* 11 (2):141-150.

Bast F, Okuda K (2010) Gametangial Ontogeny in Intertidal Green Alga: *Monostroma latissimum* (Kützing) Wittrock. *International Journal of Plant Reproductive Biology 2* (1):11-15.

Bast F, Shimada S, Hiraoka M, Okuda K (2009b) Asexual life history by biflagellate zoids in *Monostroma latissimum* (Ulotrichales). *Aquatic Botany* 91 (3):213-218.

Bast F, Shimada S, Hiraoka M, Okuda K (2009c) Seasonality and thallus ontogeny of edible seaweed *Monostroma latissimum* (Kützing) Wittrock (Chlorophyta, Monostromataceae) from Tosa Bay, Kochi, Japan. *Hydrobiologia* 630 (1):161-167.

Bhattacharyya S, Borthakur A, Dudeja PK, Tobacman JK (2008) Carrageenan induces cell cycle arrest in human intestinal epithelial cells in vitro. The Journal of nutrition 138 (3):469-475.

Bird KT, Benson PH (1987) Seaweed cultivation for renewable resources. In: *Developments in Aquaculture and Fisheries Science.* p 381.

Bischof K, Gomez I, Molis M, Hanelt D, Karsten U, Lüder U, Roleda MY, Zacher K, Wiencke C (2007) Ultraviolet radiation shapes seaweed communities. *Life in Extreme Environments:*187-212.

Blouin NA, Brodie JA, Grossman AC, Xu P, Brawley SH (2010) *Porphyra*: a marine crop shaped by stress. *Trends in plant science* 16 (1):29-37.

Bocanegra A, Bastida S, Benedi J, Rodenas S, Sanchez-Muniz FJ (2009) Characteristics and nutritional and cardiovascular-health properties of seaweeds. *J Med Food* 12 (2):236-258. doi:DOI:10.1089/jmf.2008.0151.

Bolton J, Robertson-Andersson D, Shuuluka D, Kandjengo L (2009) Growing *Ulva* (Chlorophyta) in integrated systems as a commercial crop for abalone feed in South Africa: a SWOT analysis. *Journal of applied phycology* 21 (5):575-583.

Boyer KE, Kertesz JS, Bruno JF (2009) Biodiversity effects on productivity and stability of marine macroalgal communities: the role of environmental context. *Oikos* 118 (7):1062-1072.

Boylan HJ (1971) The commercial exploitation of seaweeds, with special reference to alginate manufacture. *Planta Med* 20 (2):178-180. doi:10.1055/s-0028-1099687.

Bruton T, Lyons H, Lerat Y, Stanley M, Rasmussen MB (2009) *A review of the potential of marine algae as a source of biofuel in Ireland.* Sustainable Energy Ireland, Dublin.

Buschmann AH, López DA, Medina A (1996) A review of the environmental effects and alternative production strategies of marine aquaculture in Chile. *Aquacultural engineering* 15 (6):397-421.

Cahill PL, Hurd CL, Lokman M (2010) Keeping the water clean--Seaweed biofiltration outperforms traditional bacterial biofilms in recirculating aquaculture. *Aquaculture* 306 (1-4):153-159.

Cavalier-Smith T (1998) A revised six-kingdom system of life. *Biological reviews of the Cambridge Philosophical Society 73* (3):203-266.

Cavalier-Smith T (2004) Only six kingdoms of life. Proceedings Biological sciences / *The Royal Society* 271 (1545):1251-1262. doi:10.1098/rspb.2004.2705.

Chaoyuan W, Guangheng L (1987) Progress in the genetics and breeding of economic seaweeds in China. *Hydrobiologia* 151 (1):57-61.

Chapman VJ (1950) *Seaweeds and their uses.* Methuen, London.

Cheney DP (1990) Genetic improvement of seaweeds through protoplast fusion. *Economically Important Marine Plants of the Atlantic: Their Biology and Cultivation Connecticut Sea Grant College Program,* CT-SG-89-07, Groton, CT:15-26.

Chopin T, Buschmann AH, Halling C, Troell M, Kautsky N, Neori A, Kraemer GP, Zertuche González JA, Yarish C, Neefus C (2001) Integrating seaweeds into marine aquaculture systems: a key toward sustainability. *Journal of phycology* 37 (6):975-986.

Clifton KE (1997) Mass spawning by green algae on coral reefs. Science 275 (5303):1116.

Cohen J (2008) Microbicide Fails to Protect Against HIV. *Science* 319 (5866):1026.

Craigie JS (2010) Seaweed extract stimuli in plant science and agriculture. *Journal of applied phycology:*1-23. doi:10.1007/s10811-010-9560-4.

Cruz-Suárez LE, León A, Peña-Rodríguez A, Rodríguez-Peña G, Moll B, Ricque-Marie D (2010) Shrimp/*Ulva* co-culture: A sustainable alternative to diminish the need for artificial feed and improve shrimp quality. *Aquaculture* 301 (1-4):64-68.

Dai J, Zhang Q, Bao Z (1993) Genetic breeding and seedling raising experiments with *Porphyra* protoplasts. *Aquaculture* 111 (1-4):139-145.

Dalton R (2002) Ocean tests raise doubts over use of algae as carbon sink. *Nature* 420 (6917):722-722.

Dan A, Ohno M, Matsuoka M (1997) *Cultivation of the green alga Enteromorpha prolifera using chopped tissue for artificial seeding.* Suisanzoshoku (Japan).

Dawes CP (1995) Suspended cultivation of *Gracilaria* in the sea. *Journal of applied phycology* 7 (3):303-313.

de Paula Silva PH, McBride S, de Nys R, Paul NA (2008) Integrating filamentous 'green tide' algae into tropical pond-based aquaculture. *Aquaculture* 284 (1-4):74-80.

Doty M, *Status of marine agronomy, with special reference to the tropics*. In, 1979. pp 35-58.

Doty MS, Caddy J, Santelices B (1986) The production and use of *Eucheuma*. Case studies of seven commercial seaweed resources:123-164.

Drew KM (1949) Conchocelis-phase in the life-history of *Porphyra umbilicalis* (L.) Kütz. *Nature* 164 (4174):748-749.

Eklöf J, de La Torre Castro M, Adelsköld L, Jiddawi N, Kautsky N (2005) Differences in macrofaunal and seagrass assemblages in seagrass beds with and without seaweed farms. Estuarine, *Coastal and Shelf Science* 63 (3):385-396.

Eklöf J, Henriksson R, Kautsky N (2006) Effects of tropical open-water seaweed farming on seagrass ecosystem structure and function. *MEPS* 325:73-84.

FAO (2010) *The state of world fisheries and aquaculture.* FAO Fisheries and Aquaculture Department, United Nations, Rome.

Fitton JH (2003) Brown marine algae: a survey of therapeutic potentials. *Alternative & Complementary Therapies* 9 (1):29-33.

Fletcher RL (1995) Epiphytism and fouling in *Gracilaria* cultivation: an overview. *Journal of applied phycology* 7 (3):325-333.

Funahashi H, Imai T, Tanaka Y, Tsukamura K, Hayakawa Y, Kikumori T, Mase T, Itoh T, Nishikawa M, Hayashi H (1999) Wakame Seaweed Suppresses the Proliferation of 7, 12 Dimethylbenz (a) anthracene induced Mammary Tumors in Rats. *Cancer Science* 90 (9):922-927.

Ge L, Wang P, Mou H (2011) Study on saccharification techniques of seaweed wastes for the transformation of ethanol. *Renewable Energy* 36 (1):84-89.

Glazer AN (1994) Phycobiliproteins—a family of valuable, widely used fluorophores. *Journal of applied phycology* 6 (2):105-112.

Glenn EP, Doty MS (1990) Growth of the seaweeds Kappaphycus alvarezii, K. striatum and *Eucheuma* denticulatum as affected by environment in Hawaii. *Aquaculture* 84 (3-4):245-255.

Goldstein Ph D (1970) Gametogenesis and fertilization in *Caulerpa*. *Annals of the New York Academy of Sciences* 175 (1):660-672.

Hanaoka K, Yosida K, Tamano M, Kuroiwa T, Kaise T, Maeda S (2001) Arsenic in the prepared edible brown alga hijiki, *Hizikia fusiforme*. *Applied organometallic chemistry* 15 (6):561-565.

Hayden HS, Waaland JR (2004) A molecular systematic study of *Ulva* (Ulvaceae, Ulvales) from the northeast Pacific. *Phycologia* 43 (4):364-382.

Hehemann JH, Correc G, Barbeyron T, Helbert W, Czjzek M, Michel G (2010) Transfer of carbohydrate-active enzymes from marine bacteria to Japanese gut microbiota. *Nature* 464 (7290):908-912.

Hernández-González MC, Buschmann AH, Cifuentes M, Correa JA, Westermeier R (2007) Vegetative propagation of the carrageenophytic red alga *Gigartina skottsbergii* Setchell et Gardner: Indoor and field experiments. *Aquaculture* 262 (1):120-128.

Higashi T, Nagamori E, Sone T, Matsunaga S, Fukui K (2004) A novel transfection method for mammalian cells using calcium alginate microbeads. *Journal of bioscience and bioengineering* 97 (3):191-195.

Hiraoka M, Oka N (2008) Tank cultivation of *Ulva prolifera* in deep seawater using a new "germling cluster" method. *Journal of applied phycology* 20 (1):97-102.

Horstmann U (1983) Cultivation of the green alga, *Caulerpa racemosa*, in tropical waters and some aspects of its physiological ecology. *Aquaculture* 32 (3-4):361-371.

Hurtado-Ponce A (1990) Vertical rope cultivation of *Gracilaria* (Rhodophyta) using vegetative fragments. *Botanica Marina* 33 (6):477-482.

Hurtado-Ponce AQ, Agbayani RF, Chavoso EAJ (1996) Economics of cultivating *Kappaphycus alvarezii* using the fixed-bottom line and hanging-long line methods in Panagatan Cays, Caluya, Antique, Philippines. *Journal of applied phycology* 8 (2):105-109.

Israel A, Levy I, Friedlander M Experimental tank cultivation of *Porphyra* in Israel. In: Eighteenth International Seaweed Symposium, 2007. *Developments in Applied Phycology*. Springer, pp 9-14. doi:10.1007/978-1-4020-5670-3_2.

Jonasson A (1984) *Laminaria*-a modern cervix dilatation method with more than a 100-year history. *Jordemodern* 97 (6):187-195.

Jousson O, Pawlowski J, Zaninetti L, Meinesz A, Boudouresque C (1998) Molecular evidence for the aquarium origin of the green alga *Caulerpa taxifolia* introduced to the Mediterranean Sea. *Marine Ecology Progress Series* 172 (0):275-280.

Kalimuthu S, Kaliaperumal N, Ramalingam J (1991) Standing crop, algin and mannitol of some alginophytes of Mandapam coast. *Journal of the Marine Biological Association of India* 33 (1&2):170-174.

Kawamoto H, Miki Y, Kimura T, Tanaka K, Nakagawa T, Kawamukai M, Matsuda H (2006) Effects of fucoidan from Mozuku on human stomach cell lines. *Food science and technology research* 12 (3):218-222.

Kenicer G, Bridgewater S, Milliken W (2000) The ebb and flow of Scottish seaweed use. *Botanical Journal of Scotland* 52 (2):119-148.

Kronick M, Grossman PD (1983) Immunoassay techniques with fluorescent phycobiliprotein conjugates. *Clinical chemistry* 29 (9):1582-1586.

Kuwano K, Kono S, Jo YH, Shin JA, Saga N (2004) Cryopreservation Of The Gametophytic Cells Of Laminariales (Phaeophyta) In Liquid Nitrogen. *Journal of phycology* 40 (3):606-610.

Kylin H (1956) *Die Gattungen der Rhodophyceen.* CWK. Gleerups, Lund.

Lane CE, Mayes C, Druehl LD, Saunders GW (2006) A Multi Gene Molecular Investigation Of The Kelp (Laminariales, Phaeophyceae) Supports Substantial Taxonomic Re Organization. *Journal of phycology* 42 (2):493-512.

Lewallen E, Lewallen J (1996) Sea vegetable gourmet cookbook and wildcrafter's guide, vol 1. *Mendocino Sea Vegetable* Company (Mendocino, Calif.).

Li D, Zhou Z, Liu H, Wu C (1999) A new method of *Laminaria japonica* strain selection and sporeling raising by the use of gametophyte clones. *Hydrobiologia* 398:473-476.

Li X, Liu J, Cong Y, Qu S, Zhang Z, Dai H, Luo S, Han X, Huang S, Wang Q (2008) Breeding and trial cultivation of Dongfang No. 3, a hybrid of *Laminaria* gametophyte clones with a more than intraspecific but less than interspecific relationship. *Aquaculture* 280 (1-4):76-80.

Liu F, Pang SJ, Chopin T, Xu N, Shan TF, Gao SQ, Sun S (2010) The dominant *Ulva* strain of the 2008 green algal bloom in the Yellow Sea was not detected in the coastal waters of Qingdao in the following winter. *Journal of applied phycology* 22 (5):1-10.

Lüning K, Pang S (2003) Mass cultivation of seaweeds: current aspects and approaches. *Journal of applied phycology* 15 (2):115-119.

Maeda H, Hosokawa M, Sashima T, Funayama K, Miyashita K (2005) Fucoxanthin from edible seaweed, *Undaria pinnatifida*, shows antiobesity effect through UCP1 expression in white adipose tissues. *Biochem Biophys Res Commun* 332 (2):392-397. doi:DOI:10.1016/j.bbrc.2005.05.002.

Matsumura W, Yasui H, Yamamoto H (2001) Successful sporophyte regeneration from protoplasts of *Undaria pinnatifida* (Laminariales, Phaeophyceae). *Phycologia* 40 (1):10-20.

McCarty M (1996) *American macrobiotic cuisine.* Penguin.

McHugh DJ (1987) Production and utilization of products from commercial seaweeds. *FAO Fisheries Technical Paper* (FAO). FAO, Rome (Italy).

Miyamura S (2005) Behavior of Flagella and Eyespot during Fertilization of Anisogametes in *Caulerpa racemosa* (Forsskål) J. Agardh var. *laetevirens* (Montagne)(Ulvophyceae,

Chlorophyta) Revealed by Field Emission Scanning Electron Microscopy. *Cytologia* 70 (1):93-99.

Morand P, Merceron M (2009) Macroalgal population and sustainability. *Journal of coastal research* 21 (5):1009-1020.

Msuya FE, Neori A (2010) The Performance Of Spray Irrigated *Ulva Lactuca* (Ulvophyceae, Chlorophyta) As A Crop And As A Biofilter Of Fishpond Effluents. *Journal of phycology* 46 (4):813-817.

Mulhall M (2008) *Saving the Rainforest of the Sea: An Analysis of International Efforts to Conserve Coral Reefs.* Duke Envtl L & Pol'y F 19:321.

Naylor J (1976) Production, trade and utilization of seaweeds and seaweed products. In: FAO Fisheries Technical Paper: 159. *Food and Agricultural Organization of the United Nations.*

Neill K, Heesch S, Nelson W (2010) Diseases, pathogens and parasites of *Undaria pinnatifida.* Ministary of Agriculture and Foresty- Biosecurity, New Zealand, Wellington.

Nelson W, Brodie J, Guiry M (1999) Terminology used to describe reproduction and life history stages in the genus *Porphyra* (Bangiales, Rhodophyta). *Journal of applied phycology* 11 (5):407-410.

Neori A, Chopin T, Troell M, Buschmann AH, Kraemer GP, Halling C, Shpigel M, Yarish C (2004) Integrated aquaculture: rationale, evolution and state of the art emphasizing seaweed biofiltration in modern mariculture. *Aquaculture* 231 (1-4):361-391.

Nisizawa K, Noda H, Kikuchi R, Watanabe T (1987) The main seaweed foods in Japan. *Hydrobiologia* 151 (1):5-29.

Novaczek I, Athy A (2001) *Sea vegetable recipes for the Pacific Islands.* University of the South Pacific, Fiji.

Ohba H, Nashima H, Enomoto S (1992) Culture studies on *Caulerpa* (caulerpales, chlorophyceae) III. Reproduction, development and morphological variation of laboratory-cultured *C. racemosa* var. *peltata*. *Journal of Plant Research* 105 (4):589-600.

Ohno M, Largo DB, Ikumoto T (1994) Growth rate, carrageenan yield and gel properties of cultured kappa-carrageenan producing red alga Kappaphycus alvarezzi (Doty) Doty in the subtropical waters of Shikoku, Japan. *Journal of applied phycology* 6 (1):1-5.

Ozaki A, Uchiyama M, Tagaya H, Ohida T, Ogihara R (2007) The Japanese Centenarian Study: autonomy was associated with health practices as well as physical status. *J Am Geriatr Soc* 55 (1):95-101. doi:10.1111/j.1532-5415.2006.01019.x.

Pang SJ, Shan TF, Zhang ZH, Sun JZ (2008) Cultivation of the intertidal brown alga Hizikia fusiformis (Harvey) Okamura: mass production of zygote derived seedlings under commercial cultivation conditions, a case study experience. *Aquaculture Research* 39 (13):1408-1415.

Pickering TD, Skelton P, Sulu RJ (2007) Intentional introductions of commercially harvested alien seaweeds. *Botanica Marina* 50 (5/6):338-350.

Polne-Fuller M, Biniaminov M, Gibor A (1984) Vegetative propagation of *Porphyra* perforata. *Hydrobiologia* 116 (1):308-313.

Polne-Fuller M, Gibor A (1986) Calluses, cells, and protoplasts in studies towards genetic improvement of seaweeds. *Aquaculture* 57 (1-4):117-123.

Qi H, Rorrer GL (1995) Photolithotrophic cultivation of Laminaria saccharina gametophyte cells in a stirred tank bioreactor. *Biotechnology and bioengineering* 45 (3):251-260.

Qin S, Jiang P, Yu D, Li F, Sun G (2010) Transformation of seaweed utilizing biolistic gene transfer to spores. *US Patent,* 7,663,026 B2, Feb 16, 2010.

Raymundo Piñero E, Cadek M, Wachtler M, Béguin F (2011) Carbon Nanotubes as Nanotexturing Agents for High Power Supercapacitors Based on Seaweed Carbons. *ChemSusChem* 4 (7):943–949. doi:DOI:10.1002/cssc.201000376.

Reddy CRK, Jha B, Fujita Y, Ohno M (2008) Seaweed micropropagation techniques and their potentials: an overview. *Journal of applied phycology* 20 (5):609-617.

Reith M, Munholland J (1993) A high-resolution gene map of the chloroplast genome of the red alga *Porphyra* purpurea. *The Plant Cell Online* 5 (4):465.

Renn DW (1990) Seaweeds and biotechnology—inseparable companions. *Hydrobiologia* 204 (1):7-13.

Roberts JN, Buck CB, Thompson CD, Kines R, Bernardo M, Choyke PL, Lowy DR, Schiller JT (2007) Genital transmission of HPV in a mouse model is potentiated by nonoxynol-9 and inhibited by carrageenan. *Nature medicine* 13 (7):857-861.

Robertson-Andersson D, Leitao D, Bolton J, Anderson R, Njobeni A, Ruck K (2006) Can kelp extract (KELPAK®) be useful in seaweed mariculture? *Journal of applied phycology* 18 (3):315-321.

Rosenberg JN, Oyler GA, Wilkinson L, Betenbaugh MJ (2008) A green light for engineered algae: redirecting metabolism to fuel a biotechnology revolution. *Current opinion in biotechnology* 19 (5):430-436.

Santelices B (1999) A conceptual framework for marine agronomy*. *Hydrobiologia* 398: 15-23.

Sato M, Hosokawa T, Yamaguchi T, Nakano T, Muramoto K, Kahara T, Funayama K, Kobayashi A (2002) Angiotensin I-converting enzyme inhibitory peptides derived from wakame (Undaria pinnatifida) and their antihypertensive effect in spontaneously hypertensive rats. *Journal of agricultural and food chemistry* 50 (21):6245-6252.

Shimada S, Yokoyama N, Arai S, Hiraoka M Phylogeography of the genus *Ulva* (Ulvophyceae, Chlorophyta), with special reference to the Japanese freshwater and brackish taxa. In: *Nineteenth International Seaweed Symposium*, Kobe, 2009. Springer, pp 529-539.

Smit AJ (2004) Medicinal and pharmaceutical uses of seaweed natural products: a review. *Journal of applied phycology* 16 (4):245-262.

Smith S (1981) Marine macrophytes as a global carbon sink. *Science* 211 (4484):838.

Stachowicz JJ, Graham M, Bracken MES, Szoboszlai AI (2008) Diversity enhances cover and stability of seaweed assemblages: the role of heterogeneity and time. *Ecology* 89 (11):3008-3019.

Stiller JW, Waaland JR (1993) Molecular Analysis Reveals Cryptic Diversity in *Porphyra* (Rhodophyta) 1. *Journal of phycology* 29 (4):506-517.

Sudo Y, Araki A (2004) Research on the land culture of *Caulerpa lentillifera* using ocean deep water. *Examination on the cultivation effect of ocean deep water.* Okinawaken Kikaku Kaihatsubu Kaiyo Shinsosui Kenkyujo Kenkyu Gyomu Hokoku Heisei 14 Nendo Dai3go:76-81.

Thornber CS, Kinlan BP, Graham MH, Stachowicz JJ (2004) Population ecology of the invasive kelp *Undaria pinnatifida* in California: environmental and biological controls on demography. *Marine Ecology Progress Series* 268:69-80.

Troell M, Berg H (1997) Cage fish farming in the tropical Lake Kariba, Zimbabwe: impact and biogeochemical changes in sediment. *Aquaculture Research* 28 (7):527-544.

Trono GC (1999) Diversity of the seaweed flora of the Philippines and its utilization. *Hydrobiologia* 398:1-6.

Watanabe F, Takenaka S, Katsura H, Masumder SAMZH, Abe K, Tamura Y, Nakano Y (1999) Dried green and purple lavers (Nori) contain substantial amounts of biologically active vitamin B12 but less of dietary iodine relative to other edible seaweeds. *Journal of agricultural and food chemistry* 47 (6):2341-2343.

Watt J, Marcus R (1971) Carrageenan-induced ulceration of the large intestine in the guinea pig. *Gut* 12 (2):164.

Winterhalter P, Rouseff R Carotenoid-derived aroma compounds: an introduction. In: *ACS Symposium Series,* 2002. American Chemical Society, pp 1-19.

Wright J (2005) Differences between native and invasive Caulerpa taxifolia: a link between asexual fragmentation and abundance in invasive populations. *Marine Biology* 147 (2):559-569.

Xu Z, Dapeng L, Yiping Z, Xiaoyan Z, Zhaoling C, Wei C, Fan O (2002) Comparison of photobioreactors for cultivation of Undaria pinnatifida gametophytes. *Biotechnology letters* 24 (18):1499-1503.

Yan L, Jianzhou C, Xueyan S, Tao L, Qingli G (2006) The Application of DNA Molecular Marker Technique in Heritable Breeding of Laminaria. *Transactions of Oceanology and Limnology* 1 (107):75.

Yan X, Chuda Y, Suzuki M, Nagata T (1999) Fucoxanthin as the major antioxidant in Hijikia fusiformis, a common edible seaweed. *Biosci Biotechnol Biochem 63* (3):605-607.

Yokoyama H, Ishihi Y (2010) Bioindicator and biofilter function of *Ulva* spp.(Chlorophyta) for dissolved inorganic nitrogen discharged from a coastal fish farm-potential role in integrated multi-trophic aquaculture. *Aquaculture* 310 (1-2):74-83.

Zemke-White W, Ohno M (1999) World seaweed utilisation: An end-of-century summary. *Journal of Applied Phycology* 11 (4):369-376.

Zhang QS, Tang XX, Cong YZ, Qu SC, Luo SJ, Yang GP (2007) Breeding of an elite Laminaria variety 90-1 through inter-specific gametophyte crossing. *Journal of applied phycology* 19 (4):303-311.

Zhang X, Xu D, Mao Y, Li Y, Xue S, Zou J, Lian W, Liang C, Zhuang Z, Wang Q (2011) Settlement of vegetative fragments of Ulva prolifera confirmed as an important seed source for succession of a large-scale green tide bloom. *Limnology and oceanography* 56 (1):233-242.

Zhao FJ, McGrath SP, Meharg AA (2010) Arsenic as a food chain contaminant: mechanisms of plant uptake and metabolism and mitigation strategies. *Annual Review of Plant Biology* 61:535-559.

In: Seaweeds
Editor: Vitor H. Pomin

ISBN: 978-1-63117-571-8

Chapter 3

SEAWEED USES: HUMAN HEALTH AND AGRICULTURE

Maha Ahmed Mohamed Abdallah
National Institute of Oceanography and Fisheries,
Marine Pollution Lab, Alexandria, Egypt

ABSTRACT

Seaweeds are major coastal resources which are valuable to human consumption and environment. Also, it is a foodstuff that has been historically consumed around the globe but is only consumed in appreciable amounts in certain areas of the world today and it is also considered as a renewable natural resource with extensive distribution along the ocean coast. Biologically, seaweeds are categorized by their pigmentation, morphology, anatomy, and nutritional composition as brown (*Phaeophyta*), red (*Rhodophyta*) or green algae (*Chlorophyta*). The chemical composition of seaweeds varies with species, habitat, maturity and environmental conditions.

Seaweeds are valuable sources of protein, fiber, vitamins, polyunsaturated fatty acids, macro and trace elements, as well as important bioactive compounds. Thus, they have been recognized as being beneficial for human and animal health and may be used to the food industry as a source of ingredients with high nutritional value. Seaweeds also can be provided as a dietary alternative due to its nutritional value and its commercial value can be enhanced by improving the quality and expanding the range of seaweed – based products. In recent years, seaweeds have caused emerging interest in biomedicine and the food area, because they possess a wealth of bioactive compounds (such as sulfated polysaccharides, carotenoids, dietary fiber, protein, essential fatty acids, vitamins, minerals, terpenoids, oxylipins, phlorotannins and steroids) with potential industrial and agricultural applications.

INTRODUCTION

The marine environment representing approximately half of the global biodiversity is an enormous resource for new compounds. The huge diversity in terms of the number of different species of macro- and microalgae that coupled with the hostile environments in which these organisms live, make macro- and microalgae key targets for bioactive compound screening projects. The capacity of seaweed to produce compounds with valuable biological properties has been known for a long time and has been intensively exploited, especially for pharmaceutical purpose. In the last decades, increased demand for new drugs and food components based on natural products has placed seaweed at the center of large-scale screening programs and has led to the discovery of numerous new compounds. Algae have been the source of about 35% of newly discovered chemicals between 1977-1987 [1]. The diversity of the compounds produced by marine macroalgae is exceptional and covers a very wide range of biological activities [2]. Seaweeds have a wide range of mineral content, not found in edible land plants, and are related to factors such as seaweed phylum, geographical origin, seasonal environment and physiological variation. The mineral fraction of some seaweed even accounts for up to 40% of dry matter; however, in some cases the mineral content of some seaweed is recorded even higher than that of land plants and animal products [3]. Evaluation of minerals in any edible seaweed is important from both nutritional and toxicological point of view Abdallah 2012 book [4].

Seaweeds or marine algae are potentially prolific sources of highly bioactive primary and secondary metabolites, and their discovery has significantly expanded in the past three decades, that might represent useful leads in the development of new pharmaceutical agents. Many reports have been published about isolated compounds from algae with biological activity, demonstrating their ability to produce metabolites, with high complexity and unlimited diversity of pharmacological and/or biological properties [3]. The algae synthesize a variety of compounds such as carotenoids, terpenoids, xanthophylls, chlorophyll, vitamins, saturated and polyunsaturated fatty acids, amino acids, acetogenins, antioxidants such as polyphenols, alkaloids, halogenated compounds and polysaccharides such as agar, carrageenan, proteoglycans, alginate, laminaran, rhamnan sulfate, galactosyl glycerol, fucoidan, antimicrobial, anticancer and antiviral activity [4,5].

Antioxidants from natural sources are preferred by consumers [6] due to concerns on the toxic and carcinogenic effects of synthetic antioxidants [7]. The use of antioxidants is an effective way to minimize or prevent lipid oxidation in food products, retarding the formation of toxic oxidation products, maintaining nutritional quality and prolonging the shelf life of foods The antioxidants generally produced by plants include polyphenolic compounds such as flavonoids, cinnamic acid, benzoic acid, gallic acid, phlorotannins and quercetin [8]. Among the marine organisms, marine macroalgae or seaweed represents one of the richest sources of natural antioxidants and antimicrobials [9,10].

Brown seaweeds (e.g., *Sargassum fusiforme* and *Saccharina japonica*—formerly *Laminaria japonica*) have been used as Traditional Chinese Medicines in China for thousands of years [11]. An edible brown algae, *Sargassum naozhouense,* is widely distributed along the coasts of Zhanjiang, Guangdong province, China, is commonly consumed as a sea vegetable or as crude drugs for treating internal heat, infections, laryngitis and other ailments in locals [12].

Gracilaria Greville genus (Gracilariales, Rhodophyta) is a macroalgae group with more than 30 0 species of which 160 have been accepted taxonomically. *Gracilaria* species are important for the industrial and biotechnological uses because they have phycocolloids, the main source of agar α-(1,4)-3,6-anhydro-L-galactose and β-(1,3)-D-galactose with little esterification in cell wall [13].

The genus *Caulerpa* (class Bryopsidophceae) currently includes 86 species distributed in the tropical and warm-temperate waters of the world. Studies of biologival activities in *Caulerpa* species have led to the discovery of compounds with many different biological properties. An extraction of *Caulerpa* was shown to reduce hepatocarcinogenicity of aflatoxin, also a crude extract from *Caulerpa taxifolia* showed antiviral activity against the feline immunodeficiency virus [14]. A study of various species of *Caulerpa* (*C. racemosa, C. taxifolia, C. scalpelliformis, C. veravalensis* and *C. peltata*) showed high anticoagulant activity which was associated with their sugare, sulphate content and molecular weight of the extract [15].

Codium fragile (green seaweed) extract, showed antiviral activity by inactivation of virus particles [16]. Many reports have appeared describing the anticoagulant activity in Codium taxa [17]. Recent work showed the antigenotoxicity and antioxidative capacity of a crude extract of *Codium tomentosum* and the antibacterial activity of *C. decorticatum* [18].

The value of seaweed as fertilizer is not from the nitrogen, phosphorous, potash and organic matter but from trace elements and metabolites similar to plant growth regulators [19]. The plant growth hormone's effect of seaweed is advantageously made use of to stimulate germination and growth, thereby increasing the yield and resistance ability of many crops [20] and [21]. Seaweeds represent an alternative to conventional chemical fertilizers. Commercial use of liquid extracts, obtained from seaweeds is successfully used as foliar sprays for several crops [22]. Seaweeds show great promise as a source of seaweed liquid fertilizer [SLF] for raising food crops, it was found superior than chemical fertilizer because of the presence of high level of organic matter [23].

MINERALS

Minerals are essential elements required for human life, and typically exclude the organic building blocks of carbon, hydrogen, nitrogen and oxygen [24]. They cannot be made by the body and must therefore be obtained in the diet in adequate amounts required for proper health. Minerals are distinct from vitamins in that they are not organic molecules comprised of multiple atoms. They are classified as major minerals (daily requirement of at least 100 mg), trace minerals (daily requirement of less than 100 mg), and ultra trace minerals (typically 1 μg or less per day) [25]. Minerals are naturally present in food or may be added through food fortification or dietary supplements. Seaweeds are an extremely valuable source of a huge range of essential nutrients that our bodies must have to be healthy and function properly in terms of fighting disease and vitality. The body uses approximately 56 minerals to live and survive healthily. The seaweeds show great variation in the nutrient contents, which are related to several environmental factors as water temperature; salinity; light; nutrients; hydrochemical characteristics of ambient waters; the depth of occurrence; the morphology and structure of thalli [26]. Most of the environmental parameters vary according to season and the changes in ecological conditions can stimulate or inhibit the biosynthesis of several nutrients [27]. Seaweeds are a rich source of minerals, especially macro and micronutrients necessary for human nutrition; include boron, calcium, chloride, chromium, cobalt, copper, fluoride, iodine, iron, lithium, magnesium, manganese, molybdenum, potassium, phosphorus, selenium, silicon, sodium, sulfur and zinc, however, the nutritional properties of seaweeds are usually determined from their biochemical composition alone viz. proteins, carbohydrates, vitamins, amino acids, etc. [28].

The essential minerals and trace elements needed for human nutrition are present in seaweeds, oftentimes mineral content is generally high (8–40%), this is because seaweeds accumulate metal ions from salt water and concentrate those substances as carbonate salts in their fronds [29,30]. This wide range in mineral content, not found in edible land plants, is related to factors such as seaweed phyllum, geographical origin, and seasonal, environmental and physiological variations [29].

Important minerals, such as calcium, iron and copper accumulate in seaweeds at much higher levels than in many well-known terrestrial sources of minerals in terrestrial foodstuffs. The most common minerals found in seaweed are iodine, magnesium, calcium, phosphorus, iron, potassium copper and fluoride [31]. Minerals are very important for the biochemical reaction in the body as a co-factor of enzyme. For examples, Ca, P, and Mg build and maintain bones and teeth, whereas, Na and K help maintain balance of water, acids and bases in fluids outside of cells, an involve in acid-base balance and transfer of nutrients in and out of individual cells, respectively [31]. Researchers examined the mineral content in seaweeds harvested from Japanese beaches. The highest concentrations of potassium (2.71 g/L), magnesium (0.19 g/L) and calcium (0.16 g/L) ions were detected in extract from *Sargassum ringgoldianum* subsp. *coreanum*. *Codium fragile* was shown to be a rich source of sodium ions (1.21 g/L) [32]. *Kappaphycus alvarezii* contains high levels of magnesium and calcium ions [33]. Their concentrations in extracts were 581.20 mg/L and 460.11 mg/L, respectively. Mineral content has been shown to vary according to seaweed species, oceanic residence time and method of mineralization [34].

Undaria pinnatifida (Wakame) and *Laminaria digitata japonica* (Kombu) are an excellent source of calcium, magnesium and phosphorus. The benefit of daily intake of Wakame (3.62 g d.w.) and Kombu (3.70 g d.w.) in daily recommended doses of macrominerals such as calcium, magnesium and phosphorus [35].

The mineral composition of *Porphyra vietnamensis* was found relatively higher as compared to the land vegetables as well as to other edible seaweeds and it is in concurrence with the recent macrobiotic recommendation for western countries. It could therefore be used as food supplement as a spice to improve the nutritive value in the omnivorous diet. The selected micro nutrients (Fe + Zn + Mn + Cu) were found to be higher (45.5–309 mg/100 g d wt) than any of the land vegetables [36]. Figure (1) showed some kinds of marine seaweeds have high mineral contents.

The mineral content was also determined in several brown (*Fucus vesiculosus, Laminaria digitata, Undaria pinnatifida*) and red (*Chondrus crispus, Porphyra tenera*) edible marine sea vegetables ashes by [37]. The analysis of the ashes indicated that marine seaweeds contained higher amounts of both macrominerals (8.083–17,875 mg/100g;Na, K, Ca, Mg) and trace elements (5.1–15.2 mg/100 g; Fe, Zn, Mn, Cu), than those reported for edible land plants. Edible brown and red seaweeds could be used as a food supplement to help meet the recommended daily intake of some essential minerals and trace.

The green algae *Caulerpa lentillifera* was also rich in phosphorus, calcium, magnesium and copper while *Ulva reticulata* was rich in potassium, manganese and ferrous. Similar to other edible seaweed, *Gracilaria changgi* was reported to contain high level of calcium (651mg/100g d.w), zinc (13.8 mg/100g d.w), ferrous (95.6 mg/100g d.w) and copper (0.8 mg/100g d.w) [38].

Kappaphycus alvarezii Sargassum ringgoldianum

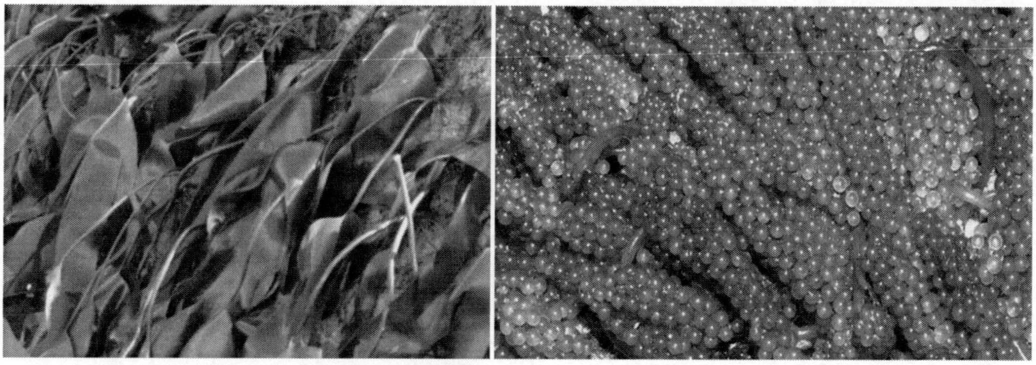

Laminaria digitata japonica Caulerpa lentillifera

Undaria pinnatifida Porphyra vietnamensis

Figure 1. Marine seaweeds have high mineral contents.

Table (1) showed mineral composition of different species of seaweeds from different sites in the world.

Table 1. Mineral composition of different marine seaweeds in μg/g

	Ca	K	Mg	Na	Cu	Fe	I	Zn
Chlorophyta								
Ulva lactuca[39]	3216 ±69.6	6500 ±86.6	13500 ±11.4	9250 ±16.1	3.4±0.07	1197±24.2	-	31.9±1.6
Ulva reticulata[40]	1400	15400	1400	-	6000	1748	11240	33
Enteromorpha compressa[41]	18900±11.1	14500±17.3	3500±7.7	744.1±7.3	13.05±0.64	1395.6±19.6	-	39.9±5.7
Caulerpa racemosa[39]	6513 ±23.6	9700 ±91.6	2040±15.4	1441 0±30.1	4.0±0.07	5388±12.5	-	15.2±1.1
Caulerpa lentillifera[40]	7800	9700	6300	-	22000	93	14240	26
Codium fragile[39]	1503 ±19.3	1073±41.6	1014 ±65.8	2948 ±26.1	2.1±0.1	185.8±8.2	-	6.3±0.17
Rhodophyta								
Hypnae musciformis[42]	10521±26.6	3088±35.2	1081±22.9	8355±87.3	6.75±0.11	1314.5±18.2	-	26.85±0.78
Pterocladia capillace[39]	7504 ±9.3	7500 ±39.6	17069 ±26.8	8550 ±21.1	2.1±0.1	568±13.4	-	18.1±1.02
Gracilaria changgi[38]	6510±5.2	-	-	-	8.0±0.1	956±3.7	-	138±4.1
Porphyra vietnamensis[36]	33.5±0.84	24.9±0.64	49.5±1.01	5230±3.2	8.3±0.02	1370±0.99	-	19.5±0.01
Porphyra tenera[37]	3900±17	35000±71	3900±17	36270±115	<5	103±0.41	-	22.1±0.17
Chondrus crispus[37]	4200±22	31840±0	7320±6	42700±62	<5	39.7±0.11	-	71.4±0.13
Phaeophyta								
Undaria pinnatifida[35]	950 ± 30	5691 ± 215	405 ± 10	6494 ± 254	0.185 ± 0.016	1.54 ± 0.07	26 ± 2.4	0.944 ± 0.038
Laminaria digitata japonica[35]	880 ± 20	5951 ± 305	550 ± 15	2532 ± 120	0.247 ± 0.076	1.19 ± 0.03	170 ± 5.5	0.886 ± 0.330
Colpomenia sinuosa[39]	10312±77.2	5400±18.1	9680±22.6	1441±30.2	6.3±0.09	1029±15.3	-	21.4±0.14
Fucus vesiculosus[37]	9380±7	43220±46	9940±13	54690±60	<5	42.0±0.17	-	37.1±0.37

[35] Kolb et al (2004) [36] Rao, et al (2007) [37] Rupe´rez (2002) [38]Norziah and Ching (2000) [39]Abdallah (2008) [40]Ratana-arporn and Chirapart (2006)[41]Abdallah (2007) [42]Formuzulhaque et al (2009)

PROTEIN

Seaweeds are valuable sources of protein, fiber, vitamins, polyunsaturated fatty acids, macro and trace elements, non-starch polysaccharides, minerals as well as important bioactive compounds [29]. As seaweed polysaccharides cannot be entirely digested by human, they are regarded as a new source of dietary fiber and food ingredient. Thus, they have been recognized as being beneficial for human and animal health [43]. Together with their low lipid content, seaweeds only provide a very low amount of energy. Consumption of seaweeds can increase the intake of dietary fiber and lower the occurrence of some chronic diseases [44].

Proteins which are the building blocks of living world also have a hidden capacity of modulating physiology by means of small peptides which are formed either during the processing or during *in vivo* digestion. Small peptides with biological function or physiological effects are called bioactive peptides. Active peptides are potential modulator of various regulatory processes in the living system. All types of seaweeds are sources of sufficient protein. Furthermore, the mean of protein content in some seaweed was higher than the concentrations found in higher plants [45].

The protein fraction of brown seaweeds is low (3-15% of the dry weight) compared with that of the green (20% of the dry weight) or red seaweeds (30-40% of the dry weight).Proteins are composed of various amino acids and their nutritional quality can be evaluated against the recommended amino acid pattern oftentimes, the protein content of marine algae depends on the seasonal period and species [46, 47]. An annual monitoring of protein level from *Palmaria palmata* collected from the French Atlantic coast showed that the protein content of this algae can vary between 9 and 25% (dry weight) [48]. The protein levels of *Ulva spp.* are in the range 15-20 % of the dry weight. Except for *Undaria pinnatifida*, which contains 11-24 % proteins, other brown seaweeds (*Laminaria digitata, Ascophyllum nodosum, Fucus vesiculosus and Himanthaliae longata*) have low protein content. Table 2 illustrated protein contents of different seaweed species reported by various authors.

However, for these seaweeds, the high phenolic content might limit protein availability in vivo and thus moderate in vitro figures. This situation is probably not found for the green and red seaweeds, which possess low levels of phenols and higher protein content [56].

For most seaweed species, aspartic and glutamic acids constitute a large part of the amino acid make-up of these proteins. Aspartic and glutamic acids are highest in brown seaweeds, with red seaweeds having lower amounts of these two amino acids. Essential amino acids such as histidine, leucine, isoleucine, and valine are present in many types of seaweed, such as *Palmaria palmata* and *Ulva*s pp. The bioavailability of seaweed proteins can sometimes be inhibited by the entrapped nature of the proteins in the cellular matrix. Increasing the bioavailability by using physical processes or fermentation to break down the fibers and liberate more protein has been studied, particularly for *Palmaria palmate* as has the use of digestive enzymes [57, 58].

Red seaweeds contain a group of proteins called biliproteins, some of which are extracted for their valuable use as fluorescent markers. Phycoethrin is a major light-harvesting pigment in red algae and is regularly used as a fluorescent probe in scientific experiments. The other members of this group of proteins known as phycobiliproteins are phycocyanins and all phycocyanins. These proteins are assembled into a structure in the cell that absorbs light where chlorophyll absorbs poorly, i.e., the light wavelength from 450 nm to 650 nm [59]. This allows some red seaweed to survive in relatively deep water, depending on opacity and other conditions.

Table 2. Protein contents of different seaweed species reported by various authors

Species	Protein content
Ulva rigida[49]	6.4
Ulva lactuca[51]	7.06
Ulva intestinalis[54]	17.9
Cauler paracemosa[50]	3.98
Caule rpalentillifera[40]	12.49
Codium fragile[39]	18.57
Codium isthmocladum[50]	3.50
Gelidium pristoides[49]	11.80
Gracilaria cornea[50]	5.47
Gracilaria cervicornis[53]	19.70
Hypnea japonica[51]	19.00
Pterocladia capillacea[39]	14.37
Prophyra tenera[52]	34.20
Chnoospora minima[55]	11.3
Sargassum vulgare[53]	13.61
Sargassum filipendula[50]	8.72
Padina pavonia[39]	12.69
Padina gymnospora[55]	10.5
Colpomenia Sinuosa[39]	42.0

Amino Acids

Scientists are studied frequently amino acid composition of seaweeds and compared it with that of other earth plants. Aspartic and glutamic acids constitute together a large part of amino acid fractions in most marine seaweeds, these two amino acids represents between 22% and 44% of the total amino acids [60]. There were some pronounced differences between the amino acid profiles of Rhodophyceae (red seaweeds) and Chlorophyceae (green seaweeds). Previous results revealed that the sulfur-containing amino acids of the red seaweeds were higher than that of green seaweeds [61].

The essential amino acids (EAAs) in algae are available throughout the year although seasonal variations in their concentrations are known to occur and represent almost a half of total amino acids and their protein profile is close to the profile of egg protein [62]. In case of non-EAAs, all three groups (green, brown, and red seaweeds) contain the similar amount. Red seaweed seems to be a good source of protein because its value reaches 47%. The issue of protein malnutrition supports the trend to find a new and cheap alternative source of protein. Seaweed could play an important role in the above-mentioned challenge because of relatively high content of nitrogen compounds. Algae may be used in the industry as a source of ingredients with high nutritional quality.

Interestingly, taurine is not a typical component of traditional European food and the taurine content represents a nutrient feature which is characteristic of red algae varieties. The abundance of taurine in red algae varieties and in *Porphyra sp.* from Japan and Korea amounted to 4 g Tau/16 g N. This concentration was significantly higher than those observed in red algae from China. Taurine was detected at low concentrations in brown algae varieties. However brown algae contained phosphoserine (P-Ser) and the content of this amino acid was significantly higher than those in red algae varieties [63].

For essential amino acids, the *Hypnea charoides*, *Hypnea japonica* and *Ulva lactuca* seaweed were rich in leucine, valine and threonine. Also, the limiting amino acids of the *Hypnea* seaweed protein concentrations were the sulphur-containing amino acids and lysine. The methionine and cystine levels of the Hypnea species in other study were above the FAO/WHO requirement (EAA score ranged from 1.90 to 1.95) With respect to the total essential amino acid in the FAO/WHO requirement pattern, all red and green seaweeds seemed to be able to contribute adequate levels of total essential amino acid [64]. According to [29], the high levels of aspartic and glutamic acids were responsible for the special flavor and taste of the seaweeds.

All of the essential amino acids i.e. methionine, leucine, lysine, cysteine, phenylalanine, tyrosine, arginine, iso-leucine, threonine and valine and six non-essential amino acids were found to be present in *Gracilaria changgii*. Among the essential amino acids assayed, lysine with a chemical score of 53% appeared to be the most limiting when compared with the essential amino acid pattern of egg protein. The total amino acids in *Gracilaria changgii* was 602 μmol amino acid/g sample (dry weight). From this total, 252 μmol amino acid/g sample is made up of essential amino acids. The ratio of essential amino acids to total amino acid is 0.4, i.e., almost half of the amino acid in *Gracilaria changgii* consist of essential amino acids [45].

The proteins of *Ulva lactuca* (green algae) contained a high level of amino acids, especially essential amino acids; valine, leucine, lysine and threonine. Furthermore, *Ulva lactuca* seaweed contained all the essential amino acids (in different proportions, excluding tryptophan), which accounted for 42% of the total amino acid content, the main limiting amino acid in *Ulva* sp.is isoleucine, followed by leucine. The non-essential amino acids, such as aspartic and glutamic acid constituted a substantial amount of the total amino acids (26% of total amino acid content) for the U. lactuca seaweeds [65]. *Ulva pertusa* was rich in leucine, arginine and valine whereas *Ulva intestinalis* was rich in leucine, threonine, valine, and arginine. The non-EAAs, which are histidine, aspartic acid, glutamic acid, serine, proline, glycine, and alanine, were present in relatively high levels. The two species contained a large amount of aspartic and glutamic acids which gave the special flavors and tastes [54].

In brown seaweed, total amino acid content is 47.1 % in *Undaria pinnatifida* and 50.7 % in *Laminaria digitata japonica,* the both contains all essential amino acids .The first limiting amino acid in the two algae is tryptophan. The amino acid scores Lysine- Methionine + Cystine, Lysine- Methionine+ Cystine-Trp and Lysine - Methionine + Cystine - Tryptophan - Threonine are in both two algae very high as well as the essential amino acid index (EAAI). EAAI is calculated as the geometric mean value of the ratios for 8 essential amino acids, and the EAAI above 100 is considered [35].

IODINE

Marine algae have long been known as an important source of iodine. Seaweed can enrich iodine due to a higher concentration factor compared with terrestrial plants. In algae, the majority of the iodine is bound with protein, part of it with pigment and polyphenol, and just a small part with polysaccharides, such as algin, fucoidan and cellulose [66]. Iodine content in seaweed varied widely depending on species, harvest location and preparation, and is typically highest in fresh cut blades and lowest in sun bleached blades [67]. Seaweeds can have 10–100 times higher mineral content, such as iodine, than land-based vegetables [68, 69]. Furthermore, concentrations of inorganic elements in marine algae, including iodine, exhibit seasonal variations. Lower concentrations in the initial and later growth stages and the highest concentrations in the middle growth stage have been shown [70]. Accordingly, Teas *et al.* [67] found about two-fold higher iodine content in growing juvenile compared with adult seaweed, with higher iodine concentrations in the meristematic tissue, the growing area at the base of the seaweed blade, than in the stripe. After the discontinuation of growth, iodine content rapidly diminishes, as shown by the markedly decreased iodine concentrations in sun-bleached blades collected from the beach or from floating drifts of seaweed. Postharvest storage conditions are additional factors that might affect the iodine content in marine algae. When seaweeds were stored in watertight bags or boxes, their iodine concentration was not affected, whereas storage in open containers or in paper bags significantly diminished iodine content, in particular under humid conditions [71]. The latter finding might be based on the high degree of water solubility of iodine in seaweed, ranging from 40% in the red alga *Sargassum kjellmanianum* to 99.2% in the brown alga *L. japonica* [72]. The high water solubility of iodine in seaweed also has an impact on food preparation and cooking. In *L. japonica*, 99% of the iodine has been found in water after 15 min of boiling [73]. Marine algae not only differ in iodine content, but also in the chemical species of iodine.

Brown algae, such as *Laminaria japonica* with mean iodine content of 2288 μg/g, were found to have higher iodine concentrations than red algae, including *Porphyra sp.* with an average of 33.8 μg/g iodine, and green algae, such as *Ulva lactuca* containing an average of 95 μg/g iodine. Different iodine concentrations reported for one species harvested in diverse regions indicate a geographic variation in iodine content. For instance, iodine concentrations in *Undaria pinnatifida* were 24-fold higher in samples collected in China compared with the mean value of samples from Tasmania, Japan and New Zealand. Causative factors for these differences might be salinity and coldness of the water, depth of the seaweed and distance from the equator [74].

In accordance with the remarkably different bioavailability of various chemical species of iodine, Aquaron *et al.* [75] found differences in the iodine bioavailability of diverse marine algae in humans. The chemical species of iodine were 80% iodide in *Laminaria hyperborean* and 80% organic iodine in *Gracilaria verrucosa.* This discrepancy between mineral and organic iodine bioavailability in red and brown seaweeds may be explained by the different nature of the carbohydrates in red and brown algae.

The brown macroalgae *Laminaria digitata* and *Fucusserratus*, and the red macroalgae *Kallymenia antarctica*, have been shown to release I^-. *Laminaria digitata* was also found to release some oxidized iodine, consistent with the release of I_2 reported by [76]. The amount of I^- released by *Laminaria digitata* is increased by simulating oxidative stress caused by attacks from alginate-degrading organisms such as bacteria and molluscs, supporting the theory that it has an antioxidant function [77]. Meanwhile, green macroalgae of the genus *Ulva* did not appear to influence dissolved inorganic iodine speciation. The influence of macroalgae on dissolved iodine speciation varied not between species, but also between specimens of the same species, presumably depending on origin, age and season [78].

ANTIOXIDANTS

Antioxidants are secondary metabolites that prevent oxidation by transforming free radicals or peroxy radicals into non-radicals by donating electrons and hydrogen, chelating transition metals, and dissolving generated peroxidation compounds. Algae are an important source of various bioactive compounds such as antioxidants, antimicrobials and antivirals [79]. These compounds are also important for protecting the algal cells against stressful conditions, e.g. ultraviolet radiation, temperature change and fluctuation in nutrient and salinity level. To enable rapid adaptation to new environmental conditions, algae produce a great variety of secondary metabolites that cannot be found in other organisms [80]. Bioactive compounds from macroalgae as fucoidans and phlorotannins (algal polyphenols) have been shown to have antiviral, antibacterial and antioxidative properties [81]. Strong antioxidant activities of marine macroalgae have been reported in Table 3, whereas those of freshwater macroalgae have been rarely investigated.

Table 3. Major groups of antioxidant compounds in macroalgae with specific examples and potential algal sources (from Cornish and Garbary, 2010) [82]

Antioxidant compounds	Example compounds	Algal source
Carotenoid	β-carotene Fucoxanthin Antheraxathin, lutein, violaxanthin	*Chondru scrispus*[83] Brown algae [84] Red algae [85]
Phenolic compounds	Stypodiol, isoepitaondiol, taondtol	*Taonia atomaria*[86]
Phycobilin pigments	Phycoerythrine, phycocyanin	Red algae in general [87,88]
Polyphenols	Catechin, epicatechin Falvonoids Phlorotannins	*Halameda sp.*[89] *Palmaria palmate* [90] *Sargassum pallidum*[91] *Fucus vesiculosus*[92]
Sulphated polysaccharides	Fucoidan, alginic acid, laminaran Fucodan Sulphated galactans Galactans Sulphated glycosaminoglycan Porphyran	*Turbinaria conoides*[93] *Laminaria japonica* [94] Some red algae [95] Most red algae [96] *Sargassum swightii*[97] *Prophyra sp.* [98]
Vitamins	Ascorbate Vitamin A	*Chondru scrispus*[99] *Kappaphycus alvarezii*[100]

A high correlation between the total phenolic content and antioxidant activity has been reported by many researchers. Seaweed polyphenols such as phlorotannins have been reported to scavenge free radicals, superoxide radicals [101], peroxyl radical, chelate ferrous ions and nitric oxide [102]. Phlorotannins such as eckol, dieckol, phlorofucofuroeckol A and 8,8′-bieckol isolated from the Japanese brown algae Eiseniabicyclis, Ecklonia cava and Eckloniakurome have shown 2–10 times more antioxidant activities as compared to catechin, α-tocopherol and ascorbic acid [103].

The antioxidant activity (total antioxidant and free radical scavenging activities) of *Porphyra sp., Laminaria sp., Undaria sp. and Hijikia sp.* have been reported for the water and ethanol extracts. In water extract, *Laminaria* sp. showed the highest total antioxidant activity of 63% compared with other samples. *Laminaria* sp., *Porphyra sp.* and *Hijikia sp.* exhibited higher radical scavenging activity. *Undaria* sp. exhibited the highest antioxidant and free radical scavenging activities in ethanolic extract with 58% and EC50 = 0.42 mg/ml respectively [104]. A water extract of *Scytosi phonlomentaria* showed strong antioxidant activity in linoleic acid peroxidation assay (about 22 mg catechin equivalents/g dry sample) [101]. Cho et al. (2007) [105] reported that ethanol extracts of *Sargassum siliquastrum* exhibited a 95% scavenging effect of DPPH radicals at 0.5 mg/ml or greater.

Four marine algae, *Cauler paracemosa*, *Gracilaria tenuistipitata*, *Sargassum* sp. and *Ulva lactuca* were extracted by boiling dried seaweed and by autoclaving each sample. The boiling extracts of the seaweeds, except *C. racemosa*, were found to have higher total phenolic contents (TPC) than those obtained from the autoclave method. The antioxidant results also showed that $O2^{\bullet-}$ scavenging activity existed only in the boiling extracts of *C. racemosa*, *G. tenuistipitata*, and *U. lactuca*. In other evidence the brown seaweed, *Sargassum* sp., showed the significantly higher TPC and antioxidant activities than the red and green algae used in this study [106].

The green seaweeds *Cauler palentillifera* and *Cauler paracemosa* and the brown seaweed *Sargassum polycystum* showed better radical-scavenging and reducing power ability, and higher phenolic content. These seaweeds could be potential rich sources of natural antioxidants [107].

Polysiphonia fucoides and the *Fucus* species (brown algae) showed highest radical scavenging activity, reducing power, inhibition of oxidation in liposome model system and in fish oil and were high in phenolic content. These seaweeds could be potential rich sources of natural antioxidants for protection of foods against oxidation [108]. Lee et al (2012) [109] found that Octaphlorethol A (OPA), a new phenolic compound isolated from the brown alga, *Ishige foliacea*, increases Glut4-mediated glucose uptake by activating PI3-K/Akt and the AMPK signaling pathway in skeletal muscle cells, a novel target for the treatment of type 2 diabetes. OPA can be further developed as a potential anti-diabetic therapy.

The complex polysaccharides from the brown, red and green seaweeds possess broad spectrum therapeutic properties. Especially, the sulfated polysaccharides, viz. fucans, carrageenans and ulvans have exhibited strong antioxidant, antitumor, immunostimulatory, anti-inflammatory, pulmonary fibrosis anticoagulant/antithrombotic, lipid lowering, antiviral, antibacterial, antiprotozoan, hyperplasia prevention, gastrointestinal, regenerative and nano medicine applications. Fucans, a family of sulfated polysaccharides present in brown seaweed, have several biological activities. Their use as drugs would offer the advantage of no potential risk of contamination with viruses or particles such as prions. A fucan prepared from *Spatoglossum schröederi* was tested as a possible inhibitor of cell-matrix interactions using wild-type [110].

Fertilizers and Soil Conditioners

Algae and consequently their extracts can be the treasure trove of biologically active compounds. Their beneficial properties for humans, animals and plants were recognized in the past and are appreciated nowadays, in the development of new biotechnological products. Products with functional properties containing organic compounds derived from natural sources, rather than being a product of heavy organic synthesis are increasingly demanded by consumers. Currently, recipients of products have higher demands, expecting that they will be on one hand "bio" and "organic", but on the other - will be a concentrate of compounds with desired properties. This poses a particular challenge for modern biotechnology [111].

People lives near the coastline using seaweeds, especially the large brown seaweeds, to fertilize nearby land. For example in a more tropical climate like the Philippines, large quantities of *Sargassum* have been collected, used wet locally, but also sun dried and transported to other areas. Part of this algal mass has been composted and then used in trials for growing tomato plants in various types of soil. In all cases, the addition of the compost increased water holding capacity and plant growth, so composting simultaneously solved environmental pollution problems and produced a useful organic fertilizer. Species of *Ascophyllum*, *Ecklonia* and *Fucus* are the common ones. They are sold as soil additives and function as both fertilizer and soil conditioner. They have a suitable content of nitrogen and potassium, but are much lower in phosphorus than traditional animal manures and the typical N:P:K ratios in chemical fertilizers. The large amounts of insoluble carbohydrates in brown seaweeds act as soil conditioners (improve aeration and soil structure, especially in clay soils) and have good moisture retention properties. Their effectiveness as fertilizers is also sometimes attributed to the trace elements they contain, but the actual contribution they make is very small compared to normal plant requirements [112].

Liquid extracts obtained from seaweeds are successfully used as foliar sprays for several crops. The value of seaweeds as fertilizers was not only due to nitrogen, phosphorus and potash content, but also because of the presence of trace elements and metabolites [19]. Seaweed liquid fertilizer was found superior than chemical fertilizer because of the presence of high level of organic matter. The amount of different macro and micro elements present in *Ulva lactuca* extract was found higher. The commercial seaweed extracts obtained from *laminariaceae* and *fucaceae* contained the plant growth regulators (PGRs) like cytokinin. The concentration of PGRs namely, auxin and cytokinin were found greater in *Ulva lactuca* [113]. Seaweed extracts are now widely accepted in the horticultural industry. When applied to fruit, vegetable and flower crops, some improvements have included higher yields, increased uptake of soil nutrients, increased resistance to some pests such as red spider mite and aphids, improved seed germination, and more resistance to frost. There have been many, many controlled studies to show the value of using seaweed extracts, with mixed results.

The red alga *Chondracanthus squarrulosus* was cultured under semi-controlled conditions to valuate growth (biomass production) with agricultural fertilizers (ammonium nitrate, ammonium sulphate and urea) vs. analytical grade inorganic salts; sodium nitrate (analytical grade) and seawater were used as controls [114].

Finally there is the question, are seaweed extracts an economically attractive alternative to NPK fertilizers? Perhaps not when used on their own, but when used with NPK fertilizers they improve the effectiveness of the fertilizers, so less can be used, with a lowering of costs. Then there are always those who prefer an "organic" or "natural" fertilizer, especially in horticulture, so seaweed extracts probably have a bright future.

REFERENCES

[1] Smit AJ . Medicinal and pharmaceutical uses of seaweed natural products: A review. *J. Appl. Phycol.* 2004,16: 245-262.

[2] Holdt, S and Kraan, S. Bioactive compounds in seaweed; functional food applications and legislation. Review Article.2011, *Journal of Applied Phycology,* 23, 543-597.

[3] Chanda, S.; Dave, R.; Kaneria, M.; Nagani ,K., Seaweeds: A novel, untapped source of drugs from sea to combat Infectious diseases. Current Research. *Technology and Education Topics in Applied Microbiology and Microbial Biotechnology*. 2010.

[4] Cardozo, K.H.M.; Guaratini, T.; Barros, M.P.; Falcão, V.R.; Tonon, A.P.; Lopes, N.P.; Campos, S.; Torres, M.A.; Souza, A.O.; Colepicolo, P.; *et al.* Metabolites from algae with economical impact. *Comp. Biochem. Physiol. C: Comp. Pharmacol.* 2006, *146*, 60–78.

[5] Paniagua-Michel, J.; Capa-Robles, W.; Olmos-Soto, J.; Gutierrez-Millan, L.E. The carotenogenesis pathway via the isoprenoid-beta-carotene interference approach in a new strain of *Dunaliella salina* isolated from Baja California Mexico. *Mar. Drugs* 2009, *7*, 45–56.

[6] Kranl K, Schlesier K, Bitsch R, Hermann H, Rohe M, Böhm V (2005) Comparing antioxidative food additives and secondary plant products - use of different assays. *Food Chem.* 93:171–175.

[7] Safer AM, Al-Nughamish AJ (1999) Hepatotoxicity induced by the antioxidant food additive butylated hydroxytoluene (BHT) in rats: An electron microscopical study. *Histol.Histopathol.* 14:391–406.

[8] Krishnaiah, D., Sarbatly, R., & Bono, A. (2007). Phytochemical antioxidants for health and medicine—A move towards nature. *Biotechnology and Molecular Biology Review*,1, 97–104.

[9] Barrow, C., & Shahidi, F. (2008). Marine nutraceuticals and functional foods. New York: Cox, S., Abu-Ghannam, N., & Gupta, S. (2010). An assessment of the antioxidant and antimicrobial activity of six species of edible Irish seaweeds. *International Food Research Journal*, 17, 205–220.

[10] Taskin, E., Ozturk, M., Taskin, E., & Kurt, O. (2007). Antibacterial activities of some marine algae from the Aegean Sea (Turkey). *African Journal of Biotechnology*, 6, 2746–2751.

[11] O'Sullivan, L.; Murphy, B.; McLoughlin, P.; Duggan, P.; Lawlor, P.G.; Hughes, H.; Gardiner, G.E. Prebiotics from marine macroalgae for human and animal health application. *Mar. Drugs* 2010, *8*, 2038–2064.

[12] Wang J., Zhang Q., Zhang Z., Zhang H., Niu X. Structural studies on a novel fucogalactan sulfated from the brown seaweed *Laminaria japonica*. *Int. J. Biol. Macromol.* 2010;47:126–131. doi: 10.1016/j.ijbiomac.2010.05.010.
[13] Wang B., Huang H., Xiong H.P., Xie E.Y., Li Z.M. Analysis on nutrition constituents of *Sargas sumnaoz houense* sp. nov. *Food Res. Dev*. 2010;31:195–197.
[14] Nicoletti E, Della Pietà F, *et al.*(1999). Antiviral properties of a crude extract from a green alga Caulerpataxifolia (Vahl) C. *Agardh. Phytother Res* 13 (3): 245-7.
[15] Shanmugam M, Mody KH (2000). Heparinoid-active sulphated polysaccharides from marine algae as potential blood anticoagulant agents. *Curr. Sci.* 79:1672-1683.
[16] Hudson J B, Kim JH, Lee MK, DeWreede RE, Hong YK (1999) Antiviral compounds in extracts of Korean seaweeds: Evidence for multiple activities. *Journal of Applied Phycology* 10: 427-434.
[17] Matsubara, K.; Matsubara, Y.; Basic, A.; Liao, M. L. and Hori, K. (2001), Anticoagulant properties of a sulfated galactan preparation from a marine green alga *Codiumcylindricum. Inter. J. Biol. Macromol.*, 28: 395-399.
[18] Sunilson, A.J., Suraj, R., Anandarajagopal, K., Rejitha, G., Vignesh, M., Promwichit, P. 2009. Preliminary Phytochemical Analysis, Elemental Determination and Antibacterial Screening of Codium decorticatum-A Marine Green Algae. *International Journal of Biological Chemistry* 3: 84-89.
[19] Booth, E. 1969. The manufacture and properties of liquid seaweed extracts. *Proc. Int. Seaweed Symp.,*6 : 622-655.
[20] Moller, M., Smith, M.L. The significance of the mineral component of seaweed suspensions on lettuce (*Lactuca sativa* L.) seedling growth. *J. Plant Physiol.*1998, 153, 658 – 663.
[21] Verkleij, F.N. *A Review of Biological Agriculture and Horticulture*. 1992, 8: 309-324.
[22] Bokil, K.K., Mehta, V.C. and Datar, D.S. Seaweeds as manure: II pot culture manorial experiments on wheat, *Phykos.*1974, 13, 1–5.
[23] Aitken, J.B. and Senn, T.L. 1965. Seaweed products as a fertilizer and soil conditioner for horticultural crops. *Bot.Mar.,*8 : 144-148.
[24] Institute of Medicine, Food and Nutrition Board. Dietary Reference Intakes for Vitamin A, Vitamin K, Arsenic, Boron, Chromium, Copper, Iodine, Iron, Manganese, Molybdenum, Nickel, Silicon, Vanadium, and Zinc. Washington: National Academy Press; 2001.
[25] Offenbacher EG, Pi-Sunyer FX, Stoecker BJ. Chromium. In: O'Dell BL, Sunde RA, eds. Handbook of Nutritionally Essential Mineral Elements. New York: Marcel Dekker; 1997:389.
[26] Dawes, C.J., 1998. Marine Botany. John Wiley & Sons, Inc., New York, p. 480.
[27] Lobban, C.S., Harrison, P.J., Duncan, M.J., 1985. The Physiological Ecology of Seaweeds. Cambridge University Press, Cambridge.
[28] Darcy-Vrillon, B., 1993. Nutritional aspects of the enveloping use of marine macro algae for the human industry. *International Journal of Food Science Nutrition*, 44: 23-35.
[29] Mabeau, S., Fleurence, J., 1993. Seaweed in food products: biochemical and nutritional aspects. *Trends in Food Science and Technology* 4,103–107.

[30] Ortega-Calvo, J.J., Mazuelos, C.,Hermosı´n,B., &Sa´iz-Jime´nez,C.(1993).Chemical composition of Spirulina and Eukaryotic algae food products marketed in Spain. *Journal of Applied Phycology*, 5, 425–435.

[31] Ensminger AH, Ensminger ME, Konlande JE, RobsonJRK. 1995. The Concisen Encyclopedia of Foods and Nutrition. CRC Press, Boca Raton Florida.

[32] Kuda, T.; Ikemori, T. Minerals, polysaccharides and antioxidant properties of aqueous solutions obtained from macroalgal beach-casts in the Noto Peninsula, Ishikawa, Japan. *Food Chem*., 2009, *112*, 575-581.

[33] Rathore, S.S.; Chaudhary, D.R.; Boricha, G.N. Effect of seaweed extract on the growth, yield and nutrient uptake of soybean (Gly-cine max) under rainfed conditions. *S. Afr. J. Bot*., 2009, *75*, 351-355.

[34] Yoshie, Y., Suzuki, T., Shirai, T., Hirano, T. Changes in the contents of dietary fibers, minerals, free amino acids and fatty acids during processing of dry Nori. *Nippon Suisan Gakkaishi*, 1994, 60: 117-123.

[35] Kolb, N., Vallorani,L., Milanovi, N., Stocchi, V. 2004. Evaluation of Marine Algae Wakame (*Undaria pinnatifida*) and Kombu (*Laminaria digitata japonica*) as Food Supplements. *Food Technol. Biotechnol. 42* : 57–61.

[36] Rao, P.V.S., Mantri, A.V., K. Ganesan, K. 2007. Mineral composition of edible seaweed *Porphyra vietnamensis. Food Chemistry* 102: 215–218.

[37] Rupérez, P. Mineral content of edible marine seaweeds. *Food Chemistry*, 2002, 79, 23-26.

[38] Norziah M.H. and C.Y. Ching. 2000. Nutritional composition of edible seaweed *Gracilaria changgi*. *Food Chemistry* 68: 69-76.

[39] Abdallah M.A.M. 2008. Chemical composition, mineral content and heavy metals of some marine seaweeds from Alexandria coast, Egypt: potential uses. *Egyptian journal of aquatic research*. vol. 34, no. 2 2008: 84-94.

[40] Ratana-arporn, P. and Chirapart, A. 2006. Nutritional Evaluation of Tropical Green Seaweeds *Cauler palentillifera* and *Ulva reticulate. Kasetsart J. (Nat. Sci.)* 40 : 75 – 83.

[41] Abdallah, M.A.M. 2007, Chemical Composition and Trace Element Concentration of Marine Algae *Entero morpha spp*. As potential food source. Third Symposium of Marine vegetation and seaweeds, UNEP/MAP/RAC-SPA. (France), March 27-29. pp. 225-228.

[42] Formuzulhaque, K.M., Chy, S.Y., Akter, S., Abdul Wahab, M.D. Nath, K.K. 2009. Collection, Identification and Biochemical Analyses of Different Seaweeds from Saint martin's island. *Bangladesh J. Agril. Res*. 34 : 59-65.

[43] Fleurence, J. 1999. Seaweed proteins: biochemical nutritional aspects and potential uses. *Trends in Food Science and Technology*. 10, 25–28.

[44] Southgate D.A.T. 1990. Dietary fiber and health. *In* D.A.T. Southgate, K. Waldron, I.T. Johnsons, and G. R. Fenwick. Dietary Fiber: Chemical and Biological Aspects. The Royal Society of Chemistry. Cambridge. pp.10-19.

[45] Norziah, M.A., Ching, C.Y., 2000. Nutritional composition of edible seaweed Gracilaria changgi. *Food Chemistry* 68, 69–76.

[46] FAO/WHO/UNU. 1985. Energy and protein requirements, Report of a joint FAO/WHO/UNU Expert Consultation, World Health Organization (WHO) Technical Report Series No. 724, WHO, Geneva, Switzerland.

[47] Matanjun, P., Mohamed, S., Mustapha, N.M. and Muham, mad, K. 2009. Nutrient content of tropical edible seaweeds, *Eucheuma cottonii*, *Cauler palentillifera* and *Sargassum polycystum*. *Journal of Applied Phycology*. 21, 75–80.

[48] Fleurence, J. Seaweeds proteins: Biochemical, nutritional aspects and protein uses, *Trends in Food Sci.&Technol.,* 1999. 10, 25-28.

[49] Foster, G.G. and Hodgson, A.N. 1998. Consummation and apparent dry matter digestibility of six intertidal macro-algae by *Turbo sarmaticus* (Mollusca: Vetigastropoda: Turbinidae). *Aquaculture* 167: 211-227.

[50] Robledo, D., Frelie Pelegrín, Y. 1997. Chemical and Mineral Composition of six Potentially Edible Seaweed Species of Yucatán. *Botanica Marina*.40: 301-306.

[51] Wong, K.H. and Cheung, C.K. 2000. Nutritional evaluation of some subtropical red and green seaweeds Part I: proximate composition, amino acid profiles and some physicochemical properties. *Food Chemistry* 71: 475-482.

[52] Arasaki, S. And Arasaki, T. 1983. Vegetables from the Sea. Japan Publ. Inc., Tokyo.

[53] Marinho-Soriano, E., Fonseca, P.C., Carneiro, M.A.A., Moreira, W.S.C.2006. Seasonal variation in the chemical composition of two tropical seaweeds. *Bioresource Technology* 97 (2006) 2402–2406.

[54] Benjama, O., Masniyom, P. 2011. Nutritional composition and physicochemical properties of two green seaweeds (*Ulvapertusa* and *U. intestinalis*) from the Pattani Bay in Southern Thailand. *Song klanakarin J. Sci. Technol.* 33 : 575-583.

[55] Rameshkumar, S., Ramakritinan, CM., Eswaran, K., Yokeshbabu, M. 2012. Proximate composition of some selected seaweeds from Palk bay and Gulf of Mannar, Tamilnadu, India. *Asian Journal of Biomedical and Pharmaceutical Sciences* 3: 1-5.

[56] Munda, I.M., Differences in amino acid composition of estuarine and marine fucoids. *Aquatic Botany*, 1977. 3,273-280.

[57] Marrion O, Schwertz A, Fleurence J, Gueant JL, Villaume C. 2003. Improvement of the digestibility of the proteins of the red alga *Palmaria palmate* by physical processes and fermentation. *Nahrung/Food*. 2003;47:339–344.

[58] Marrion O, Fleurence J, Schwertz A, et al. Evaluation of protein in-vitro digestibility of *Palmaria palmate* and *Gracilaria verrucosa*. *J. Appl. Phycol*. 2005; 17:99–102.

[59] Galland-Iarmouli AV, Lons L, Lucon M, et al. 2009. One step purification of R-hycoerythrin from the red macroalga *Palmaria palmate* using preparative polyacrylamide gel lectrophoresis. *J. Chromatogr B*. 2000;739:117–123.

[60] Munda, I.M., Differences in amino acid composition of estuarine and marine fucoids. Aquatic Botany, 1977. 3,273-280.

[61] Qasim, R. (1991). Amino acids composition of some seaweeds. *Pakistan Journal of Pharmaceutical Science*. 4, 49-54.

[62] Galland-Irmouli, A. V., Fleurence, J., Lamghari, R., Luc_on, M., Rouxel, C., Barbaroux, O., et al. (1999). Nutritional value of proteins from edible seaweed Palmariapalmata (Dulse). *Journal of Nutritional Biochemistry*. 10, 353–359.

[63] Dawczynski, C., Schubert, R., Jahreis, G. 2007. Amino acids, fatty acids, and dietary fiber in edible seaweed products. *Food Chemistry*. 103 ,891–899.

[64] Wong, K.H., Cheung, C.K.P. Nutritional evaluation of some subtropical red and green seaweeds Part I Ð proximate composition, amino acid profiles and some physico-chemical properties. *Food Chemistry*.2000. 71, 475-482.

[65] Yaich, H., Garna, H., Besbes, S., Paquot, M., Blecker, C., Attia, H. Chemical composition and functional properties of Ulvalactuca seaweed collected in Tunisia. *Food Chemistry* 128 (2011) 895–901.

[66] Hou, X., Yan, X., Chai, C., 2000. Chemical species of iodine in some seaweeds II. Iodine bound macromolecules. *J. Radioanal. Nucl. Chem.* 245, 461–467.

[67] Teas J, Pino S, Critchley AT, Braverman LE: Variability of iodine content in common commercially available edible seaweeds. *Thyroid* 2004, 14:836-841.

[68] Arasaki S, Arasaki T (1983) Low calorie, high nutrition vegetables from the sea to help you look and feel better. Japan Publications, Tokyo, 196 pp (cited in Holdt&Kraan (2011).

[69] Nisizawa K (2002) Seaweed Kaiso, bountiful harvest from the seas. Sustenance for health and well-being by preventing common lifestyle related diseases. Kochi University, Kochi, 106 pp (cited in Holdt&Kraan (2011)).

[70] Hou, X., Yan, X. (*1998*) Study on the concentration and seasonal variation of inorganic elements in 35 species of marine algae. Sci. Total Environ. *222:141–156.*

[71] M., Ninane, L., Schwartz, D., Ménager, M., Delange, F. and Aquaron, R. (2000). In: (ed. R.M. Geertman), *8th World Salt Symposium*. Elsevier Inc., Amsterdam, The Netherlands, pp. 1015–1020.

[72] Hou, X., Chai, C., Qian, Q., Yan, X., Fan, X. Determination of chemical species of iodine in some seaweeds (I). Sci. Total Environ. 1997. *204,215–221.*

[73] Aquaron, R.; Delange, F.; Marchal, P.; Lognone, V.; Ninane, L. 2002: Bioavailability of seaweed iodine in human beings. *Cell Mol. Biol.* 48, 563-569.

[74] Saenko GN, Kravtsova YY, Ivanenko VV, Sheludko SI (1978) Concentration of Iodine and bromine by plants in the seas of Japan and Okhotsk. *Mar Biol* 47,243–250.

[75] Ishizuki Y, Yamauchi K, Miura Y 1989 Transient thyrotoxicosis induced by Japanese kombu. *Folia Endocrinol* 65,91–98.

[76] Palmer, C.J., Anders, T.L., Carpenter, L.J., Ku¨ pper, F.C., McFiggans, G.B., 2005. Iodine and halocarbon response of Laminaria digitata to oxidative stress and links to atmospheric new particle production. *Environmental Chemistry* 2, 282–290.

[77] Küpper, F.C., Carpenter, L.J., McFiggans, G.B., Palmer, C.J., Waite, T.J., Boneberg, E.-M., Woitsch, S., Weiller, M., Abela, R., Grolimund, D., Potin, P., Butler, A., Luther, G.W., Kroneck, P.M.H., Meyer-Klaucke, W., Feiters, M.C. Iodide accumulation provides kelp with an inorganic antioxidant impacting atmospheric chemistry. *Proceedings of the National Academy of Sciences of the United States of America* 2008, 105, 6954–6958.

[78] Chance, R., Alex R. Baker, R.A., Frithjof C. Küpper, C.F., Claire Hughes, C., Kloareg, B., Malin, G. Release and transformations of inorganic iodine by marine macroalgae. *Estuarine, Coastal and Shelf Science*.2009. 82, 406–414.

[79] Pulz, O., Gross, W. Valuable products from biotechnology of microalgae, *Appl. Microbiol. Biotechnol.*, 2004, 65, 635-648.

[80] Rodriguez-Meizoso, I., Jaime, L., Santoyo, S. , Señoráns, F. J., Cifuentes, A., Ibáñez, E. Subcritical water extraction and characterization of bioactive compounds from *Haematococcus pluvialis* microalga, *J. Pharm. Biomed. Anal.*, 2010, *51*, 456-463.

[81] Heo, S.J., Park, E. J., Lee, K. W. , Jeon, Y. J. Antioxidant activities of enzymatic extracts from brown seaweeds, *Bio resour. Technol.*, 2005, *96*, 1613-1623.

[82] Cornish, M.L., Garbary, D.J. Antioxidants from macroalgae: potential applications in human health and nutrition. 2010. *Algae.* 25, 155-171.

[83] Lohrmann, N.L., Logan, B.A., Johnson, A.S. Seasonal acclimatization of antioxidants and photosynthesis in chondruscripus and mastocarpus stellatus, two co-occurring red algae with differing stress tolerance. *Boil. Bulletin.* 2004. 207, 225-232.

[84] Sachindra, N.M., Sato, E., Maeda, H., Hosokawa, M., Niwano, Y., Kohno, M., Mjyashita, K. Radical scavenging and single oxygen quenching activity of marine carotenoid fucoxanthin and its metabolites. *Journal of Agriculture and Food Chemistry.* 2007. 55, 8516-8522.

[85] Schubert, N., Garcia-mendoza, E., Pacheco-Ruiz, I. Carotenoid composition of marine red algae. *Journal of Psychology.* 2006. 42, 1208-1216.

[86] Nahas, R., Abatis, D., Anagnostopoulou, M.A., Kefalas, P., Vaglas, C., Roussis, V. Radical-scavenging activity of Agean Sea marine algae. *Food Chemistry.* 2007. 102, 577-581

[87] Romany, G., González, R., Ledón, N., Remirez, D., Rimbau, V. C-phycocyanin: a biliprotien with antioxidant, anti-inflammatory and neuroprotective effects. *Curr. Protien Rept. Sci.* 2003. 4, 207-216.

[88] Sekar, S., Chandramohan, M. Phycobiliprotiens as a commodity: trends research, patents and commercialization. *Journal appl. Phyco.* 2008. 20, 113-136.

[89] Devi, K.P., Suganthy, N., Kesika, P. &Pandian, S. K. Bioprotective properties of seaweeds: *in vitro* evaluation of antioxidant activity and antimicrobial activity against food borne bacteria in relation to polyphenolic content. *BMC Complement. Altern. Med.* 2008, 8:38.

[90] Yuan, Y. V., Bone, D. E. & Carrington, M. F. 2005. Antioxidant activity of dulse (*Palmaria palmata*) extract evaluated *in vitro*. *Food Chem.* 91:485-494.

[91] Ye, H., Wang, K., Zhou, C., Liu, J. & Zeng, X. 2008. Purification, antitumor and antioxidant activities *in vitro* of polysaccharides from the brown seaweed *Sargassum pallidum*. *Food Chem.* 111:428-432.

[92] Díaz-Rubio, M. E., Pérez-Jiménez, J. &Saura-Calixto, F. 2009. Dietary fiber and antioxidant capacity in *Fucus vesiculosus* products. *Int. J. Food Sci. Nutr.* 60 Suppl. 2:23-34.

[93] Chattopadhyay, N., Ghosh, T., Sinha, S., Chattopadhyay, K., Karmakar, P. & Ray, B. 2010. Polysaccharides from *Turbinaria conoides*: structural features and antioxidant capacity. *Food Chem.* 118:823-829.

[94] Luo, D., Zhang, Q., Wang, H., Cui, Y., Sun, Z., Yang, J., Zheng, Y., Jia, J., Yu, F., Wang, X. & Wang, X. 2009. Fucoidan protects against dopaminergic neuron death *in vivo* and *in vitro*. *Eur. J. Pharmacol.* 617:33-40.

[95] Rocha de Souza, M.C., Marques, C.T., Dore, C.M.G., Ferreira da Silva, F.R., Rocha, H.A.O., Leite, E.L. 2007. Antioxidant activities of sulfated polysaccharides from brown and red seaweeds. *J. Appl. Phycol.* 19:153-160.

[96] Costa, L.S., Fidelis, G. P., Cordeiro, S.L., Oliveira, R.M., Sabry, D.A., Câmara, R. B. G., Nobre, L.T.D.B., Costa, M.S.S.P., Almeida-Lima, J., Farias, E.H.C., Leite, E.L. & Rocha, H.A.O. Biological activities of sulfated polysaccharides from tropical seaweeds. *Biomed. Pharm.* 2020. 64:21-28.

[97] Josephine, A., Nithya, K., Amudha, G., Veena, C. K., Preetha, S. P. & Varalakshmi, P. 2008. Role of sulphated polysaccharides from *Sargassum Wightii* in Cyclosporine A-induced oxidative liver injury in rats. *BMC Pharmacol.* 8:4.

[98] Athukorala, Y., Kim, K.N., Jeon, Y.J. Antiproliferative and antioxidant properties of an enzymatic hydrolysate from brown alga, *Ecklonia cava. Food Chem. Toxicol.* 2006. 44:1065-1074.

[99] Lohrmann, N. L., Logan, B. A. & Johnson, A.S. Seasonal acclimatization of antioxidants and photosynthesis in *Chondru scrispus* and *Mastocarpus stellatus*, two co-occurring red algae with differing stress tolerances. *Biol. Bull.* 2004. 207,225-232.

[100] Kumar, K. S., Ganesan, K., Subba Rao, P.V. Antioxidant potential of solvent extracts of *Kappaphycusalvarezii*(Doty) Doty: an edible seaweed. *Food Chem.* 2008. 107,289-295.

[101] Kuda, T., Tsunekawa, M., Hishi, T., Araki, Y. Antioxidant properties of dried 'kayamonori', a brown alga Scytosiphonlomentaria (Scytosiphonales, Phaeophyceae). *Food Chemistry*, 2005. 89, 617–622.

[102] Valentão, P., Trindade, P., Gomes, D., de Pinho, P. G., Mouga, T., Andrade, P. B. Codium tomentosum and Plocamium cartilagineum: Chemistry and antioxidant potential. *Food Chemistry*, 2010. 119, 1359–1368.

[103] Shibata, T., Ishimaru, K., Kawaguchi, S., Yoshikawa, H., Hama, Y. Antioxidant activities of phlorotannins isolated from Japanese Laminari aceae. *Journal of Applied Phycology,* 2008. 20, 705–711.

[104] Amin Ismail, A., Hong, T.S. Antioxidant Activity of Selected Commercial Seaweeds. 2002. *Mal. J. Nutr.*8(2): 167-177.

[105] Cho, S.H., Kang, S.E., Cho, J.Y., Kim, A.R., Park, S.M., Hong, Y.K. The antioxidant properties of brown seaweed (Sargassum siliquastrum) extracts. *Journal of Medicinal Food,* 2007. 10, 479–485.

[106] Yangthong, M., Hutadilok-Towatana, N., Phromkunthong, W. Antioxidant activities of four edible seaweeds from the southern coast of Thailand. *Plant Foods Hum. Nutr.* 2009. 64:218-223.

[107] Matanjun, P., Mohamed, S., Mustapha, M.N., Muhammad, K., Ming, C.H. Antioxidant activities and phenolics content of eight species of seaweeds from north Borneo. *J. Appl.Phycol.* 2008. 20,367–373.

[108] Sabeena Farvin, K.H., Jacobsen, C. Phenolic compounds and antioxidant activities of selected species of seaweeds from Danish coast. *Food Chemistry* 138 (2013) 1670–1681.

[109] Lee, S., Kang, S., Ko, S., Lee, D., Jeon, Y. Octaphlorethol A, a novel phenolic compound isolated from a brown alga, Ishige foliacea, increases glucose transporter 4-mediated glucose uptake in skeletal muscle cells. *Biochemical and Biophysical Research Communications* 420 (2012) 576–581.

[110] Rochal, H.A.O., Franco, C.R.C., Trindade, E.S., Carvalho, L.C.M., Veiga, S.S., Leite, E.L., Dietrich, C.P., Nader, H.B. A fucan from the brown seaweed *Spato glossum schröederi* inhibits Chinese hamster ovary cell adhesion to several extracellular matrix proteins. *BArnatzi-ialidanhe Jsoiuvern faulcoafn Medical and Biological Research* (2001) 34, 621-626.

[111] Chojnacka, K., Saeid, A., Witkowska, Z., Tuhy, L. Biologically Active Compounds in Seaweed Extracts - the Prospects for the Application. *The Open Conference Proceedings Journal*, 2012, 3, (Suppl 1-M4) 20-28.

[112] McHugh, D.J. A guide to the seaweed industry. FAO Fisheries Technical Paper. No. 441. Rome, FAO. 2003. 105p.
[113] Sridhar, S., Rengasamy, R. Effect of seaweed liquid fertilizer on growth, pigment concentration and yield of Amaranthus roxburghinus and Amaranthus tricolor under field trial. *International Journal of Current Research*. 2011, 3, 131-134.
[114] Pacheco-ruiz, I. Zertuche-Gonza´lez, J.A., Arroyo-Ortega, E., Valenzuela-Espinoza, E. Agricultural fertilizers as alternative culture media for biomass production of *Chondra canthus squarrulosus* (Rhodophyta, Gigartinales) under semi-controlled conditions. *Aquaculture,* 2004, 240, 201-209.

In: Seaweeds
Editor: Vitor H. Pomin
ISBN: 978-1-63117-571-8

Chapter 4

SCIENTIFIC EVIDENCES OF THE HEALTH-PROMOTING EFFECT OF SEAWEED INTAKE IN HUMANS

Antonio Jiménez-Escrig*
Spanish National Research Council, CSIC, Madrid, Spain

ABSTRACT

There is a long tradition in East-Asian countries of consuming seaweed, although its intake has recently increased in Western countries. Nutritional aspects of seaweeds as foods are well established, as it is considered a valuable food source. In addition, an extensive body of evidence of the associated health effects of seaweeds measured *in vitro* is described. However, knowledge of the health effects of seaweed based upon human intervention studies is very limited. The purpose of this chapter is to describe the scarce substation of the scientific and health-promoting effects attributed to brown and red edible seaweeds. Different health aspects will be described, e.g. protection of postmenopausal breast cancer, modulation of postpandrial plasma glucose and insulin levels, or modulation of serum estrogen hormone levels.

INTRODUCTION

Epidemiological studies have shown that diet plays a meaningful role in health, more specifically; consumption of fruits and vegetables is associated with reduced risk of chronic diseases. As a consequence, an increased consumption of products from vegetable origin, which contain high levels of dietary fiber and phytochemical constituents, has been recommended (Kris-Etherton et al., 2002; Willet, 1994). On this basis, nutrition and health claims have been demonstrated to be strong marketing incentives for the food industry, providing opportunities for product differentiation based on a health-related positioning (Verbeke et al., 2009). However, consumers should be able to make choices based on clear

* Email-a.jimenez.escrig@csic.es

and accurate information (Richardson, 2012). In this sense, there is a good understanding on the basis of nutrient requirements for health maintenance but the next steps towards substation of scientific-evidence of the relationships between nutrition and health have proven to be difficult (Van Ommen et al. 2010). It is a current opinion that the design of well-controlled nutrient studies is essential for the former substation (Kris-Etherton and Dietschy, 1997). Therefore, clear guidelines based on consistent clinical trials need to be developed to provide the evidence-based information for the consumer (Jefferey and Keck, 2008). Specifically, demonstration of health effect of diets through human intervention studies is required (Gallagher et al., 2011). Thus, it is clear that substation of scientific of the health-promoting effects attributed to edible seaweeds need to include intervention studies. In this sense, certain factors of critical importance to the success of these studies are well-defined subject eligibility criteria, taking into account the response of different target groups (men and women, and older and younger subjects). Another factor is the need to precisely define experimental diets in which the only variable is the nutrient of interest, to unambiguously establish the effect of a specific nutrient. Also it is necessary to develop and implement an appropriate experimental design including the number of subjects, which it is important for the statistical power, and the collection and analysis of end point measurements (Kris-Etherton, 1997). In this sense, randomized controlled clinical trials (RCT) are essential for establishing cause-and-effect relations (Roussel and Kris-Etherton, 2013), providing this type of human design study the most persuasive evidence of efficacy (Gallagher et al., 2011).

Since ancient times, green, brown, and red seaweeds have been part of the diet in Asian countries, especially China, Japan, and Korea (Nisizawa et al. 1987). In the last three decades the discovery of metabolites with biological activities from macroalgae has increased significantly. However, despite the intense research effort by academic and corporate institutions, very few products with real potential have been identified or developed. Moreover, regarding the health benefits of constituents of edible seaweeds and their role in nutrition and disease prevention it is stated that much research remains to be done before science-based dietary recommendation can be made (Smit, 2004).

The purpose of this chapter is to describe certain intervention human studies of the scarce substation of the scientific and health-promoting effects attributed to brown and red edible seaweeds.

INTERVENTION HUMAN STUDIES

A single-blind placebo-controlled study assessed the impact of introducing seaweed- to non-seaweed consuming American postmenopausal women. Fifteen healthy postmenopausal women are recruited for a 3-month single blinded placebo controlled clinical trial; five have no history of breast cancer (controls) and ten are breast cancer survivors. Participants ingested ten 500-mg capsules daily of the placebo for 4 weeks, seaweed [(*Undaria pinnatifida* (Harvey) Suringar, so-called Japanese Wakame] for 4 weeks, then placebo for another 4 weeks (Teas et al., 2013). Brown seaweeds have no land equivalents in terms of their specific components of dietary fiber (alginate), primary carotenoid (fucoxanthin), sulfated polysaccharide (fucoidan and laminarin), and polyphenol defense compounds, each of which has been reported to have strong anti-cancer activity. Justification of the study is based upon the preliminary relationship between the relative breast cancer risk and seaweed intake among humans, which is only now unfolding. A small body of *in vivo* research suggests seaweed may be useful in breast cancer prevention. In addition, seaweeds are specifically used to treat tumors in Traditional Chinese Medicine and Japanese folk medicine (Teas et al., 2013). As a result of this intervention study, urinary human urokinase-type plasminogen activator receptor concentrations (uPAR) changed significantly between groups, decreasing by about half following seaweed supplementation while placebo returned to pre-seaweed concentration. This study concludes that by lowering uPAR, dietary brown seaweed may help explain lower breast cancer incidence and mortality among postmenopausal women in Japan. In many ways, it is difficult to imagine how a merely 5 g / day could have such a profound effect on systemic concentrations of uPAR and specific protein expression. One possibility is that seaweed, like other dietary fibers, is known to increase the mucus bilayer that protects the colonic mucosa, stimulating mucus secretion and thickness. Other suggested consequences of dietary seaweed fiber include increased stool bulk which would provide protection from dietary toxins and carcinogens and reduction in amylase activity leading to decreased glycemic response (Teas et al., 2013).

Another recent study examined the impact of brown seaweed on post-load plasma glucose and insulin concentrations in men and women. Twenty-three participants (11 men, 12 women) aged 19-59 years are recruited in this double-blind, randomized, placebo-controlled crossover study. The test product consisted of a commercially available blend of brown seaweed (*Ascophyllum- nodosum and Fucus vesiculosus*) with known inhibitory action on α-amylase and α-glucosidase activities. Two 250 mg seaweed capsules and 2 placebo capsules are consumed on each occasion 30 min prior to the consumption of 50 g of carbohydrates from bread. Subjects participated in two 3-h meal tolerance tests that are separated by a 1-week ishout period. Participants are recruited from the general population. Inclusion criteria are women and men aged between 18 and 60 years, nonsmokers, with a body mass index (BMI) between 20-30 kg x m^{-2}. Criteria of exclusion of potential volunteers are the following: Consume of more than 2 alcoholic drinks per day, or more than 9 alcoholic drinks per week, as well as those using natural health products are criterion of exclusion for participating in the study. Also volunteers suffering diseases that could compromise their safety, such as iodine allergy, gastrointestinal problems, diabetes, thyroid or liver dysfunction, or other diseases are excluded. In addition, volunteers with untreated hypertension (systolic blood pressure > 140 mm Hg and (or) diastolic blood pressure > 90 mm Hg) are excluded from the study. Breastfeeding or pregnant women as well as those using unrecognized contraceptive methods are not eligible. Volunteers consuming more than 2 alcoholic drinks per day or more than 9 alcoholic drinks per week for the duration of the project as well as those using natural health products are also excluded. Plasma glucose and insulin concentrations are measured over a period of 3 h postcarbohydrate ingestion at predetermined time points. Compared with placebo, consumption of seaweed is associated with a 12.1% reduction in the insulin incremental area under the curve. The single ingestion of 500 mg of brown seaweed has no significant effect on the glucose response. Glucose and insulin responses are similar between men and women. Consumption of the seaweed capsules is not associated with any adverse event. This data suggests that brown seaweed may alter the insulin homeostasis in response to carbohydrate-rich meal (Paradis et al., 2011).

The edible brown kelp, *Fucus vesiculosus* (bladderwrack), as well as other brown kelp species, can lower plasma cholesterol levels. Since cholesterol is a precursor to sex hormone biosynthesis, kelp consumption may alter circulating sex hormone levels and menstrual cycling patterns. In particular, dietary kelp may be beneficial to women with or at high risk for estrogen-dependent diseases. To test this, bladderwrack is administered to three pre-menopausal women with abnormal menstrual cycling patterns and/or menstrual-related disease histories. As a result, intake of bladderwrack is associated with significant increases in menstrual cycle lengths, ranging from an increase of 5.5 to 14 days. In addition, hormone measurements ascertained for one woman revealed significant anti-estrogenic and progestagenic effects following kelp administration. These pilot data suggest that dietary bladderwrack may prolong the length of the menstrual cycle and exert anti-estrogenic effects in pre-menopausal women. Further, these studies also suggest that seaweed may be another important dietary component apart from soy that is responsible for the reduced risk of estrogen-related cancers observed in Japanese populations. However, these studies will need to be performed in well-controlled clinical trials to confirm these preliminary findings (Skibola, 2004).

Seaweed and soy foods are consumed daily in Japan, where breast cancer rates for postmenopausal women are significantly lower than in the West. Likely mechanisms include differences in diet, especially soy consumption, and estrogen metabolism. Fifteen healthy postmenopausal women participated in a double-blind trial of seaweed supplementation with soy challenge. Participants were randomized to 7 week of either 5 g/d seaweed (Alaria) or placebo (maltodextrin). During week 7, participants also consumed a daily soy protein isolate (2 mg isoflavones/kg body weight). After a 3-week washout period, participants were crossed over to the alternate supplement schedule. There was an inverse correlation between seaweed dose (mg/kg body weight) and serum estradiol, which is linear across the range of weights. Soy supplementation increased urinary daidzein, glycitein, genistein, and O-desmethylangolensin and decreased matairesinol and enterolactone. Soy and seaweed plus soy (SeaSoy) increased urinary excretion of 2-hydroxyestrogen (2-OHE) and the ratio of 2-OHE:16a-hydroxyestrone. For the 5 equol excretors, soy increased urinary equol excretion; the combination of SeaSoy further increased equol excretion by 58%. Equol producers also had a 315% increase in 2:16 ratio with SeaSoy. Seaweed favorably alters estrogen and phytoestrogen metabolism and these changes likely include modulation of colonic bacteria (Teas et al., 2009).

More intervention studies dealing with the substation of the scientific evidence of health-promoting effects of edible seaweeds through well-designed randomized clinical trials are necessary.

REFERENCES

Gallagher, A. M., Meijer, G. W., Richardson, D. P., Rondeau, V., Skarp, M., Stasse-Wolthuis, M., Tweedie, G. C. & Witkamp, R. (2011). A standardised approach towards PROving the efficacy of foods and food constituents for health CLAIMs (PROCLAIM), Providing guidance. *British Journal of Nutrition*, *106*, S16–S28.

Jeffery, E. H. & Keck, A. S. (2008). Translating knowledge generated by epidemiological and *in vitro* studies into dietary cancer prevention. *Molecular Nutrition Food Research*, *52*, S7-S17.

Kris-Etherton, P. M. & Dietschy, J. (1997). Design criteria for studies examining individual fatty acid effects on cardiovascular disease risk factors: human and animal studies. *American Journal of Clinical Nutrition*, *65*, 1590S-1596S.

Kris-Etherton, P. M., Hecker, K. D., Bonanome, A., Coval, S. M., Binkoski, A. E., Hilpert, K. F., Griel, A. E. & Etherton, T. D. (2002). Bioactive compounds in foods: Their role in the prevention of cardiovascular disease and cancer. *American Journal of Medicine*, *113*, 71S-88S.

Nisizawa, K., Noda, H., Kikuchi, R. & Watanabe, T. (1987). The main seaweed foods in Japan. *Hydrobiologia*, 151-152, 5-29.

Paradis, M. E., Couture, P. & Lamarche, B. (2011). A randomised crossover placebo-controlled trial investigating the effect of brown seaweed (*Ascophyllum nodosum* and *Fucus vesiculosus*) on postchallenge plasma glucose and insulin levels in men and women. *Applied Physiology, Nutrition and Metabolism*, *36*, 913-919.

Roussell, M. & Kris-Etherton, P. (2012). Reply to AM Bernstein and WC Willett. *American Journal of Clinical Nutrition*, *95*, 1295-1296.

Skibola, C. F. (2004). The effect of *Fucus vesiculosus*, an edible brown seaweed, upon menstrual cycle length and hormonal status in three pre-menopausal women: A case report. *BMC Complementary and Alternative Medicine*, *4*, 1-9.

Smit, A. J. (2004). Medicinal and pharmaceutical uses of seaweed natural products: a review. *Journal of Applied Phycology*, *16*, 245–262.

Teas, J., Vena, S., Cone, D. L. & Irhimeh, M. (2013). The consumption of seaweed as a protective factor in the etiology of breast cancer: proof of principle. *Journal of Applied Phycology*, *25*, 771-779.

Teas, J., Hurley, T. G., Hebert, J. R., Franke, A. A., Sepkovic, D. W. & Kurzer, M. S. (2009). Dietary seaweed modifies estrogen and phytoestrogen metabolism in healthy postmenopausal women. *Journal of Nutrition*, *139*, 939-944.

Verbeke, W., Scholderer, J. & Lähteenmäki, L. (2009). Consumer appeal of nutrition and health claims in three existing product concepts. *Appetite*, *52*, 684-692.

Willett, W. C. (1994). Diet and health: What should we eat? *Science*, *264*, 532-537.

In: Seaweeds
Editor: Vitor H. Pomin
ISBN: 978-1-63117-571-8

Chapter 5

EVALUATION OF BIOACTIVE COMPOUNDS IN SEAWEEDS: A REVIEW

***Julia López-Hernández*[*] *and Ana Rodríguez-Bernaldo de Quirós*[†]**
Department of Analytical Chemistry, Nutrition and Food Science University of Santiago de Compostela, Campus Vida s/n, Santiago de Compostela (A Coruña), Spain

ABSTRACT

Several studies have demonstrated that seaweeds besides their nutritional properties such as high content in dietary fibre, proteins, polyunsaturated fatty acids and low content in saturated fats are an excellent source of compounds with biological activity. Among the bioactive substances, antioxidants have attracted the attention of the scientific community due to the positive effects against serious disorders. Several compounds, of different nature, with antioxidant activity have been identified in macroalgae including, carotenoids (i.e. fucoxanthin), polyphenols (i.e. phlorotannins) or vitamins such as α-tocopherol (Vitamin E) and ascorbic acid (Vitamin C).

This study reports an up-to-date overview on the analysis of bioactive compounds in seaweeds. Relevant features of chromatographic methods are commented on. In addition methods to evaluate the antioxidant activity are also reviewed.

Keywords: Analytical methods, antioxidant activity, bioactive compounds, seaweeds

1. INTRODUCTION

Seaweeds are considered an excellent source of compounds with biological activity such as, antioxidants, polysaccharides with antiviral action or for example ω-3 fatty acids with beneficial effects against coronary diseases (Plaza et al. 2008). Within the bioactive compounds natural pigments present in macroalgae like carotenoids and chlorophylls have

[*] E-mail: julia.lopez.hernandez@usc.es.
[†] E-mail: ana.rodriguez.bernaldo@usc.es.

shown to have possible effects on the health for example antioxidant, anti-inflammatory and so on (Pangestuti and Kim, 2011).

Another family of compounds with interesting properties are the polyphenols, many research works have reported their biological activities, particularly their antioxidant activity. Phlorotannins are the major phenolic compounds present in algae, however other phenolic compounds such as gallic acid, catechins and their galloylated esters (gallates) have been also identified in different species of seaweeds (Rodríguez-Bernaldo de Quirós et al. 2010; Yoshie et al. 2000).

Besides, they contain important quantities of vitamins, including vitamin C, which is involved in biochemical processes, such as collagen formation, iron absorption and in immune responses (Martínez, 1998), vitamin E, vitamin B complex or folic acid. The vitamin composition of each specie depends on several factors such as the environmental conditions.

Polyunsaturated fatty acids (PUFAs), particularly ω-3 fatty acids have attracted the attention of the scientists due to their positive effects, these compounds are believed to be implicated in reducing the incidence of cardiovascular disease (Kinsella, 1988) and to have beneficial effects on hypertension, inflammation, arrhythmias, and so on (Lee et al. 1985).

Seaweeds also have in their composition sulfated polysaccharides, that have many biological activities, for example anti-coagulant, antioxidant, anti-thrombotic, anti-platelet aggregation and anti-viral properties among others (Karmakar et al. 2009).

In the present chapter a review of the methods to determine natural pigments, polyphenols, vitamins, PUFAs and sulfated polysaccharides is presented. In addition, a summary of the antioxidant activity exhibited for the different species of macroalgae indicating the method of determination is reported.

2. Pigments

As it has been comment on previously, among the bioactive compounds present in seaweeds, the pigments have been studied in depth and attracted the interest of the scientific community due to their beneficial physiological and biological functions on health for example antioxidants, enhancers of the immune response and in addition some of them have been shown to be effective against certain types of cancer, cardiovascular diseases and age-related macular degeneration (Pangestuti and Kim, 2011; Rodríguez-Bernaldo de Quirós and Costa, 2006; Machlin, 1995; Bone et al. 2000).

Four types of chlorophylls have been identified in seaweeds chlorophylls a, b, c, and chlorophyll d, the last one has been found in red algae (Pangestuti and Kim, 2011). Regarding the carotenoids, from the chemical point of view are isoprenoid polyenes and can be classified in two groups: carotene hydrocarbons containing only carbon and hydrogen atoms and xantophylls which carry at least one oxygen function. Structurally, the carotenoids are characterized by a conjugated double bond system that confers to this class of compounds special spectroscopic properties (Nollet, 2000; Rodríguez-Bernaldo de Quirós and López-Hernández, 2009).

Different carotenoids have been identified in seaweeds, among them fucoxanthin is the main carotenoid present in brown seaweeds; this pigment belongs to the group of xanthophylls and has exhibited a potent antioxidant activity (Yuan, 2007; Kumar et al. 2008).

Hegazi et al. (1998) developed a high performance liquid chromatographic method to study the pigment composition of three seaweeds belonging to three different groups *Caulerpa prolifera* (green algae), *Jania rubens* (red algae) and *Padina pavonica* (brown algae). A LiChroCART (250 x 4 mm i.d. packed with LiChrospher 100, RP-18 e (5-μm spherical particles) with a precolumn ODS-Hypersil (C_{18}), 5 μm, 20 x 4.0 mm (Hewlett-Packard) was used as a stationary phase and a gradient composed by methanol, acetone and ammonium acetate solution (1 M) at a flow rate of 1 mL/min as mobile phase. The wavelengths were set at 430 nm for carotenoids identification and at 660 nm for chlorophylls identification. The injection volume was 20 μL. The pigments identified in the species studied were the follows: in *Caulerpa prolifera* 18 pigments were separated chlorophyllide a, siphonein, neoxanthin, neoxanthin-like, violaxanthin, microxanthin, micronone, micronone-like, lutein-5,6-epoxide, siphonoxanthin, lutein, chlorophyll b, chlorophyll b´, chlorophyll a, chlorophyll a´, α-carotene, β-carotene, and phaeophytin a. The most characteristic compounds of green algae were chlorophyll a and b. In *Jania rubens* 16 pigments were separated: chlorophyllide a, chlorophyll d, fucoxanthin, violaxanthin, fucoxanthol, α-cryptoxanthin-like, α-cryptoxanthin, β-cryptoxanthin, zeaxanthin, lutein-5,6-epoxide, lutein, chlorophyll a, chlorophyll a´, α-carotene, β-carotene, and phaeophytin a. And in *Jania rubens* 14 pigments were separated: chlorophyll c_1 , chlorophyll c_2 , fucoxanthin, violaxanthin, flavoxanthin, fucoxanthol, antheraxanthin, 9-cis-neoxanthin, diatoxanthin, zeaxanthin, chloro phyll a, chlorophyll a´, β-carotene and phaeophytin a.

Total carotenoid as well as the chlorophylls a, b and c in different algae belonging to Chlorophyta, Phaeophyta and Rhodophyta were determined in a work carried out by Dere et al. (2003). Total carotenoids were extracted by using 90% acetone as solvent. The values obtained ranged from 0.01 to 0.18 mg/g fresh weight.

Rodríguez-Bernaldo de Quirós et al. (2010) evaluated the pigment composition of *Himanthalia elongata*, *Laminaria* sp., *Laminaria saccharina* and *Undaria pinnatifida*. The pigments were separated and identified by on line HPLC coupled with UV–Vis and fluorescence detectors and confirmed by liquid chromatography coupled to mass spectrometry (LC–MS APCI). The separation was performed on a Tracer Extrasil C18 ODS 2 (25 x 0.4 cm, 5 μm) column and using a gradient of three eluents (A) MeOH, (B) water and (C) dichloromethane: hexane (50:50, v/v). The detection was done at 450 nm and the chlorophyllic pigments were also detected by fluorescence (λex 410 nm and λem 670 nm). Fucoxanthin, β-carotene, chlorophyll a and phaeophytin a were identified in all the samples.

Marquardt and Hanelt (2004) studied the carotenoids present in sixteen different red algae from polar and temperate habitats. The analysis of the pigments were performed by thin layer chromatography (TLC) and high-performance liquid chromatography (HPLC) coupled to a photodiode array detector. All species exhibited the typical carotenoid profile of the rhodophytes except *Delesseria lancifolia*. The characteristic profile was composed by a xanthophyll mainly lutein and in some cases α- and β-carotene. *Pantoneura baerii* and *Polysiphonia urceolata* had in their composition zeaxanthin instead lutein. Some species also contained another xanthophyll α-cryptoxanthin or β-cryptoxanthin although in low concentrations. In the case of *Delesseria lancifolia* the major xanthoplhylls were violaxanthin, antheraxanthin and zeaxanthin.

In another study conducted by Schubert et al. (2006) the determination of carotenoids in red algae collected on different coastal locations near Ensenada, Baja California was carried out. The carotenoids were separated on a reversed-phase Nucleosil C-18 (250 mm x 4.6mm i. d., 5 μm particle size) column protected by a guard column Eclipse XDB-C8 (12.5 mm x 4.6 mm i. d. 5 μm particle size) and using as mobile phase binary solvent gradient consisting of ethylacetate and acetonitrile: water (9:1, v/v). Chromatograms were monitored at 447 nm. Different carotenoids were identified in the species studied, such as lutein, zeaxanthine, violaxanthin, antheraxanthin or α- and β-carotene.

Suzuki and Ishimaru (1992) identified the pigments of *Gymnodinium mikimotoi*; the analysis were performed by HPLC with a photo-diode array detetector and using a Merk Lichrosphere, 100 RP-18e octadecyl silica (125 x 4 mm, 5 μm particle size) as a stationary phase and methanol-water (88:12 v/v) and methanol as mobile phase at flow-rate of 1 mL/min. The wavelength was set at 440 nm. Pigments were extracted from the samples with 100% methanol. The major pigments identified were cholorophyll c_3, fucoxanthin, 19'-hexanoyloxyfucoxanthin, and 19'-butanoyloxyfucoxanthin among others.

A reversed-phase high-performance liquid chromatographic (HPLC) method was used to determine the photopigments in *Symbiodinium* spp. Methanol was used to extract the analytes from the samples. The chromatographic conditions were as follows: a Waters Spherisorb S5 ODS2 (250 x 4 mm) column protected by a guard-column Waters Spherisorb S% ODS2 (10 x 4.6 mm); ternary gradient elution consisting of: eluent (A) methanol-0.5M ammonium acetate (pH 7.2) 80:20 (v/v), eluent (B) acetonitrile-water 80:20 (v/v) and eluent (C) ethyl acetate. The flow rate was 0.76 mL/min and the injection volume 50 μL. Pigments were detected at 436 nm and chlorophylls and chlorophyll break down products were identified by fluorescence (λ_{ex} 424 nm λ_{em} 640 nm). The main photopigments identified were chlorophyll c, peridinin, diadinoxanthin, and chlorophyll a (Rogers and Marcovich, 2007).

The chlorophyll and carotenoid contents in seaweeds *Kappaphycus alvarezii* were investigated in a study carried out by Thirumaran et al. (2009). Pigments were extracted using 80% acetone and were determined spectrometrically; chlorophylls were detected at 645 and 663 nm and carotenoids at a wavelength of 480 nm. The chlorophyll content ranged from 0.998±0.073 to 0.941±0.086 and the carotenoid content from 0.337±0.036 to 0.440±0.05 mg/g fr.wt.

3. Polyphenols

Among the components with biological activity present in the macroalgae the polyphenols are one of the most important substances with antioxidant properties present in the aqueous fraction (Plaza et al. 2008; Kumar et al. 2008). Phlorotannins are the main group of phenolic compounds in marine brown algae, chemical and structurally are oligomers of 1,3,5-trihydroxybenzene (phloroglucinol) they are synthesized via the acetatemalonate pathway (Li et al. 2011; Gupta and Abu-Ghannam, 2011). It can be distinguished four classes of phlorotannins according the linkage fuhalols and phlorethols (phlorotannins with an ether linkage), fucols (with a phenyl linkage), fucophloroethols (with an ether and phenyl linkage), and eckols (with a dibenzodioxin linkage) (Li et al. 2011). The concentration of polyphenols in seaweeds depends on environmental factors (Parys, et al. 2007).

Koivikko et al. (2007) developed a high-performance liquid chromatographic method to analyze phlorotannins from the brown alga *Fucus Vesiculosus*. In the study they evaluated different methods with normal and reversed stationary phases. The reversed-phase method involved the use of a gradient elution system consisted of eluent (A) (1% formic acid in water) and eluent (B) (acetonitrile) at a flow rate of 1 mL/min and a Superspher 100 RP-18 (250 × 4 mm i.d.; 4 μm; Merck, Germany) column. Besides formic acid, 0.05 M phosphoric acid and acetic acid were assayed in the composition of the eluent A.

A LiChrospher Si 60 (250 × 4 mm i.d.; 5 μm; Merck) protected with a guard cartridge (LiChrospher Si 60; 4 × 4 mm i.d.; 5 μm; Merck) was used as normal-phase and a binary solvent gradient composed by solvent (A) dichloromethane-methanol-water-acetic acid (82:14:2:2, v/v) and solvent (B) methanol-water-acetic acid (96:2:2, v/v) at a flow rate of 1 mL/min as mobile phase. Polyphenols were detected at 280 nm.

With the NP-HPLC method a better separation and resolution of the chromatographic peaks were achieved. Due the lack of standards, in order to identify the compounds, the analysis were also performed by NP-HPLC-ESI/MS operated in negative mode.

The same authors applied a colorimetric method to determine the total phlorotannin content in order to study the variation of phlorotannins among three populations of *Fucus vesiculosus*; quantification was performed with a modification of the Folin–Ciocalteu method and using phloroglucinol as the standard (Koivikko et al. 2008).

Chkhikvishvili and Ramazanov (2000) proposed a gas-liquid chromatography to determine phlorotannins after their acetylation in brown algae. A capillary column Supercowax 10 (SUP, USA) was used to separate the acetylated derivatives of phlorotannins. *Cystoseira compressa* and *Sargassum furcatum* were the species that showed the highest values of the phenolic compounds.

Parys et al. (2007) evaluated four different methods for the determination of polyphenols in algal extracts of *Ascophyllum nodosum* and *Fucus vesiculosus*. Two of them were spectrophotometric determinations based on the reaction of phenolic compounds with Folin-Ciocalteu's reagent and on the reaction of phlorotannins with 2, 4-dimethoxybenzaldehyd. A gravimetric method that involves the linkage of phlorotannins to polyvinylpyrrolidone and a ^{1}H NMR based method were also investigated. A statistical study were carried out and it might conclude that the Folin-Ciocalteu's method and the ^{1}H NMR were the most precise.

Glombitza and Schmidt (1999) isolated and identified nonhalogenated and halogenated phlorotannins from the brown seaweed *Carpophyllum angustifolium*. For that purpose mass spectrometry and ^{13}C NMR were employed. Two new halogenated phlorotannins were identified. Other classes of polyphenols such as phenolic acids, catechins and their galloylated esters (gallates) have been also identified in edible red and brown seaweeds. In a study conducted by Rodríguez-Bernaldo de Quirós et al. (2010) catechin, epicatechin, epicatechin gallate, catechin gallate, epigallocatechin, epigallocatechin gallate and gallic acid were found in the seaweeds collected on the Atlantic coastal region in Galicia (NW Spain). The analysis were performed by high-performance-liquid-chromatography with UV detector (HPLC-UV), a TEKNOKROMA, Tr-010065 Mediterranean sea18 (15 cm x 0.4 cm, i.d. 3 μm) was used as column and a gradient of two eluents (A) water-acetic acid (99:1, v/v) and (B) (water-acetonitrile- acetic acid) (67:32:1, v/v/v) as mobile phase. The wavelength was set at 280 nm. The flow rate was 1.2 ml/min during the first 12 min and 1 ml/min until the end of the run. The species of seaweeds studied were the following: *Palmaria* spp., *Porphyra* spp., *Himanthalia elongata*, *Laminaria ochroleuca* and *Undaria pinnatifida*. Epigallocatechin was the predominant catechin, and was detected in four of the five species studied in the range of concentration of 252.5 ± 9.5-760.2 ± 5.2 μg/g dry weight. LC-MS in the negative ion electrospray ionization (ESI) mode was used as a confirmatory technique.

Yoshie et al. (2000) investigated the distribution of catechins including, catechin, epicatechin, epigallocatechin, catechin gallate, epicatechin gallate and epigallocatechin gallate, in green, red and brown seaweeds collected in different areas of Japan. The determination of the phenolic compounds was carried out by liquid chromatography. Regarding chromatographic conditions an ODS (Intersil ODS-2 250 x 4.6 mm i.d., 5μm GL Science Inc., Tokyo Japan) thermostatted at 40ºC was used as column and acetonitrile-ethyl acetate-0.1% phosphoric acid (85:20:895 v/v/v) as mobile phase, the wavelength was set at 280 nm. Epigallocatechin was the polyphenol found in the highest concentration. Catechin gallate, epicatechin gallate and epigallocatechin gallate were only detected in brown and red algae. The total phenolic content has been determined in several species of seaweeds by means of colorimetric determinations based on Folin-Ciocalteu's method. Like this, Rodríguez-Bernaldo de Quirós et al. (2010) evaluated the total polyphenol content in four species of brown algae included *Himanthalia elongata*, *Laminaria* sp., *Laminaria saccharina* and *Undaria pinnatifida* the values expressed as g phloroglucinol/ kg wet weight ranged from 1.3 to 10.

Matanjun et al. (2008) evaluated the total phenolic contents in red, green and brown seaweeds from Borneo including *Eucheuma cottonii, E. spinosum, Halymenia durvillaei, Caulerpa lentillifera, C. racemosa, Dictyota dichotoma, Sargassum polycystum* and *Padina* sp. and found values between 15.82 and 45.16 mg phloroglucinol equivalents (PGE)/ g dry extract.

Yuan and Walsh (2006) studied the polyphenol content in *Palmaria palmata*, *Laminaria setchellii, Macrocystis integrifolia* and *Nereocystis leutkeana* by using gallic acid as standard, the values varied between 1.84 and 12.8 μg gallic acid equivalents.

Yuan et al. (2005) evaluated the antioxidant properties of Palmaria palmata (dulse) and found that total polyphenol content was 10.3 μg gallic acid.

The Folin–Ciocalteu method was used to measure the total polyphenol content in *Padina antillarum, Caulerpa racemosa* and *Kappaphycus alvarezzi*. The results obtained ranged between 28.4 and 2430 mg gallic acid equivalents(GAE) /100 g dried samples (Chew et al. 2008).

4. VITAMINS

Seaweeds can be considered an excellent source of vitamins, watersoluble vitamins such as vitamin C, folic acid or vitamin B complex as well as liposoluble vitamins including α-tocopherol and vitamin A have been identified in different species of seaweeds. The content of vitamins in seaweeds depends on several factors such as environmental conditions (light, temperature), season, and the specie of the algae among others (Gupta and Abu-Ghannam, 2011; Kumar et al. 2008).

Ferraces-Casais et al. (2012) evaluated the content of vitamin C in four species of seaweeds, *Laminaria* spp., *Porphyra umbilicalis*, *Himanthalia elongata*, and *Palmaria palmata*. To determine the vitamin the authors used a chromatographic method. The stationary phase was a Kinetex C18 (150 × 4.6 mm, 2.6 μm; Phenomenex®, USA) and the mobile phase was composed by acetic acid–Milli-Q water (0.1% v/v) at a flow rate of 0.7 mL/min. The detection was done at 245 nm. The antioxidant was extracted from the matrix by using 4.5% metaphosphoric acid as extraction solvent. The vitamin content in the species studied ranged from 0.61±0.02 to 46.66±5.94 mg/100g for *Palmaria palmata* and *Himanthalia elongata*, respectively.

Norziah and Ching (2000) used the titrimetric method AOAC no:967.21 to determine the vitamin C content in *Gracilaria changgi*. The value obtained was 28.5±0.1 mg vitamin C/100 g sample (wet weight), which is similar to some vegetables. The authors also evaluated the vitamin A activity by means of the analysis of β-carotene. For that, they used a reversed-phase high-performance liquid chroamtographic method witha μ-Bondapak C_{18} column (30 cm x 3.9 mm i.d.) thermostatted at 30°C. The column was protected by a guard column (Waters Guard-Pak pre-column). A mixture of acetonitrile- methanol, dichloromethane (75:20:5 v/v/v) containing 0.1% BHT and 0.05% trie- thylamine (TEA) at a flow rate of was 1 mL/min was used as the mobile phase. The estimated vitamin A activity was 865 μg rectinol equivalents /100 g of sample.

McDermid and Stuercke (2003) evaluated the vitamin content (Vitamin C, and B complex) in Hawaiian macroalgae including Chlorophyta, Phaeophyta, Rhodophyta species. The analysis of the vitamins were performed by high-performance liquid chromatography. *Enteromorpha flexuosa* showed the higher content of vitamin C with 3.0 mg/g. The riboflavin content was very low in the species studied *Ulva fasciata*, presented 0.010 mg/g. Niacin contents ranged from 0.06 to 2.00 mg/g for *Chondrus ocellatus* and *Caulerpa lentillifera*, respectively.

A reversed-phase high-performance liquid chromatography with fluorescence detection was employed to determine thiamin and riboflavin in dried and canned macroalgae. The dried samples were *Himanthalia elongata*, *Laminaria ochroleuca*, *Undaria pinnatifida*, *Palmaria* sp., and *Porphyra* sp. and the canned samples, *H. elongata* and *Saccorhiza polyschides*. The extraction of the analytes from the matrix involves an enzymatic incubation with Clara-Diastase. Thiamin was determined as thiochrome after its oxidation. A 25- × 0.4-cm i.d. Kromasil 100 C_{18} column packed with 5-μm particles (Teknokroma, Barcelona, Spain) was used as the stationary phase, and it was thermostatted at 35°C and 72:28 (v/v) mixture of 0.005M ammonium acetate–methanol at a flow rate of 1.35 mL/min was used as mobile phase. *Porphyra* sp. presente the highest levels of both vitamins, 2.02 ± 0.15 g/g dry weight of thiamine and 6.15 ± 0.49 g/g dry weight.

Folic acid and folates have also been found in seaweeds, Rodríguez-Bernaldo de Quirós et al. (2004) proposed a method that involves the extraction of folates from seaweed by heat treatment followed by deconjugation of folate polyglutamates to their respective monoglutamate forms by incubation with hog kidney conjugase, then the seaweed extracts were concentrated and purified with strong anion-exchange (SAX) solid-phase extraction (SPE) cartridges. The vitamins were separated on a Tracer Extrasil ODS2 column (250mm × 4.6 mm, 5 μm) protected by a guard column TR-C160-1 ODS (15 mm × 4.0 mm, 5 μm). The mobile phase consisted of a gradient of acetonitrile and 30 mM potassium phosphate buffer (pH 2.2) at a flow rate of 0.8 mL/min. UV was used as a detection system for folic acid at a wavelength of 290 nm whereas for 5-methyltetrahydrofolate, tetrahydrofolic acid and 5-formyltetrahydrofolate fluorescence detection was employed (λ_{em} 356 nm λ_{ex} 290 nm). The method was applied to estimate the vitamin contents in several species of seaweeds,

Himanthalia elongata, Laminaria ochroleuca, Palmaria spp., *Undaria pinnatifida Porphyra* spp. and *Saccorhiza polychides*). The amount of folate (as folic acid) in the studied species ranged from 61.4 to 161.6μg per 100 g dry mass.

Several studies reported in the literature have evaluated the vitamin B_{12} contents in seaweeds. Watanabe et al.(1999) determine the vitamin B_{12} in *Enteromorpha* sp. and *Porphyra* sp. by a microbiological and a chemiluminescence method, the values obtained were about 65 μg/100 g of dry wt for *Enteromorpha* sp. and about 30 μg/100 g of dry wt for *Porphyra* sp. Both methods lead to the same results. Miyamoto et al. (2009) also analyzed the vitamin B_{12} in *Porphyra* sp. by a microbiological assay, and they found a value of 133.8 μg/100 g. Seaweeds also content liposoluble vitamins, among them α-tocopherol. Sánchez-Machado et al. (2002) developed a liquid chromatographic method to determine the vitamin in macroalgae. The detection was performed by UV-Vis and fluorescence. With respect to chromatographic conditions a Kromasil 100 C_{18} (25 x 0.4 cm I.D., 5 μm particle size) and a mobile phase consisted of methanol–acetonitrile (50:50 v/v) at a flow-rate of 1.3 ml/min was used in the HPLC-FLD system. The wavelength selected was (λ_{em} 329 nm λ_{ex} 288 nm). In the UV-Vis system the column and mobile phase used were a Tracer Extrasil ODS2 (25 x 0.4 cm I.D., 5 μm) and a mixture of methanol–acetonitrile (30:70 v/v), flow-rate 1.0 ml/min, respectively. The detection was done at 472 nm. LC-MS in positive atmospheric pressure chemical inonization (APCI +) was used as a confirmatory technique. The vitamin was extracted after saponification. Two species of dehydrated seaweeds (*Himanthalia elongata, Laminaria ochroleuca*) and two canned seaweeds (*Himanthalia elongata, Saccorhiza polychides*) were analyzed. The values obtained were: 33.3 ± 4.2 and 8.96±2.1 μg/g dry mass for the dehydrated species *H. elongata* and *L. ochroleuca* and 12.0±2.0 and 5.7±1.3 μg/g dry mass for the canned species *H. elongata* and S. *polychides.*

A study performed by Ortiz et al. (2006) reported the determination of tocopherols and tocotrienols in the edible seaweeds *Durvillaea antarctica* and *Ulva lactuca* by HPLC-FLD. The stationary phase used was a LichroCART Superspher Si 60 column (25 cm x 4 mm i.d., particle size 5 μm) and the mobile phase was propan-2-ol in hexane (0.5:99.5 v/v) at a flow rate of 1 mL/min. Chromatograms were monitored at λ_{em} 330 nm λ_{ex} 290 nm. *D. antarctica* presented the highest values of α- β- and δ-tocopherol and *U. lactuca* showed the highest values of α-tocopherol and γ-tocopherol and α- and γ-tocotrienol.

In another study these authors (Ortiz et al. 2009) evaluated the tocopherols and tocotrienols in the Chilean seaweeds *Codium fragile*, *Gracilaria chilensis* and *Macrocystis pyrifera*. The contents of tocotrienols and tocopherols ranged from 25.2 ± 1.4 to 365.9± 2.3 μg/g lipid and from 7.7 ± 0.5 to 1327.7 ± 4.4 μg/g lipid, respectively.

5. Other Bioactive Compounds

Besides the active compounds reported earlier, other substances with biological activities such as polyunsaturated fatty acids (PUFAs) or sulphated polysaccharides have also been identified in different species of macroalgae. Within the lipidic fraction, the analysis of polyunsaturated fatty acids has gained great interest in recent years due to the beneficial effects of these compounds against certain types of diseases.

Norziah and Ching (2000) evaluated the fatty acids profile in the edible seaweed *Gracilaria changgi*. The analysis was performed by gas chromatography-flame ionization detector (GC-FID), the column used was a fused silica capillary column Omegawax 320 (30 m x 0.32 mm, film thickness 0.25 μm), helium was the carrier gas. The temperature program started at 110 °C and increased until 220 °C at a rate of 8°C/min. The detector temperature was 260 °C. The seaweed analyzed presented high levels of unsaturated fatty acids (74%) particularly omega fatty acids, eicosapentaenoic acid and docosahexanoic.

The fatty acid profile of several species of seaweeds (*Saccorhiza polyschides, Himanthalia elongata, Laminaria ochroleuca, Undaria pinnatifida, Palmaria* sp. Porphyra sp.) collected in the northwest Iberian coast was investigated by Sánchez-Machado et al. (2004) by using GC-FID. The compounds were separated on a Supelco-Wax capillary column (30 m x 0.32 mm, stationary phase thickness, 0.25 μm). The injector and detector temperature were 240 °C and 260 °C, respectively. The was as follows: initially at 160 °C (1 min), then raised at a rate of 3.5 °C/min to 230 °C and held 14 min. The percentage of PUFAs ω3 ranged from 7.20±1.48 to 44.70±5.05. Gas chromatography coupled to mass spectrometry has also been widely used for the determination of fatty acids. Matanjun et al. (2009) studied the fatty acid contents in three seaweeds *Eucheuma cottonii* (Rhodophyta), *Caulerpa lentillifera* (Chlorophyta) and *Sargassum polycystum* (Phaeophyta).A Supelco-Wax (30m x 0.32 mm x 0.25 mm) was used as stationary phase. E. *cottonii* presented the highest levels of PUFAs; all the species studied had in their composition ω-3 fatty acids.

Van Ginneken et al. (2011) evaluated the fatty acids in different species of *seaweeds from north Atlantic and tropical seas including Ulva lactuca, Laminaria* hyperborea, *Palmaria palmata, Ascophyllum nodosum, Sargassum natans,* and *Caulerpa taxifolia,* the ω-3 fatty acids percentages ranged from 8 to 63% of the total fatty acids.

Another interesting group of compounds with beneficial effects on the health present in algae are the sulfated polysaccharides. Guerra Dore et al. (2013) characterized the polysaccharides obtained from *Sargassum vulgare* by means electrophoresis and purified by gel filtration. The monosaccharide composition was also studied by high-performance liquid chromatography. Four heteropolysaccharide fractions were determined in the seaweed, and regarding the monosaccharides, were identified fucose, galactose, xylose, glucuronic acid and mannose. Karmakar et al. (2009) determined sulfated fucan from *Padina tetrastromatica*. For that purpose after the extraction anion exchange chromatography and size exclusion chromatography were applied. Thin layer chromatographic and gas–liquid chromatography–mass spectrometry were used to identified the following monosaccharides, rhamnose, fucose, xylose, mannose, galactose and glucose.

Kolender and Matulewicz (2002) studied the sulfated polysaccharides from *Georgiella confluens*. Seven fractions were isolated. The structural analysis were carried out by using ^{13}C NMR spectroscopy among other techniques.

6. ANTIOXIDANT ACTIVITY

Numerous works are focused on the determination of the antioxidant activity in natural products, a summary of the antioxidant activity exhibited by different species of seaweeds is reported in Table 1, the method used for the determination is also included.

Table 1. Antioxidant activity of different species of seaweeds

Seaweeds	Method	Antioxidant activity	Reference
Fucus vesiculosus, Laminaria ochroleuca, Undaria pinnatifida, Chondrus crispus, Porphyra umbilicalis	Free radical (DPPH) scavenging, ferric-reducing antioxidant power (FRAP)	DPPH: 3.07-91.85 dry matter (DM) g^{-1} DPPH FRAP: N.D-66.13 μmol Trolox g^{-1} DM	Jiménez-Escrig et al. (2001)
Red seaweeds *(Eucheuma cottonii, E. spinosum, Halymenia durvillaei)* green seaweeds *(Caulerpa lentillifera and C. racemosa)* and brown seaweeds (*Dictyota dichotoma, Sargassum polycystum, Padina* sp.)	TEAC (trolox equivalent antioxidant capacity) and FRAP (ferric reducing antioxidant power)	TEAC: 1.49-2.16 mM Trolox equivalents mg^{-1} dry extract FRAP: 153.97-366.69 μM mg^{-1} dry extract	Matanjun et al. (2008)
Red seaweeds (*Gracilaria birdiae, Gracilaria cornea)*	DPPH Radical-Scavenging and β-Carotene-Linoleic Acid Assay	DPPH: IC_{50} methanolic and ethanolic extracts for *Gracilaria birdiae* was in both cases 0.76 mg mL^{-1}. IC_{50} methanolic and ethanolic extracts for *Gracilaria cornea* were 0.86 and 0.77 mg mL^{-1} β-carotene oxidation inhibition of up to 40%	Souza et al. (2011)
Palmaria palmata, Laminaria setchellii, Macrocystis integrifolia, Nereocystis leutkeana	Reducing activity using L-Ascorbic acid as standard.	Reducing activity: 2.23-4.48 μg ascorbic acid equivalents)	Yuan and Walsh (2006)
Palmaria palmata (dulse)	Radical scavenging activity (DPPH), radical cation scavenging activity (ABTS)	DPPH (1,1-diphenyl-2-picrylhydrazyl): EC_{50} 12.5 mg/mL ABTS (2,20-azino-bis(3-ethyl-benzthiazoline-6-sulfonic acid) cation free radical): EC_{50} 29.5 mg/mL	Yuan et al. (2005)
Padina antillarum, Caulerpa racemosa, Kappaphycus alvarezzi	1,1-diphenyl-2-picrylhydrazyl (DPPH) radical scavenging, ferric reducing antioxidant power (FRAP)	DPPH: IC_{50}: 0.337-14.3 mg/mL FRAP: 0.561-15.7 mg GAE/g	Chew et al. (2008)
Sargassum marginatum, Padina tetrastomatica, Turbinaria conoides	DPPH radical scavenging	DPPH radical scavenging activity (%) of total extract: *S. marginatum* 11.00 %; *P. tetrastomatica* 14.78%; *T. conoides* 17.23 %	Chandini et al. (2008)
Halimeda incrassata	β-carotene-linoleate assay	Hight antioxidant activity in polar fractions containing phenolic acids	Vidal Novoa et al. (2011)
Enteromorpha prolifera	DPPH radical scavenging activity	DPPH: radical scavenging activity (% mg/mL) crude extract: 23.1-96.3% at concentrations ranging from 0.06 to1.0 mg/mL	Cho et al. (2011)

Seaweeds	Method	Antioxidant activity	Reference
Rhodomela confervoides, Symphyocladia latiuscula, Grateloupia filicina, Scytosiphon lomentarius, Chondrus ocellatus, Polysiphonia urceolata, Laminaria japonica, Porphyra yezoensis, Ceramium kondoi, Corallina officinalis, Chondria tenuissima, Porphyra haitanensis, Gloiopeltis furcata, Undaria pinnatifida, Chondria crassicaulis, Plocamium telfairiae	β-carotene-linoleate assay	Antioxidant activities (%) of the seaweeds varied from 2.9 to 68.8 at a concentration of 100 ppm and from 14.8 to 79.2 at a concentration of 200 ppm	Huang and Wang (2004)

CONCLUSION

In brief, in the present chapter a review about the analytical methods to determine bioactive compounds, particularly, natural pigments, polyphenols, vitamins, polyunsaturated fatty acids, and sulfated polysaccharides and the antioxidant activity in seaweeds is reported. The chapter is focused mainly in chromatographic techniques. The analytical conditions as well as other interesting features are commented on.

REFERENCES

Bone, R. A.; Landrum, J. T.; Dixon, Z.; Chen, Y.& Llerena, C. M. (2000). Lutein and zeaxanthin in the eyes, serum and diet of human subjects. *Experimental Eye Research,* 71, 239–245.

Chandini, S. K., Ganesan, P. & Bhaskar, N. (2008). In vitro antioxidant activities of three selected brown seaweeds of India. *Food Chemistry,* 107, 707–713.

Chew, Y. L. , Lim, Y. Y., Omar, M. & Khoo, K. S. (2008). Antioxidant activity of three edible seaweeds from two areas in South East Asia. *LWT*, 41, 1067–1072.

Chkhikvishvili, I. D. & Ramazanov, Z. M. (2000). Phenolic substances of brown algae and their antioxidant activity. *Applied Biochemistry and Microbiology*, 36, 289-291.

Cho, M., Lee, H., Kang, I., Wond, M. & You, S. (2011). Antioxidant properties of extract and fractions from Enteromorpha prolifera, a type of green seaweed. *Food Chemistry*, 127, 999–1006.

Dere, Ş.; Dalkiran, N.; Karacaoğlu, D.; Yildiz, G. & Dere, E. (2003). The determination of total protein, total soluble carbohydrate and pigment contents of some macroalgae collected from Gemlik-Karacaali (Bursa) and Erdek-Ormanlı (Balıkesir) in the Sea of Marmara, Turkey. *Oceanologia,* 45, 453-471

Ferraces-Casais, P., Lage-Yusty, M. A., Rodríguez-Bernaldo de Quirós, A. & López-Hernández, J. (2012). Evaluation of bioactive compounds in fresh edible seaweeds. *Food Analytical Methods,* 5, 828-834.

Glombitza, K. W. & Schmidt, A. (1999). Nonhalogenated and Halogenated Phlorotannins from the Brown Alga Carpophyllum angustifolium. *Journal of Natural Products,* 62, 1238-1240.

Guerra Dore C. M. P., Faustino Alves, M. G. das C., Pofírio Will, L. S. E., Costa, T. G., Sabry, D. A., de Souza Rêgo, L. A.R., Accardo, C. M., Rocha, H. A. O., Luciana Guimarães A. Filgueira, L. G. A. & Lisboa Leite, E. (2013). A sulfated polysaccharide, fucans, isolated from brown algae Sargassum vulgare with anticoagulant, antithrombotic, antioxidant and anti-inflammatory effects. *Carbohydrate Polymers,* 91, 467-475.

Gupta, S. & Abu-Ghannam, N. (2011). Bioactive potential and possible health effects of edible brown seaweeds. *Trends in Food Science & Technology*, 22, 315-326.

Hegazi, M. M.; Pérez-Ruzafa, A.; Almela, L. & Candela, M. E. (1998). Separation and identification of chlorophylls and carotenoids from Caulerpa prolifera, Jania rubens and Padina pavonica by reversedphase high-performance liquid chromatography. *Journal of Chromatography A*, 829, 153-159.

Huang, H. & Wang, B. (2004). Antioxidant capacity and lipophilic content of seaweeds collected from the Qingdao Coastline. *Journal of Agricultural and Food Chemistry*, 52, 4993-4997.

Jiménez-Escrig, A., Jiménez-Jiménez, I., Pulido, R. & Saura-Calixto, F. (2001). Antioxidant activity of fresh and processed edible seaweeds. *Journal of the Science of Food and Agriculture,* 81, 530-534.

Karmakar, P., Ghosh, T., Sinha, S., Saha, S., Mandal, P., Ghosal, P. K. & Ray, B. (2009). Polysaccharides from the brown seaweed Padina tetrastromatica: Characterization of a sulfated fucan. *Carbohydrate Polymers,*78, 416-421.

Kinsella, J.E. (1988) Food lipids and fatty acids: importance in food quality, nutrition and health. *Food Technology*, 42, 124-145

Koivikko, R., Loponen, J., Pihlaja, K. & Jormalainen, V. (2007). High-performance liquid chromatographic analysis of phlorotannins from the brown alga Fucus Vesiculosus. *Phytochemical Analysis*, 18, 326-332.

Koivikko, R.; Eränen , J. K., Loponen, J. & Jormalainen V. (2008). Variation of Phlorotannins Among Three Populations of Fucus vesiculosus as Revealed by HPLC and Colorimetric Quantification. *Journal of Chemical Ecology*, 34, 57-64.

Kolender, A. A. & Matulewicz, M. C. (2002). Sulfated polysaccharides from the red seaweed Georgiella confluens. *Carbohydrate Research, 337,* 57–68.

Kumar, C. S.; Ganesan, P.; Suresh, P. V. & Bhaskar, N. (2008). Seaweeds as a source of nutritionally beneficial compounds – A review. *Journal of Food Science and Technology, 45,* 1-13.

Lee, T. H., Hoover, R.L., Williams, J.D., Sperling, R. I., Ravalese, J., Spur, B. W., Robinson, D.R., Corey, E.J., Lewis, R. A. & Austin, K. F. (1985). Effect of dietary enrichment with eicosapentaenoic and docosahexaenoic acids on in vitro neutrophil and monoyte leukotriene generation and neutrophil function. *New England Journal of Medicine, 312,* 1217-1224.

Li, Y. X., Wijesekara, I., Li, Y. & Kim, S. K. (2011). Phlorotannins as bioactive agents from brown algae. *Process Biochemistry*, 46, 2219–2224.

Machlin, L. J.(1995). Critical assessment of epidemiological data concerning the impact of antioxidant nutrients on cancer and cardiovascular disease. *Critical Reviews in Food Science and Nutrition,* 35, 41–50.

Marquardt, J. & Hanelt, D. (2004). Carotenoid composition of Delesseria lancifolia and other marine red algae from polar and temperate habitats. *European Journal of Phycology*, 39, 285-292.

Martínez, J. A. (1998). *Fundamentos teórico-prácticos de Nutrición y dietética*. España: McGraw-Hill, Interamericana.

Matanjun, P., Mohamed, S., Mustapha, N. M., Muhammad, K. & Ming, C. H. (2008). Antioxidant activities and phenolics content of eight species of seaweeds from north Borneo. *Journal of Applied Phycology*,20, 367-373.

Matanjun, P., Mohamed, S., Mustapha, N. M. & Muhammad, K. (2009). Nutrient content of tropical edible seaweeds, Eucheuma cottonii, Caulerpa lentillifera and Sargassum polycystum. *Journal of Applied Phycology*,21, 75-80.

McDermid, K. J. & Stuercke, B. (2003). Nutritional composition of edible Hawaiian seaweeds. *Journal of Applied Phycology*, 15, 513-524.

Miyamoto, E., Yabuta, Y., Kwak, C. S., Enomoto, T. & Watanabe, F. (2009). Characterization of vitamin B12 compounds from korean purple laver (Porphyra sp.) products. *Journal of Agricultural and Food Chemistry*, 57, 2793–2796.

Nollet, L. M. L. (2000). *Food analysis by HPLC*. (Second Edition). New York: Marcel Dekker, Inc.

Norziah, M. H. & Ching, C. Y. (2000). Nutritional composition of edible seaweed Gracilaria changgi. *Food Chemistry*, 68, 69-76.

Ortiz, J., Romero, N., Robert, P., Araya, J., Lopez-Hernández , J., Bozzo, C., Navarrete, E., Osorio, A. & Rios, A. (2006). Dietary fiber, amino acid, fatty acid and tocopherol contents of the edible seaweeds Ulva lactuca and Durvillaea antarctica. *Food Chemistry*, 99, 98–104.

Ortiz, J., Uquiche, E., Robert, P., Romero, N., Quitral, V. & and Llantén, C. (2009). Functional and nutritional value of the Chilean seaweeds Codium fragile, Gracilaria chilensis and Macrocystis pyrifera. *European Journal Lipid Science and Technolgy*, 111, 320-327.

Pangestuti, R.& Kim, S. (2011). Biological activities and health benefit effects of natural pigments derived from marine algae. *Journal of Functional Foods*, 3, 255-266.

Parys, S., Rosenbaum, A., Kehraus, S., Reher, G., Glombitza, K. W. & König, G. M. (2007). Evaluation of Quantitative Methods for the Determination of Polyphenols in Algal Extracts. *Journal of Natural Products*, 70, 1865–1870.

Plaza, M., Cifuentes, A. & Ibáñez, E. (2008). In the search of new functional food ingredients from algae. *Trends in Food Science & Technology*, 19, 31-39.

Rodríguez-Bernaldo de Quirós, A., Castro de Ron, C., López-Hernández, J. Lage-Yusty, M. A. (2004). Determination of folates in seaweeds by high-performance liquid chromatography. *Journal of Chromatography A*, 1032, 135–139.

Rodríguez-Bernaldo de Quirós, A. & Costa, H.(2006). Analysis of carotenoids in vegetable and plasma samples: A review. *Journal of Food Composition and Analysis, 19,* 97–111.

Rodríguez Bernaldo de Quirós, A. & López Hernández, J. (2009). Seafood: A natural source of carotenoids. In Leiv Haugen and Terje Bjornson (Editors), *Beta Carotene: Dietary Sources, Cancer and Cognition* (First Edition, 329-335). New York: Nova Science Publishers, Inc.

Rodríguez-Bernaldo de Quirós, A., Lage-Yusty, M .A. & López-Hernández, J. (2010). Determination of phenolic compounds in macroalgae for human consumption. *Food Chemistry, 121*, 634–638.

Rodríguez-Bernaldo de Quirós, A.; Frecha-Ferreiro, S.; Vidal-Pérez, A. M. &. López-Hernández, J. (2010). Antioxidant compounds in edible brown seaweeds. *European Food Research Technology*, 231, 495-498.

Rogers, J. E. & Marcovich, D. (2007). A simple method for the extraction and quantification of photopigments from Symbiodinium spp. *Journal of Experimental Marine Biology and Ecology,* 353, 191-197.

Sánchez-Machado, D. I., López-Hernández, J. & Paseiro-Losada, P. (2002). High performance liquid chromatographic determination of α-tocopherol in macroalgae. *Journal of Chromatography A, 976,* 277-284.

Sánchez-Machado, D.I. , Ló pez-Cervantes, J. , López-Herná ndez, J. Paseiro-Losada, P. (2004). Fatty acids, total lipid, protein and ash contents of processed edible seaweeds. *Food Chemistry*, 85, 439–444.

Sánchez-Machado, D. I., López-Cervantes, J., López-Hernández, J. & Paseiro-Losada, P. (2004). Simultaneous determination of thiamine and riboflavin in edible marine seaweeds by high-performance-liquid chromatography. *Journal of Chromatographic Science*, 42, 117-120.

Schubert, N.; García-Mendoza, E. & Pacheco-Ruiz, I. (2006). Carotenoid composition of marine red algae. *Journal of Phycology*, 42, 1208-1216.

Souza, B. W. S., Cerqueira, M. A., Martins, J. T., Quintas, M. A. C., Ferreira, A. C. S., Teixeira, J. A. & Vicente, A. A. (2011). Antioxidant potential of two red seaweeds from the Brazilian coasts. *Journal of Agricultural and Food Chemistry, 59,* 5589-5594.

Suzuki, R. & Ishimaru, T. (1992). Characteristics of photosynthetic pigment composition of Gymnodinium mikimotoi Miyake et Kominami ex Oda. *Journal of Oceanography*, 48, 367-375.

Thirumaran, G., Manivannan, K., Karthikai-Devi, G., Anantharaman, P. & Balasubramanian, T. (2009). Photosymthetic pigments of different colour strains of the cultured seaweeds Kappaphycus alvarezii (Doty) Doty ex. P.Silva in Vella Estuary. *Academic Journal of Plant Sciences,* 2, 150-153.

van Ginneken1, V. J.T., Helsper, J.P.F.G ., de Visser, W., van Keulen, H. & Brandenburg, W. A. (2011). Polyunsaturated fatty acids in various macroalgal species from north Atlantic and tropical seas. *Lipids in Health and Disease,* 10, 104-111.

Vidal Novoa, A., Andrade-Wartha, E. R. S., Fallarero Linares, A., Mara de O. e Silva, A., Genovese, M. I. González, A. E. B., Vuorela, P., Costa, A. & Mancini-Filho, J. (2011). Antioxidant activity and possible bioactive components in hydrophilic and lipophilic fractions from the seaweed Halimeda incrassata. *Brazilian Journal of Pharmacognosy,* 21, 53-57.

Watanabe, F., Takenaka, S., Katsura, H., Masumder, S. A. M. Z. H., Abe, K., Tamura, Y. & Nakano, Y. (1999). Dried green and purple lavers (nori) contain substantial amounts of biologically active vitamin B12 but less of dietary iodine relative to other edible seaweeds. *Journal of Agricultural and Food Chemistry, 47,* 2341-2343.

Yoshie, Y., Wang, W., Petillo, D. & Suzuki, T. (2000). Distribution of catechins in Japanese seaweeds. *Fisheries Science,* 66, 998-1000.

Yuan, Y. V., Bone, D. E. & Carrington, M. F. (2005). Antioxidant activity of dulse (Palmaria palmata) extract evaluated in vitro. *Food Chemistry*, 91, 485–494.
Yuan, Y. V. & Walsh, N. A. (2006). Antioxidant and antiproliferative activities of extracts from a variety of edible seaweeds. *Food and Chemical Toxicology*, 44, 1144–1150.
Yuan, Y. V. (2007). Antioxidants from edible seaweeds. *ACS Symposium Series*, 956, 268-301.

In: Seaweeds
Editor: Vitor H. Pomin

ISBN: 978-1-63117-571-8

Chapter 6

POPULATION STUDIES AND CARRAGEENAN PROPERTIES IN EIGHT GIGARTINALES (RHODOPHYTA) FROM IBERIAN PENINSULA

Leonel Pereira*, Filipa Meireles and Rui Gaspar
IMAR-CMA (Institute of Marine Research),
Department of Life Sciences, Faculty of Sciences and Technology,
University of Coimbra, Apartado 3046, 3001-401 Coimbra, Portugal

ABSTRACT

Eight carrageenophytes, representing seven genera and three families of the Gigartinales (Florideophyceae, Rhodophyta), were studied for 15 months. Reproductive status, dry weight and carrageenan content have been monitored by monthly random sampling. The highest carrageenan yields were found in *Chondracanthus acicularis* (61.1%), *Gigartina pistillata* (59.7%) and *Chondracanthus teedei* var. *lusitanicus* (58.0%). Cultivated *Chondrus crispus* from an integrated multi-trophic aquaculture (IMTA) system and wild *C. crispus* were compared regarding carrageenan yield and quality.

Species from the Cystocloniaceae family predominantly produce iota-carrageenans; Gigartinaceae family produces hybrid kappa-iota carrageenans (gametophytic plants) and lambda-family carrageenans (sporophytic plants); the Phyllophoraceae family produces kappa-iota hybrid-carrageenans.

Quadrate (10x10 cm) destructive sampling method was used to determine the carrageenophytes biomass in two periods: autumn/winter and spring/summer. The highest biomass was found in *C. crispus* (570 g/m^2), *C. acicularis* (99 g/m^2), *C. teedei* var. *lusitanicus* (207.5 g/m^2) and *Mastocarpus stellatus* (520 g/m^2).

Quadrate (50x50 cm) non-destructive sampling method, applied along perpendicular transects to the shoreline, was also used to determine the carrageenophytes coverage during spring/summer and along exposed and shelters *facies*. Considering the overall

---

* Corresponding author phone: +351 239 855 241, fax: +351 239 855 211, e-mail: leonel.pereira@uc.pt, URL: http://macoi.ci.uc.pt/

coverage at both exposed and sheltered areas, *C. teedei* var. *lusitanicus* (26.39%) is the most abundant species.

Keywords: Carrageenan, carrageenophytes, IMTA, vibrational-spectroscopy, NMR

1. Introduction

1.1. Carrageenans

Carrageenans are industrially important hydrocolloids that are found in various red seaweeds (Gigartinales, Rhodophyta) [1,2]. Carrageenans are a family of water soluble, linear, sulfated galactans. They are composed of alternating 3-linked β-D-galactopyranose (G-units) and 4-linked α-D-galactopyranose (D-units) or 4-linked 3,6-anhydro-α-D-galactopyranose (DA-units), forming the disaccharide repeating unit of carrageenans. The most common types of carrageenan are traditionally identified by a Greek prefix. The three commercially most important carrageenans are the kappa (κ), iota (ι) and lambda (λ) carrageenans. Their corresponding IUPAC-inspired names and letter codes, developed by Knutsen and collaborators [3], are carrageenose 2,4'-disulfate (G4S-DA2S), carrageenose 4'-sulfate (G4S-DA) and carrageenan 2,6,2'-trisulfate (G2SD2S,6S) [4]. Kappa-carrageenan is present, overall, in the species pertaining to the *Hypneae* (Hypneaceae) and *Kappaphycus* genera (Solieriaceae) and the species belonging to the *Eucheuma* genus (Solieriaceae) are the principal source of iota-carrageenan [5, 6,7]. Kappa/iota hybrid carrageenans are found in the gametophytic life phases of several species in the family Gigartinaceae and Phyllophoraceae [8]. Mu (μ) - and nu (ν)-carrageenans, existing in the native phycocolloid samples, are the biologic precursors of, respectively, kappa and iota-carrageenan [9]. *In vivo*, iota- and kappa-carrageenan are formed enzymatically from their precursor carrageenans by a sulfohydrolase [10,11]. *In vitro*, these precursor residues are converted into the corresponding gelling carrageenan on treatment with alkalis. Alkali extraction is commonly used in the commercial production of kappa and iota-carrageenan to increase 3,6-anhydro-D-galactose content, as it results in a product with enhanced gelling properties [12,13]. The tetrasporic life phase of Gigartinaceae contains carrageenans of the lambda-family. In general, carrageenan serves as gelling (kappa-family carrageenans), stabilising and viscosity-building agent (lambda-family carrageenans) in food products, pharmaceutical formulations, cosmetics and oil well drilling fluid [1,14].

The species *Chondrus crispus* and *Mastocarpus stellatus* were the first seaweeds used for carrageenan extraction. Due to the fact that both species are morphologically very similar and live in similar habitats, both are commonly harvested together. Although *C. crispus* may occur in considerable quantities to a depth of 12 m in the infralitoral zone of the maritime provinces of Canada, in Europe it is found mainly in intertidal zones along with *M. stellatus* [15,16].

In North-East Atlantic, both species are found from North Cape (Norway) to Mauritania [16].

Most coastal countries have had, at one or another moment of their history, an intense harvest. In some of them this activity has decreased and almost disappeared, due to the success of carrageenophyte culture in Asia. An example is the Philippines, where Maxwell S.

Doty pioneered the marine culture of *Eucheuma* and *Kappaphycus* species [17]. Seedlings of varieties which grow rapidly and have good resistance to diseases are tied to lines. The lines are placed at the inside of coral reefs, where the seedlings can grow without floating away [18].

1.2. World Carrageenophyte Production

Carrageenan processors have fuelled the development of *Kappaphycus alvarezii*, formerly known as *Eucheuma cottonii*, (which goes by the name "cottonii" to the trade) and *Eucheuma denticulatum* (commonly referred to as "spinosum" in the trade) farming in several countries including the Philippines, Indonesia, Malaysia, Tanzania, Kiribati, Fiji, Kenya, and Madagascar. Primarily wild-harvested genera such as *Chondrus*, *Furcellaria*, *Gigartina*, *Chondracanthus*, *Sarcothalia*, *Mazzaella*, *Mastocarpus* and *Tichocarpus* are also mainly produced as carrageenan raw materials. Producing countries include Argentina, Canada, Chile, Denmark, France, Japan, Mexico, Morocco, Portugal, North Korea, South Korea, Spain, Russia and the USA [19]. According to Bixler and Porse [20], carrageenan production exceeded 50,000 tonnes in 2007/8 with a value of over US$527 million.

2. Iberian Carrageenophytes

Nowadays the carrageenan obtained from Galician (Spain) natural beds is limited and unable to provide for all industrial requirements, and the raw material is mainly supplied by imported algae. Galician natural seaweed beds exploited for carrageenan production are dominated by *Chondrus crispus* and *Mastocarpus stellatus*. As the morphologies of these two species are very similar, they are frequently collected and sold as a mixture under the names "carrageen", "Irish moss", or "liquen". Galician carrageenan production began in the 1960's when CEAMSA, Marine Algae Company, was established. At start, supplies came mostly from the local resources of *C. crispus* and *M. stellatus* combined with carrageenophytes imported from Canada and the USA. Maximum level of exploitation was attained in the 1970's but in the 80's the activity declined mainly due to competition with developing tropical countries that produced *Eucheuma* and *Kappaphycus*. The drop in production could be also a consequence of the depletion of the natural resource caused by over-exploitation. The impairment of the resource resulted in the prohibition of *C. crispus* and *M. stellatus* harvesting during the year of 1987 [21].

In Portugal, *C. crispus* and *M. stellatus* are mainly harvested in Viana do Castelo, but not at significant industrial level. Generally, the harvest is made within a period depending on the local customs, climate and sea behaviour. *C. crispus* and *M. stellatus* have been exploited, especially in Galicia and northern Portugal (Minho, Douro and Beira Litoral) [22,23]. The carrageenophytes are plucked from the intertidal zone during summer, they are sun-dried, sold to concentrators and then exported [23,24]. Portuguese seaweed harvesting regulations, which first date from 1909, specify that the seaweed harvesting period is from July to November, they require the issuing of licences for each area of the coastline, and they impose price control [15,23].

In 1955, all carrageenans were obtained from *C. crispus* and *M. stellatus*, but today these two species do not represent more than 10% of the total raw material for extraction. Most of the currently used seaweeds in the world are cultivated species belonging to the genus *Eucheuma* and *Kappaphycus* as source of iota and kappa-carrageenan, respectively [1,25,26]. It is worthy to search for new sources of kappa/iota hybrid, iota-carrageenan and other sources of particular types of carrageenans from some underutilized carrageenophytes.

This chapter is focused on the study of population and phycocolloid ecology of several underutilized Gigartinales from the Iberian Peninsula: *Chondracanthus teedei* var. *lusitanicus* (Rodrigues) Bárbara *et* Cremades (Gigartinaceae), *Chondracanthus acicularis* (Roth) Fredericq, *Gigartina pistillata* (S.G. Gmelin) Stackhouse (Gigartinaceae), *Calliblepharis jubata* (Goodenough *et* Woodward) Kützing (Cystocloniaceae), *Gymnogongrus crenulatus* (Turner) J. Agardh (Phyllophoraceae), *Ahnfeltiopsis devoniensis* (Greville) P.C. Silva *et* DeCew (Phyllophoraceae) and to compare them with the traditionally harvested carrageenophytes *C. crispus* Stackhouse (Gigartinaceae) and *M. stellatus* (Stackhouse) Guiry (Phyllophoraceae).

For this intention, a natural population of mixed carrageenophytes located in Buarcos bay (North-West Portuguese Atlantic coast, Iberian Peninsula) was studied during 15 months. Quadrate (10x10 cm) destructive sampling method was used to determine the carrageenophytes biomass in two periods: autumn/winter and spring/summer. To determine the nature of the produced phycocolloid we examined and quantified the native and alkali-modified carrageenan, extracted from different phases of the life history of the studied carrageenophytes (tetrasporophyte, female gametophyte and non-fructified thalli) with FTIR-ATR, FT-Raman and ^{1}H-NMR. The nature of the native polysaccharides (i.e., without extraction) present in these seaweeds was determined by FTIR-ATR and FT-Raman analysis of the ground dry seaweed samples [27,28].The composition and abundance of the carrageenophytes species present along different exposed and sheltered *facies* were study during spring and summer (3 months) at the intertidal rocky shores of Buarcos bay (recent unpublished data). 50x50 cm photo-quadrate replicates (non-destructive sampling method) were taken along perpendicular transects to the shoreline. Coverage values (expressed in %) were calculated *per* species by drawing polygons on 50x50 cm photo-quadrate replicates, using the Quantum GIS (QGIS) 1.8.0 Lisboa software.

2.1. Physical and Chemical Data

In Buarcos bay the average water temperature varies between 12 °C in autumn/winter and 22 °C in spring/summer. The air average temperature varies between 10 °C, in the first period, and 23 °C in the second one. pH and salinity are relatively constant throughout the year, with an average value of 8.3 and 32.8 S‰, respectively [29].

2.2. Coverage

Different carrageenophyte species exhibit different patterns of coverage along the exposed and sheltered *facies* gradient (Figure 1). In general, carrageenophytes are more abundant in exposed *facies*, but a clear exception can be made for the species *C. teedei* var. *lusitanicus*, which is much more abundant at sheltered areas.

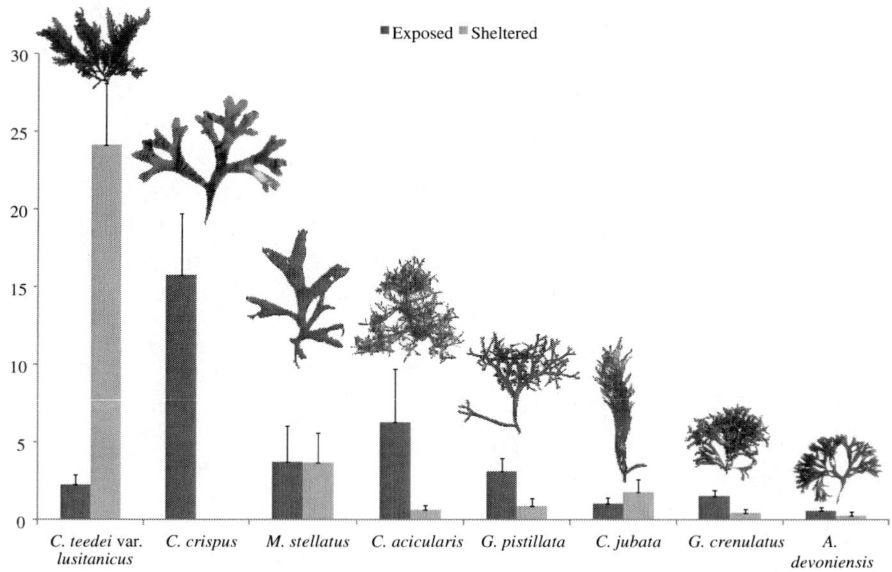

Figure 1. Carrageenophytes coverage along exposed and sheltered *facies* during spring/summer (average + standard error, n= 24).

In fact, considering the overall coverage at both exposed and sheltered areas, *C. teedei* var. *lusitanicus* (26.39%) is the more abundant species, followed by *C. crispus* (15.75%), *M. stellatus* (7.41%), *C. acicularis* (6.93%), *G. pistillata* (4.02%), *C. jubata* (2.82%), *G. crenulatus* (1.96%) and *A. devoniensis* (0.90%). *C. crispus* prefers exposed areas, avoiding completely sheltered habitats. *M. stellatus* is apparently indifferent in terms of abundance along exposure gradients. *C. acicularis, G. pistillata, G. crenulatus* and *A. devoniensis* seem to prefer exposed habitats and *C. jubata* seem to prefer sheltered habitats.

2.3. Biomass and Plant Size

Regarding coverage (see Figure 1) and the available biomass for harvesting, *C. teedei* var. *lusitanicus*, *C. crispus* and *M. stellatus* are the dominant species. The highest biomass value (570 ± 1.8 gm^{-2}, n = 10) and carrageenan content (see Table 1) of *C. crispus* was registered in spring/summer. The maximum average length was 13.8 ± 1.2 cm (n = 100) in the summer and a minimum of 8.2 ± 0.5 cm (n = 100) in the winter [29,30].

Although *M. stellatus* comes only in third place in seaweed coverage, it shows high biomass (520 ± 2.0 gm^{-2}, n = 8) in spring/summer. The average length of this species was 6.3 ± 0.5 cm (n = 13), with a maximum of 9.5 ± 1.2 cm (n = 100) in summer and a minimum of 4.1 ± 0.8 cm (n = 100) in winter [29,30]. Data on seasonal variation length show statistical significance (one-way ANOVA, $p < 0.001$).

Not being a harvested seaweed, it is surprising that, among all the carrageenophytes studied, *C. teedei* var. *lusitanicus* has one of the highest values of average carrageenan yields

(Figure 2), coverage (Figure 1) and biomass (594 ± 10.5 gm^{-2}, n = 10) in spring/summer. The maximum average length was 7.7 ± 0.4 cm (n = 100) in summer and a minimum of 2.9 ± 0.2 cm (n = 100) in winter. The biomass values ranged between 110 ± 1.9 gm^{-2} (n = 8), in summer and 594 ± 10.5 gm^{-2} (n = 8), in late spring. Biomass and length had low values in autumn and winter, as a small increase occurred in early spring and the highest values of these parameters were recorded in early summer [29,30].

The highest carrageenan yields were recorded in *G. pistillata* with an average of the 38.7% (Figure 2). Nevertheless coverage (Figure 1) and biomass are relatively low. *C. acicularis* has the fourth highest coverage value (Figure 1), even though this carrageenophyte has a low biomass, especially in spring/summer (22 ± 0.2 gm^{-2}, n = 8). This carrageenophyte has an isomorphic triphasic lifecycle, but the non-fertile thalli and the tetrasporophytes are difficult to distinguish from one another, and consequently these reproductive phases were not separated. Due to the thalli nature, very thin and tangled, plant size data were not collected. Coverage (Figure 1) and biomass are very low for the remaining carrageenophytes *C. jubata*, *G. crenulatus* and *A. devoniensis* [29,30].

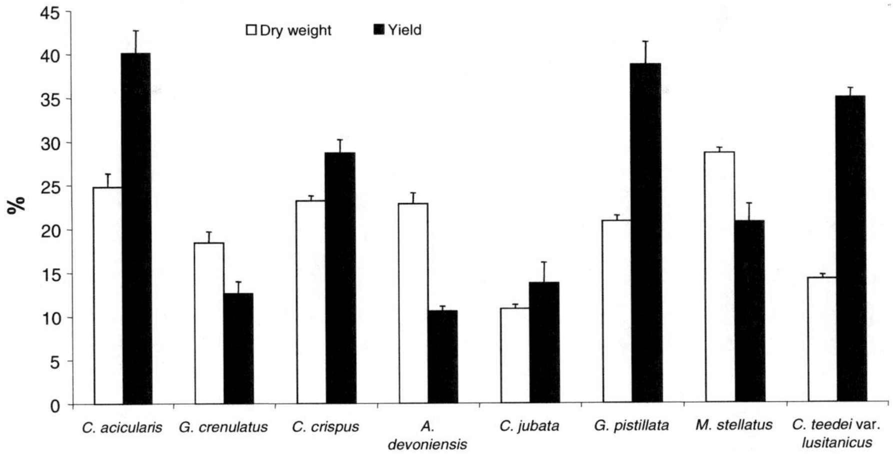

Figure 2. Carrageenophytes dry weight, expressed as percentage of fresh weight and carrageenan yields, expressed as percentage of dry weight (average + standard error, n = 13) (adapted from Pereira 2004, 2013 [29,30]).

2.4. Lifecycle Phase

C. crispus has an isomorphic triphasic lifecycle [31,32,33], although in the studied population, the non-fructified thalli were dominant in most samples (see Figure 3) and the percentage varied between 15.4% (summer) and 66.7% (winter). The female gametophytes were present in all samples, varying the proportion from 12.5% (late winter) to 44.5% (spring). The tetrasporophytes were also present in all samples, with a maximum of 50% in the spring and a minimum of 11.1% in the summer. Compared with female gametophytes (cystocarpic thalli) (29.4 ± 2.9%, n = 14), the tetrasporophytes (tetrasporic thalli) are usually less abundant (24.5 ± 3.4%, n = 14); the predominance of gametophytes in *C. crispus* populations was also found in other works [33,34,35]. The average percentage of non-fructified thalli was 46.1 ± 3.9% (n = 14). The data on seasonal variation in the percentage of individuals of each generation have statistical significance (one-way ANOVA, p <0.001).

Table 1. Yield and carrageenan composition (adapted from Pereira 2004, 2013 [29,30])

Family	Species	Lifecycle phase	Harvest Season	Carrageenan			
				Yield [1]	Letter code [2]	Alkali-extracted [3] (%mol)	Native [4]
Cystocloniaceae	*Calliblepharis jubata*	NF	Spring	20.0	G4S-DA2S	100 ι	ι (ν)
	C. jubata	T	Spring	28.4	G4S-DA2S, G4S-DA	98 ι, 2 κ	ι - κ (ν)
	C. jubata	FG	Spring	40.4	G4S-DA2S	100 ι	ι (ν)
Gigartinaceae	*Chondracanthus acicularis*	NF	Summer	61.1	G4S-DA, G4S-DA2S	-	κ - ι (μ / ν)
	C. acicularis	T	Summer	36.6	G2S-D2S, G2S-DA2S	59 ξ, 41 θ	ξ - θ
	C. acicularis	FG	Late Summer	49.8	G4S-DA, G4S-DA2S	61 κ, 34 ι, 5 pyruvate	κ - ι (μ / ν)
	Chondracanthus teedei var. *lusitanicus*	NF	Late Summer	41.0	G4S-DA, G4S-DA2S	56 κ, 44 ι	κ - ι (μ / ν)
	C. teedei var. *lusitanicus*	T	Summer	58.0	G2S-D2S, G2S-DA2S	67 ξ, 33 θ	ξ - θ
	C. teedei var. lusitanicus	FG	Late Summer	57.4	G4S-DA, G4S-DA2S	58 κ, 42 ι	κ - ι (μ / ν)
	Chondrus crispus	NF	Late Summer	36.0	G4S-DA, G4S-DA2S	64 κ, 36 ι	κ - ι (μ)
	C. crispus	FG	Summer	36.8	G4S-DA, G4S-DA2S	70 κ, 30 ι	κ - ι (μ)
	C. crispus	T	Summer	58.0	G2S-D2S,6S	100 λ	λ
	Gigartina pistillata	T	Late Summer	57.0	G2S-D2S, G2S-D2S,6S	-	ξ - λ
	G. pistillata	H (⊕)	Summer	58.5	G2S-D2S, G2S-D2S,6S	-	ξ - λ
	G. pistillata	FG	Summer	59.7	G4S-DA, G4S-DA2S	48 κ, 45 ι, 7 pyruvate	κ - ι (μ / ν)
	G. pistillata	H (♀)	Summer	58.5	G4S-DA, G4S-DA2S	49 κ, 45 ι, 6 pyruvate	κ - ι (μ / ν)

Table 1. (Continued)

Family	Species	Lifecycle phase	Harvest Season	Carrageenan			
				Yield (1)	Letter code (2)	Alkali-extracted (3) (%mol)	Native (4)
Phyllophoraceae	*Ahnfeltiopsis devoniensis*	G	Summer	13.6	G4S-DA2S, G4S-DA	82 ι, 18 κ	ι - κ (ν)
	Gymnogongrus crenulatus	TB	Late Spring	23.3	G4S-DA, G4S-DA2S	64 κ, 29 ι, 7 pyruvate	κ - ι (μ / ν)
	Mastocarpus stellatus	NF	Late Summer	41.4	G4S-DA, G4S-DA2S	62 κ, 36 ι, 2 pyruvate	κ - ι (μ / ν)
	M. stellatus	FG	Late Summer	41.0	G4S-DA, G4S-DA2S	59 κ, 41 ι	κ - ι (μ / ν)

T – Tetrasporophyte; FG – Female Gametophyte; G – Gametophyte; NF – Non-fructified thalli; TB – Tetrasporoblastic thalli; H – Heterosporic thalli: (♀ cistocarpic branch; ⊕ tetrasporic branch); 1 – Yields (maximum value found) expressed as percentage of dry weight; 3 – composition determined by ^{1}H-NMR; 4 – composition determined by FTIR-ATR and FT-Raman analysis of ground seaweed samples; the carrageenans are identified according to the Greek lettering system and letter code (2) proposed by Knutsen et al. (1994); the letters between parentheses ( ) correspond to the biological precursors of the carrageenans, present in native carrageenan samples (or ground seaweed). Kappa (κ), Mu (μ), Iota (ι), Nu (]), Theta (θ), Lambda (λ), and Xi (ξ).

G. pistillata is the type species of the genus *Gigartina* [36,37]. Despite possessing an isomorphic triphasic lifecycle [38], *G. pistillata* has heterosporic thalli (i.e., producing both tetraspores and carpospores) [39,40], which were found in all studied samples. The tetrasporophytes were dominant in most samples, and the percentage varied between 10% (spring) and 81.8% (late autumn). The female gametophytes were present in all samples, varying the proportion from 6.8% (late autumn) to 90% (spring). The heterosporic thalli were present in all samples, with a maximum of 30.8% in early winter and a minimum of 1.9% in the summer. The data on seasonal variation in the percentage of individuals of each generation have statistical significance (one-way ANOVA, $p < 0.001$).

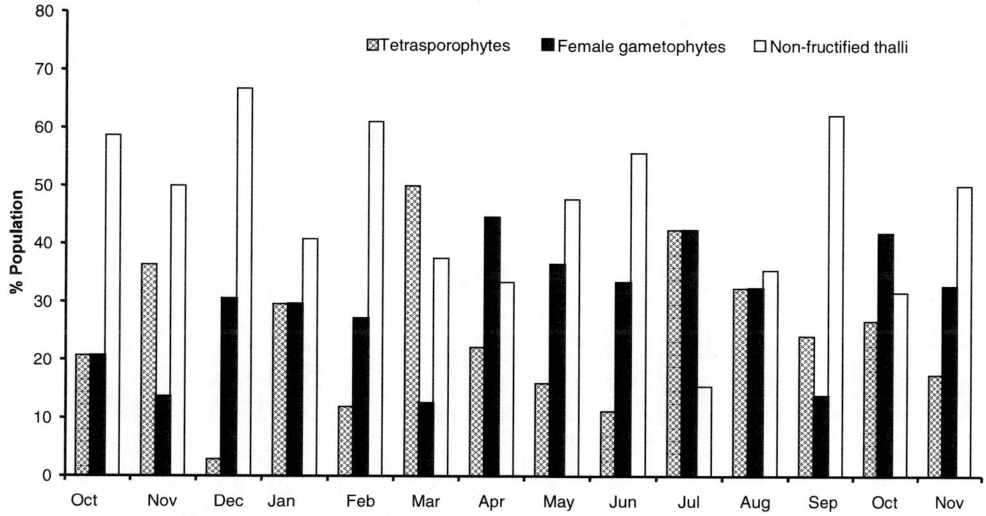

Figure 3. Population structure of *Chondrus crispus* (after Pereira 2013 [30]).

C. acicularis has an isomorphic triphasic lifecycle [37], but the non-fructified thalli and the tetrasporophytes are difficult to distinguish from one another, and consequently these reproductive phases were not separated.

M. stellatus, formerly *Petrocelis cruenta*, has a heteromorphic triphasic lifecycle with a sporophytic crust [41]. Only the gametophytic phase was studied. The studied population had a predominance of non-fructified plants in the winter and a predominance of female gametophytes in spring/summer (Figure 4).

C. teedei var. *lusitanicus* has an isomorphic triphasic lifecycle [42] and in the Buarcos bay population the non-fructified thalli were dominant in all samples (see Figure 5), the percentage varied between 43% (early autumn) and 82.5% (early summer). The female gametophytes were present in all samples, varying the proportion from 3% (late autumn) to 29% (late summer). The tetrasporophytes were also present in all samples, with a maximum of 32.5% in autumn and a minimum of 4% in the summer. As compared to fructified thalli, namely the female gametophytes bearing cystocarps (9.6 ± 1.7%, n = 17), the tetrasporophytes are generally more abundant (21 ± 1.7%, n = 17). The average percentage of non-fructified thalli was 69.4 ± 2.2% (n = 17). The data on seasonal variation in the percentage of individuals of each generation have statistical significance (one-way ANOVA, $p < 0.001$).

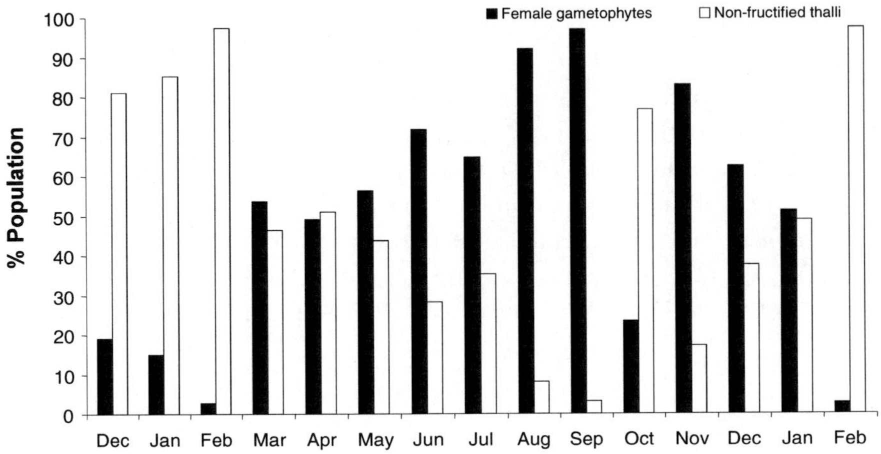

Figure 4. Population structure of *Mastocarpus stellatus* (after Pereira 2013 [30]).

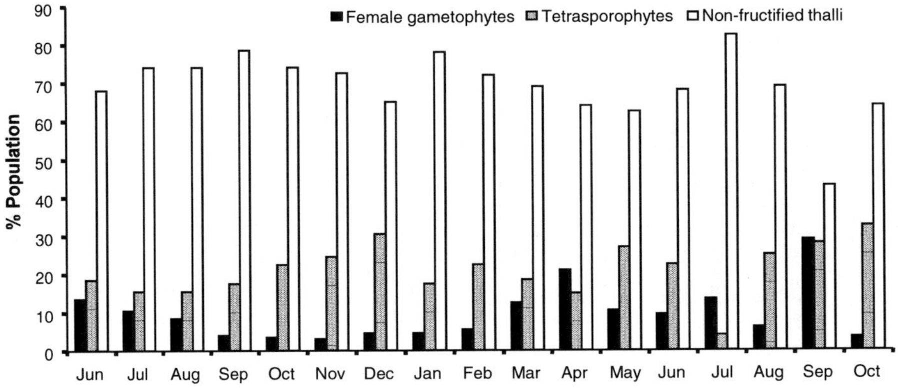

Figure 5 – Population structure of *Chondracanthus teedei* var. *lusitanicus* (adapted from Pereira 2004, 2013 [29,30]).

The *C. jubata* shows an isomorphic triphasic lifecycle [43] and the non-fructified plants were dominant in all samples, except in a summer sample (August); the percentage varied between 28.6% (summer) and100.0% (autumn and early winter) (Figure 6). The female gametophytes were present in 11 out of 13 samples, with a maximum of 17.3% in the spring and a minimum of 5.3% in the winter. Comparing the fructified plants, tetrasporophytes are always less abundant (11.3 ± 1.5%, n = 13) against the female gametophytes (20.3 ± 5.3%, n = 13). The tetrasporophytes were present in 4 out of 13 samples, with a maximum of 17.3% in the spring and a minimum of 5.3% in winter. Comparing the fructified plants, particularly female gametophytes (20.3 ± 5.3%, n = 13), tetrasporophytes are always less abundant (11.3 ± 1.5%, n = 13). The average percentage of non-fructified thalli was 20.8 ± 5.6% (n = 13). The data on seasonal variation in the percentage of individuals of each generation have statistical significance (one-way ANOVA, $p < 0.001$).

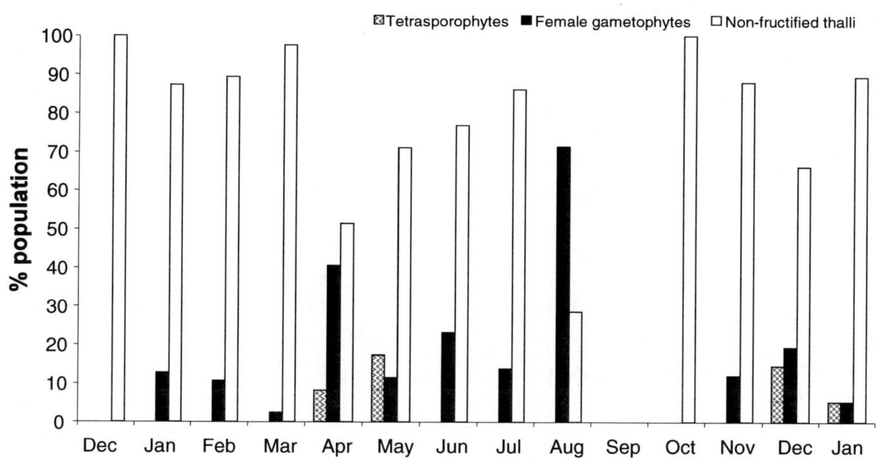

Figure 6. Population structure of *Calliblepharis jubata* (adapted from Pereira 2004, 2013 [29,30]).

The female gametophytes of *G. crenulatus* (a species with a digenetic lifecycle) have reproductive structures known as tetrasporoblasts [44,45]; these structures appear as external wart-like excrescences. All collected thalli showed tetrasporoblasts. *A. devoniensis* has a heteromorphic triphasic lifecycle with crustose tetrasporophytes [46]. Only the gametophytic phase was studied. The cystocarps were present in the gametophytes sampled in December, January and March.

2.5. Variation in Dry Weight and Carrageenan Content

Average dry weight and carrageenan content values are given in Figure 2. Maximum carrageenan content in *C. crispus* (see Table 1) was found in a tretrasporic thalli sample, collected in the summer, with 53.2% of dry weight; a sample of female gametophytes collected in late autumn had the minimum value, with 10% of dry weight. The data on seasonal variation of carrageenan content show statistical significance (one-way ANOVA, $p < 0.01$).

Maximum carrageenan content in *G. pistillata* (Table 1) was found in a sample of female gametophytes, with 59.7% of dry weight in late spring; a sample of heterosporic thalli collected in late autumn had the minimum value, with 22.7% of dry weight. The data on seasonal variation of carrageenan content show statistical significance (one-way ANOVA, $p < 0.001$).

The population of *M. stellatus* is composed of about 50% female gametophytes and 50% non-fructified thalli, both showing a similar maximum carrageenan content (see Table 1) in summer samples, with about 41% dry weight. However, the average carrageenan content ($20.7 \pm 2.0\%$, $n = 14$) is relatively low when compared to other carrageenophytes from Buarcos bay (Figure 2). The data on seasonal variation of carrageenan content show statistical significance (one-way ANOVA, $p < 0.001$).

Dry matter in *C. teedei* var. *lusitanicus* (Figure 2) varied between $12.1 \pm 1.5\%$ ($n = 3$) in late autumn and $17.5 \pm 0.8\%$ ($n = 3$) in the winter. The maximum carrageenan content (58%) was found in a tetrasporophyte sample collected in the summer (Table 1) and the minimum content was found in a non-fructified sample, with 23% in the winter. Overall phycocolloid content was minimum (26.4 ± 2.5%, n = 3) in the winter and maximum (43.6 ± 12.5%, n = 3) in early summer. The data on seasonal variation of dry weight and carrageenan content have statistical significance (one-way ANOVA, $p < 0.001$ and $p < 0.05$, respectively).

The average dry weight in *C. acicularis* ranged from 18.3 ± 0.2% (n = 3) in late spring to33.0 ± 2.7% (n = 3) in autumn. The average dry weight was 24.8 ± 2.7 (n = 13). Maximum carrageenan content (61.1%) was found in a non-fructified thalli sample collected in the summer (Table 1) and the minimum content was found in a female gametophyte sample, with 21.7% in autumn. Overall phycocolloid content was minimum (29.4 ± 4.3%, n = 3) in autumn and maximum (55.5 ± 4.6%, n = 3) in summer. The data on seasonal variation of dry weight and carrageenan content have statistical significance (one-way ANOVA, $p < 0.001$).

All three remaining species, *C. jubata*, *G. crenulatus* and *A. devoniensis*, are producers of iota-carrageenan and have been studied in other countries (Mathieson et al. 1984, Cosson et al. 1990, Zinoun and Cosson 1996). In general, our results are in accordance with those studies. The carrageenan content of these species is relatively low, varying from 10.1% in *A. devoniensis* to 13% in *C. jubata*, with 12.4% in *G. crenulatus*. Furthermore, the coverage (Figure 1) and biomass are very low.

2.6. Carrageenan Composition

The ground seaweed samples FTIR-ATR spectra (data not shown) of *C. crispus*, *M. stellatus*, *G. pistillata* (female gametophytes) and non-fructified plants of *C. crispus* exhibit strong absorption bands in the region of 930 cm^{-1} (DA) and in the region of 845 cm^{-1} (G4S), typical of the kappa-carrageenan. These spectra have low absorbance in the 805 cm^{-1} region (DA2S), related to the presence of iota-carrageenan [28]. The female gametophytes of *C. crispus*, *M. stellatus*, *G. pistillata* and the non-fructified plants of *C. crispus* ground seaweed FT-Raman spectra show two bands in the region of 807 cm^{-1} (DA2S) and 850 cm^{-1} (G4S), typical of kappa/iota hybrid carrageenans. The occurrence of additional peaks at 821 cm^{-1} (G/D6S), 830 cm^{-1} (G/D2S) and 870 cm^{-1} (G/D6S) indicates the presence of mu- and nu-carrageenan, the biological precursors of kappa- and iota-carrageenan, respectively [28]. Our results agree with those obtained in other studies with *C. crispus* [8,47,48,49], *M. stellatus* [8,50,51] and *G. pistillata* [52,53,54].

In female gametophytes and non-fructified thalli of *C. teedei* var. *lusitanicus*, the FTIR-ATR spectra show strong absorption at 930 cm^{-1} (DA) and 845 cm^{-1} (G4S) and median absorption in the 805 cm^{-1} band (DA2S). Additional peaks at 867 cm^{-1} (G/D6S), 825 cm^{-1} (G/D2S) and 820 cm^{-1} (G/D6S), with small intensity, indicate the presence of carrageenan precursors (mu and nu). The presence of bands at 820 cm^{-1}, 825 cm^{-1} and 867 cm^{-1}, corresponding to the existence of precursors, is more evident in the FT-Raman spectra. These results agree with those obtained in other studies with *C. teedei* [8,28,42,55].

For the species *A. devoniensis*, *G. crenulatus* and *C. jubata*, the FTIR-ATR spectra show absorption bands at 930, 845 and 805 cm^{-1}, which represents the characteristic triplet of the fraction iota, when dominant in a hybrid carrageenan. The FT-Raman spectra of *A. devoniensis* and *G. crenulatus* show two bands in the region of 807 cm^{-1} (DA2S) and 850 cm^{-1} (G4S), typical of hybrid kappa/iota carrageenans. The peaks related to the carrageenan precursors, mu and nu, are 821 cm^{-1} (G/D6S), 830 cm^{-1} (G/D2S) and 870 cm^{-1} (G/D6S) [28].

The intensities of the resonances in ^{1}H-NMR spectra [4,13,56] were used in order to quantify the different carrageenan fractions (see Table 1). The alkali-extracted carrageenans showed lower sulphate content and a decrease in galactose to the benefit of 3,6-anhydrogalactose. This corresponds to the conversion of the 4-linked galactose-6-sulfate in native samples to anhydrogalactose in the alkali-extracted carrageenans. Thus the carrageenan precursor's mu and nu were converted into kappa- and iota-carrageenan, respectively [42].

3. SEAWEED AQUACULTURE

Natural seaweed beds are limited and unable to supply the world industrial requirements for agar, alginates and carrageenan. According to FAO's 2012 Aquaculture and Fisheries Report, seaweed production increased from 3.8 million tonnes in 1990 to 19 million tonnes in 2010, when cultivation had already overshadowed production of seaweed collected from the wild, which accounted for only 4.5 percent of total seaweed production in 2010 [57]. *In fact,* aquaculture continues to be the fastest growing animal food producing sector worldwide, and in 2006 it had already equaled wild fisheries in the world's fish supply. Intensive aquaculture activities, particularly in marine waters, generate an excess of nutrients/organic matter, and, in this context, the integration of different trophic levels has emerged to relieve some of this stress. Integrated multitrophic aquaculture (IMTA) is an ecological engineering tool to convert monoculture into an ecological and, thus, more sustainable aquaculture. In IMTA, seaweeds assimilate the fish-excreted ammonia, phosphate and CO_2, converting them into potentially valuable biomass.

In Portugal, seaweeds are being integrated in aquaculture farming by the hand of scientific researchers, who were already working with these organisms, and interested in studying and implementing the concept of IMTA. In seaweeds from aquaculture, particularly land-based systems, conditions such as culture density and water flux in the tanks, among others, may affect morphological and/or histological aspects of the alga, as average frond size/area seems to be higher in wild specimens [58]. A comparative histological analysis of IMTA-cultivated *Chondrus crispus* specimens from a land-based IMTA system at Aveiro estuary (ALGAPlus) and specimens of the same alga collected at Buarcos bay did not unveil significant differences (Figure 7). As in wild *C. crispus*, IMTA-cultivated *C. crispus* tetrasporophytes exhibit tetrasporangial sori with an appearance of dark red spots, prominent in the thallus, on the main axis and lateral branches. In the female gametophytes, the cystocarps, usually in large numbers on the pinnules, margins of the branches and thalli surface in wild specimens, have similar density and distribution as in the cultivated individual fronds (Figure 7) [60].

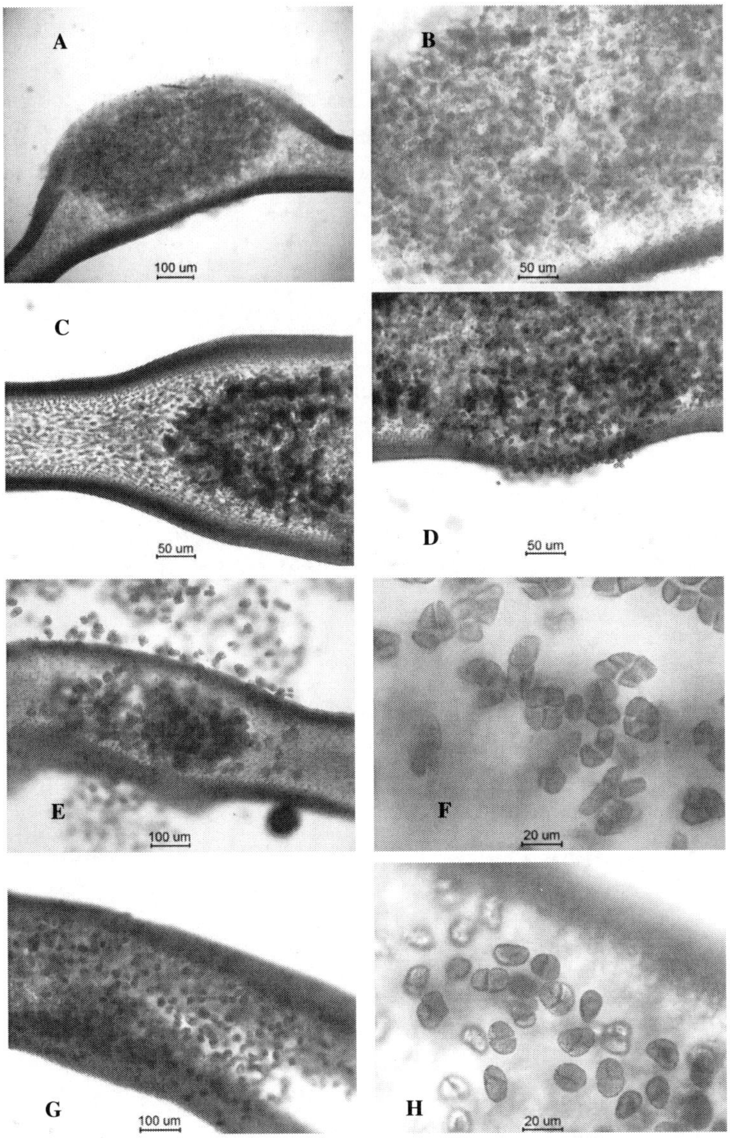

Figure 7. Cross sections of *C. crispus* thalli specimens at optical microscope. A, B Cross sections of female gametophyte thalli of wild specimens of *C. crispus* harvested in Buarcos. C, D Cross sections of female gametophyte thalli of IMTA-cultivated *C. crispus* specimens. A-D: cystocarp cross section, with carpospores within. E, F Cross sections of tetrasporophyte thalli of wild specimens of *C. crispus* harvested in Buarcos. G, H Cross sections of tetrasporophyte thalli of IMTA-cultivated *C. crispus* specimens. E-G: tetrasporangial sori cross section. F and H: detail of tetraspores (after Meireles 2013 [60]).

CONCLUSION

It can be commonly acknowledged that macroalgae species have seasonal fluctuations, being spring and summer seasons the periods of maximum development, in temperate seas, for most macroalgae species; also, that species vary along different spatial gradients in their habitats such the ones created by different wave exposures. Carrageenophyte species are apparently more abundant in exposed *facies*, but a clear exception can be made for the species *C. teedei* var. *lusitanicus*, which is much more abundant in sheltered areas.

Several investigators [25,45,48,52,59] have described seasonal fluctuations on the carrageenan quantity and composition in some Gigartinales. Most of the seaweeds described in this chapter have high carrageenan contents in summer (*C. crispus*, *C. teedei* var. *lusitanicus*, *G. pistillata*, *C. jubata* and *G. crenulatus*); however, the maximum carrageenan content for *C. acicularis* and *A. devoniensis* is found in autumn/winter.

In *C. crispus*, the carrageenan content was low in autumn and winter, a small increase occurred in early spring (April), and the largest carrageenan content was found in samples collected in the summer (July). In *G. pistillata*, the carrageenan content was low in autumn and winter, with a large increase occurring in early spring (March), and the highest carrageenan content was recorded in samples collected in spring (June) and summer (July and August). In *C. acicularis*, carrageenan content was low in autumn and winter, with a small increase occurring in early spring (March), and the highest carrageenan content was recorded in samples collected in the summer (July). In *M. stellatus*, the carrageenan content was low in winter and spring, a small increase occurred in early summer (June), and the highest carrageenan content was obtained with samples collected at the end of summer (September). In *A. devoniensis*, the carrageenan content was low throughout the studied period, with a small increase occurring only in the summer (June). In *G. crenulatus*, the carrageenan content was low during autumn, and small increase occurred in the winter; the spring samples and particularly the summer samples have higher carrageenan content. Finally, in *C. jubata*, the yield was low in autumn and winter, and the highest carrageenan content was obtained with samples collected in the spring (May).

Therefore, by the combination of high biomass and carrageenan content available in summer, we can conclude that this is the best season for harvesting the Iberian dominant carrageenophytes, with exception made for *C. acicularis*, which should be harvested in autumn/winter. Other studies carried out in North Atlantic coasts indicate an increase in carrageenan content during summer and a decrease in winter, namely in *C. crispus* [48], *C. jubata* [25,59] and *G. crenulatus* [45].

Considering the nature of the phycocolloids, our vibrational and resonance spectroscopic analysis showed that the studied Portuguese carrageenophytes indicate a similar composition to that found in other species of Cystocloniaceae, Gigartinaceae and Phyllophoraceae families [6].

In conclusion, some species found in the North-West coast of the Iberian Peninsula could be used for industrial applications. Kappa, kappa-iota hybrid and lambda fractions can be provided by harvesting *C. crispus*, *M. stellatus*, *C. teedei* var. *lusitanicus* and C. *acicularis*. However, responsible harvesting of natural populations must be always the norm, as non-sustainable procedures can have severe economic and environmental impacts. Under this perspective, land-based IMTA systems are promising and more sustainable sources of seaweed biomass. Due to its limited stock in the West coast, *G. pistillata* (source of hybrid kappa-iota and xi-lambda carrageenans), *C. jubata*, *G. crenulatus* and *A. devoniensis* (sources of iota-carrageenan) should be, in the future, objects of culture assays in order to research their potentialities in biomass production.

ACKNOWLEDGMENTS

The authors would like to thank ALGAPlus (Ílhavo, Portugal) for IMTA-cultivated *Chondrus crispus* samples and particularly the support of Dr.Helena T. Abreu.The authors also acknowledge the financial support of a PhD scholarship attributed to Rui Gaspar and the financial support of the Institute of Marine Research (IMAR-CMA), provided by the Portuguese Foundation for Science and Technology (FCT).

REFERENCES

[1] van de Velde F, de Ruiter GA. Carrageenan. In: Vandamme EJ, Baets SD, Steinbèuchel A, editors. Biopolymers V6, Polysaccharides II, *Polysaccharides from Eukaryotes*. Weinheim: Wiley; 2002. p. 245-274.

[2] Therkelsen GH. 1993. Carrageenan. In: Whistler RL, BeMiller JN, editors. Industrial gums: *Polysaccharides and their derivatives*. San Diego: Academic Press Inc.; 1993. p. 145-180.

[3] Knutsen SH, Myslabodski DE, Larsen B, Usov AI. A modified system of nomenclature for red algal galactans. *Botanica Marina*, 1994, 37, 163-169.~

[4] van de Velde F, Pereira L, Rollema HS. The revised NMR chemical shift data of carrageenans. *Carbohydrate Research*, 2004, 339, 2309-2313.

[5] Craigie JS 1990. Cell walls. In: Cole KM, Sheath RG, editors. *Biology of the red algae*. Cambridge: Cambridge University Press; 1990. p. 221-257.

[6] Chopin T, Kerin BF, Mazerolle R. Phycocolloid chemistry as taxonomic indicator of phylogeny in the Gigartinales, Rhodophyceae: A review and current developments using Fourier transform infrared diffuse reflectance spectroscopy. *Phycological Research*, 1999, 47, 167-188.

[7] Rudolph B. Seaweed product: red algae of economic significance. In: Martin RE, Carter EP, Davis LM, Flich GJ, editors. *Marine and freshwater products handbook*. Lancaster, USA: Technomic Publishing Company, Inc.; 2000. p. 515-529.

[8] Pereira L, Mesquita JF. Carrageenophytes of occidental Portuguese coast: 1-spectroscopic analysis in eight carrageenophytes from Buarcos bay. *Biomolecular Engineering*, 2003, 20, 217-222.

[9] Falshaw R, Bixler HJ, Johndro K. Structure and performance of commercial kappa-2 carrageenan extracts I. Structure analysis. *Food Hydrocolloids*, 2001, 15(4-6), 441-452.

[10] Wong KF, Craigie JS. 1978. Sulfohydrolase activity and carrageenan biosynthesis in *Chondrus crispus* (Rhodophyceae). *Plant Physiology*, 1978, 61, 663-666.

[11] de Ruiter GA, Richard O, Rudolph B, Genicot S, Kloareg B, Penninkhof B, Potin P. Sulfohydrolases, corresponding amino acid and nucleotide sequences, sulfohydrolase preparations, processes, and products thereof. *World patent application*, 2000, WO0068395.

[12] Aguilan JT, Broom JE, Hemmingson JA, Dayrit FM, Montaño MNE, Dancel MCA, Niñonuevo MR, Furneaux RH. Structural analysis of carrageenan from farmed varieties of Philippine seaweed. *Botanica Marina*, 2003, 46, 179-192.

[13] van de Velde F, Knutsen SH, Usov AI, Rollema HS, Cerezo AS. ^{1}H and ^{13}C high resolution NMR spectroscopy of carrageenans: Application is research and industry. *Trends in Food Science & Technology*, 2002, 13, 73-92.

[14] Pereira L. Chapter 2 – A Review of the Nutrient Composition of Selected Edible Seaweeds. In: Pomin VH, editor. *Seaweed: Ecology, Nutrient Composition and Medicinal Uses*. New York: Nova Science Publishers Inc.; 2011, p. 15-47.

[15] Briand X. Chapter 10 – Seaweed harvesting in Europe. In: Guiry MD, Blunden G, editors. *Seaweed, resources in Europe – uses and potential*. Chichester: John Wiley & Sons, Ltd.; 1991, p. 259-308.

[16] Perez R, Kaas R, Campello F, Arbault S, Barbaroux O. *La Culture des algues marines dans le monde*. Institut Françaíse de Recherche pour L'Explotation de la Mer, Centre de Brest, IFREMER, Plouzane ; 1992, p. 223-292.

[17] Santelices B. A conceptual framework for marine agronomy. *Hydrobiologia*, 1999, 398/399, 15-23.

[18] McHugh DJ. A guide to the seaweed industry. *FAO Fisheries Technical Paper*, 2003, 441, 52-72.

[19] Guiry MD. The Seaweed Site: information on marine algae – *Carrageenans*, Galway: National University of Ireland; 2009, URL: http://www.seaweed.ie/uses_general/carrageenans.php.

[20] Bixler HJ, Porse H. A decade of change in the seaweed hydrocolloids industry. *Journal of Applied Phycology*, 2011, 23(3), 321-335.

[21] Tasende MG, Cid M, Fraga MI. Spatial and temporal variations of *Chondrus crispus* (Gigartinaceae, Rhodophyta) carrageenan content in natural populations from Galicia (NW Spain). *Journal of Applied Phycology*, 2012, 24, 941-951.

[22] Palminha FP 1971. Exploração e utilização de algas marinhas na plataforma portuguesa e nas ilhas do Arquipélago dos Açores. *Junta Nacional do Fomento das Pescas*, 1971,7, 25-36.

[23] Sousa-Pinto I, Araújo R. The Seaweed Resources of Portugal. In: Critchley AT, Ohno M, Largo DB, editors. *World seaweed resources: an authoritative reference system*. Wokingham, Berkshire: ETI Information Services Ltd.; 2006.

[24] Santos R, Duarte P. Marine plant harvest in Portugal. *Journal of Applied Phycology*, 1991, 3, 11-18.

[25] Zinoun M, Cosson J. Seasonal variation in growth and carrageenan content of *Calliblepharis jubata* (Rhodophyceae, Gigartinales) from the Normandy coast, France. *Journal of Applied Phycology*, 1996, 8, 29-34.

[26] Pereira L, Critchley AT, Amado AM, Ribeiro-Claro PJA. A comparative analysis of phycocolloids produced by underutilized *versus* industrially utilized carrageenophytes (Gigartinales, Rhodophyta). *Journal of Applied Phycology*, 2009, 21, 599-605.

[27] Pereira L. Identification of phycocolloids by vibrational spectroscopy. In: Critchley AT, Ohno M, Largo DB, editors. *World Seaweed Resources - An authoritative reference system*. Wokingham, Berkshire: ETI Information Services Ltd.; 2006.

[28] Pereira L, Amado AM, Critchley AT, van de Velde F, Ribeiro-Claro PJA. Identification of selected seaweed polysaccharides (phycocolloids) by vibrational spectroscopy (FTIR-ATR and FT-Raman). *Food Hydrocolloids*, 2009, 23, 1903-1909.

[29] Pereira L. Estudos em macroalgas carragenófitas (Gigartinales, Rhodophyceae) da costa portuguesa - aspectos ecológicos, bioquímicos e citológicos, *PhD Thesis*. Coimbra: Departamento de Botânica - FCTUC, Universidade de Coimbra, 2004, p 293.

[30] Pereira L. Population Studies and Carrageenan Properties in Eight Gigartinales (Rhodophyta) from Western Coast of Portugal. *The Scientific World Journal, Marine Biology*, 2013, vol. 2013, Article ID 939830, 11 pages.

[31] Chen LCM, McLachlan J. The life history of *Chondrus crispus* in culture. *Canadian Journal of Botany*, 1971, 50, 1055-1060.

[32] Guiry MD. *Chondrus crispus* Stackh. 'T4' is a male clone (Rhodophyta). *Phycologia*, 1981, 20, 439-449.

[33] Brown MT, Neish A, Harwood D. Comparison of three techniques for identifying isomorphic phases of *Chondrus crispus* (Gigartinaceae). *Journal of Applied Phycology*, 2004, 16, 447-450.

[34] Scrosati R, Garbary DJ, McLachlan J. Reproductive ecology of *Chondrus crispus* (Rhodophyta, Gigartinales) from Nova Scotia, Canada. *Botanica Marina*, 1994, 37, 293-300.

[35] Scrosati R, Mudge B. Persistence of gametophyte predominance in *Chondrus crispus* (Rhodophyta, Gigartinaceae) from Nova Scotia after 12 years. *Hydrobiologia*, 2003, 519, 215-218.

[36] Kim DH. A study of the development of cystocarps and tetrasporangial sori in Gigartinaceae (Rhodophyceae, Gigartinales). *Nova Hedwigia* 1976, 27, 1-146.

[37] Hommersand MH, Guiry MD, Fredericq S, Leister GL. New perspectives in the taxonomy of the Gigartinaceae (Gigartinales, Rhodophyta). *Hydrobiologia*,1993, 261, 105-120.

[38] Hommersand M, Fredericq S, Cabioch J. Developmental morphology of *Gigartina pistillata* (Gigartinaceae, Rhodophyta). *Phycologia*, 1992, 31, 300-325.

[39] Isaac WE, Simons SM. Some observations on *Gigartina pistillata* (Gmel.) Stackh. from Port Alfred with a record of plants bearing both tetraspores and carpospores. *J. S. Afr. Bot.* 1954, 20, 117-125.

[40] Pereira L, Mesquita JF, Dias JDS. Optical and electron microscope study of heterosporic thalli (carpospores/tetraspores) in *Gigartina pistillata* (Gmel.) Stackh. (Rhodophyta), *Proceedings of ICEM 15, vol. 2 - Biology and Medicine*, Durban, South Africa; 2002, p. 747-748.

[41] Guiry MD, West JA, Kim DH, Masuda M. Reinstatement of the Genus *Mastocarpus* Kutzing (Rhodophyta). *Taxon*, 1984, 33, 53-63.

[42] Pereira L, Mesquita JF. Population studies and carrageenan properties of *Chondracanthus teedei* var. *lusitanicus* (Gigartinaceae, Rhodophyta). *Journal of Applied Phycology*, 2004, 16, 369-383.

[43] Dixon PS, Irvine LM. *Seaweeds of the British Isles, vol. I – Rodhophyta, Part 1 - Introduction, Nemaliales, Gigartinales*. London: The Natural History Museum/B. P. Society; 1995, p. 167-251.

[44] Gayral P. *Les Algues de côtes françaises (Manche et Atlantique), notions fondamentales sur l'écologie, la biologie et la systématique des algues marines*. Paris: Éditions Doin, Deren et Cie.; 1966, p. 116.

[45] Mathieson AC, Penniman CE, Tvetergallagher E. Phycocolloid ecology of underutilized economic red algae. *Hydrobiologia*, 1984, 116, 542-546.

[46] Maggs CA, Douglas SE, Fenety J, Bird CJ. A molecular and morphological analysis of the *Gymnogongrus devoniensis* (Rhodophyta) complex in the North Atlantic. *Journal of Phycology*, 1992, 28, 241-232.

[47] Kopp J. Contribuition à l'étude de l'algue rouge *Chondrus crispus* Stackh. Biochimie des carraghénanes. *PhD Thesis,* Université Bretagne Occidentale; 1975, p. 93.

[48] Kopp J, Perez J. Contribuition à l'étude de l'álgue rouge *Chondrus crispus* Stackh. Relation entre la croissance, la potentialité sexuelle, la quantité e la composition de carraghénanes. *Revue des Travaux de l'Institut des Peches Maritimes*, 1979, 42, 291-324.

[49] Chopin T, Floch JY. Ecophysiological and biochemical-study of two of the most contrasting forms of *Chondrus crispus* (Rhodophyta, Gigartinales). *Marine Ecology-Progress Series*, 1992, 81, 185-195.

[50] Hilliou L, Larotonda FDS, Abreu P, Ramos AM, Sereno AM, Gonçalves MP. Effect of extraction parameters on the chemical structure and gel properties of κ/ι-hybrid carrageenans obtained from *Mastocarpus stellatus*. *Biomolecular Engineering*, 2006, 23, 201-208.

[51] Hilliou L, Larotonda FDS, Sereno AM, Gonçalves MP. Thermal and viscoelastic properties of κ/ι-hybrid carrageenan gels obtained from the Portuguese seaweed *Mastocarpus stellatus*. *Journal of Agricultural and Food Chemistry,* 2006, 54, 7870-7878.

[52] Amimi A, Mouradi A, Bennasser L, Givernaud T. Seasonal variations in thalli and carrageenan composition of *Gigartina pistillata* (Gmelin) Stackhouse (Rhodophyta, Gigartinales) harvested along the Atlantic coast of Morocco. *Phycological Research*, 2007, 55, 143-149.

[53] Amimi A, Mouradi A, Givernaud T, Chiadmi N, Lahaye M. Structural analysis of *Gigartina pistillata* carrageenans (Gigartinaceae, Rhodophyta). *Carbohydrate Research*, 2001, 333: 271-279.

[54] Pereira L, Mesquita JF. Cytochemical studies on Portuguese carrageenophytes (Gigartinales, Rhodophyta). In: Long CA, Anninos P, editors. *ISI Book of Proceedings of the 3rd WSEAS International Conference on Cellular and Molecular Biology, Biophysics and Bioengineering*. Athens: World Scientific and Engineering Acad. and Soc.; 2007, p. 66-71.

[55] Zinoun M, Cosson J, Deslandes E. Influence of Culture Conditions on Growth and Physicochemical Properties of Carrageenans in *Gigartina teedii* (Rhodophyceae, Gigartinales). *Botanica Marina*, 1993, 36, 131-136.

[56] Pereira L, van de Velde F. Portuguese carrageenophytes: Carrageenan composition and geographic distribution of eight species (Gigartinales, Rhodophyta). Carbohydrate Polymers, 2011, 84, 614-623.

[57] FAO - Fisheries and Aquaculture Department. *The State of World Fisheries and Aquaculture.* Rome, Italy: Food and Agriculture Organization of the United Nations. World-wide electronic publication; 2012, http://www.fao.org/docrep/016/i2727e/i2727e.pdf

[58] Abreu MH, Pereira R, Mata L, Nobre A, Sousa-Pinto I. Aquacultura Multi-Trófica Integrada em Portugal. In: Vázquez Ferreiro U, Incera Filgueira M, Fernández Otero R, Maroto Leal J, editors. *Macroalgas en la Acuicultura Multitrófica Integrada*

Peninsular: Valorización de su Biomasa. Vigo, Spain: Centro Tecnológico del Mar – Fundación CETMAR; 2011, p. 54-77.

[59] Cosson J, Deslandes E, Braud JP. Preliminary approach to the characterization and seasonal-variation of carrageenans from four Rhodophyceae on the Normandy coast (France). *Hydrobiologia*, 1990, 204, 539-544.

[60] Meireles F. Carrageenans yield and quality from *Chondrus crispus* Stackhouse (Rhodophyta) cultivated in an Integrated Multi-Trophic Aquaculture (IMTA) System. *MSc Thesis*. Coimbra: DCV, FCTUC, University of Coimbra; 2013, p. 30.

In: Seaweeds
Editor: Vitor H. Pomin
ISBN: 978-1-63117-571-8

Chapter 7

D-AMINO ACIDS AND AMINO ACID RACEMASES IN SEAWEEDS

***Takehiko Yokoyama*[1] *and Koji Mikami*[2]**
[1]School of Marine Biosciences, Kitasato University, Sagamihara, Japan
[2]Faculty of Fisheries Sciences, Hokkaido University, Hakodate, Japan

ABSTRACT

The composition of free amino acids is responsible for the flavor and nutritional value of seaweeds such as *Pyropia yezoensis*, *Saccharina japonica*, *Undaria pinnatifida*, and *Sargassum fusiforme*, which are edible foods that are popular in Japan. Although only L-amino acids were thought to occur naturally, recent advances in analytical techniques have led to the identification of various free D-amino acids in eukaryotic organisms, including seaweeds. We detected high levels of D-aspartate in *S. fusiforme*, and the functional significance of this amino acid is supported by its specific localization in the medullary layer of blades and quantitative variation due to changes in environmental conditions. Despite the presence in various kinds of seaweeds, the biological function of D-alanine remains unclear. Based on the genome-wide survey, we have found gene candidates in seaweeds that encode putative serine racemase [EC 5.1.1.18] involving in the racemization of D- and L-serine, while alanine racemase [EC 5.1.1.1] and aspartate racemase [EC 5.1.1.13] were not found. Seaweed serine racemases are homologous to the human enzyme, suggesting the functional enzymatic activity of seaweed proteins. However, D-serine was not detected in any seaweed analyzed. Thus, the analysis of the enzymatic activity of putative serine racemases encoded by the genes discussed herein provides new insight on the biological significance of D-amino acids in seaweeds.

1. INTRODUCTION

The composition of free amino acids of seaweeds has been studied by several researchers. Although the data cannot be simply summarized, 20 proteinogenic amino acids have been detected [1-3]. Each seaweed differs in the composition of free amino acids with season and growth stage [4]. Most seaweeds have high aspartate, glutamate, and alanine levels (Table 1), except for the red alga *Lophocladia lallemandii*, in which proline comprises 48.2% of total amino acids [5]. In addition, several species of red algae contain high levels of nonproteinogenic amino acids, such as taurine and citrulline, compared to those in green and brown algae (Table 1). Taurine and citrulline may play fundamental roles in maintaining cellular processes in red algae.

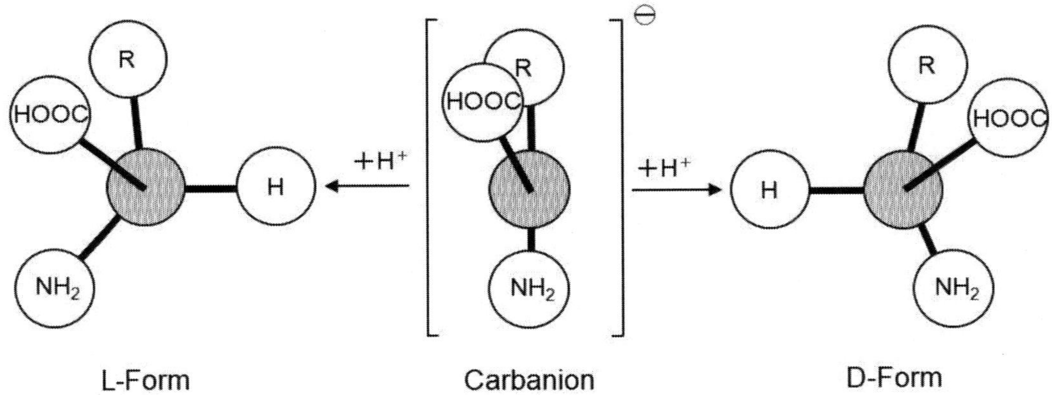

Figure 1. Amino acid enantiomer [6].

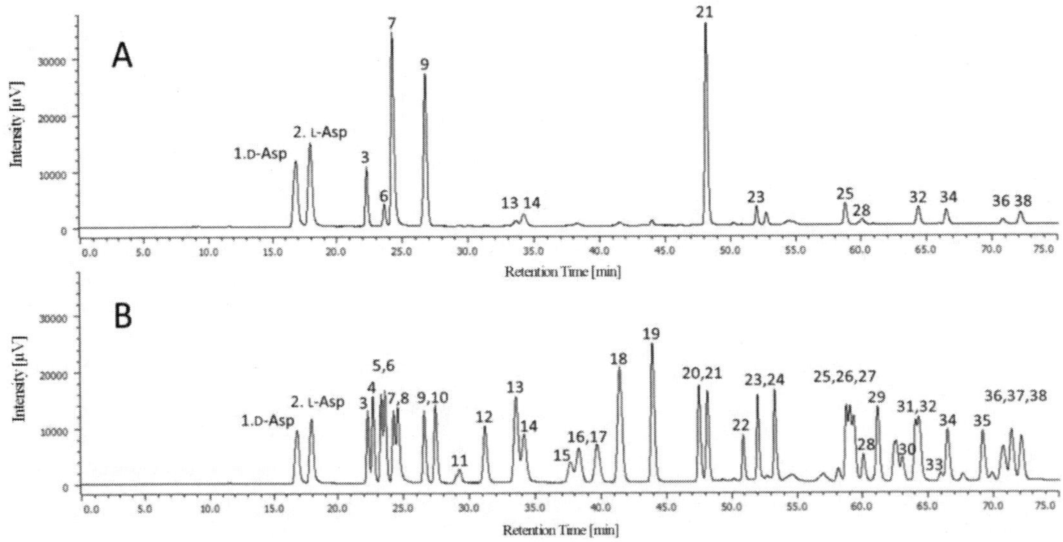

Figure 2. Typical HPLC chromatogram of (A) *Sargassum fusiforme* extract and (B) derivatized D- and L-enantiomers of standard amino acids. The following chromatographic conditions were used: column, ODS-80T$_S$ (5 µm particle size; 250×4.6 mm) at 40°C; flow rate, 1 mL/min; excitation and emission wavelengths of 340 nm and 450 nm, respectively. The peak numbers correspond to the following amino acids: *1*, D-Asp; *2*, L-Asp; *3*, L-Asn; *4*, D-Asn; *5*, D-Ser; *6*, L-Ser; *7*, L-Glu; *8*, D-Glu; *9*, L-Gln; *10*, D-Gln; *11*, DL-His; *12*, D-Thr; *13*, Gly; *14*, L-Thr; *15*, L-citrulline; *16*, L-Arg; *17*, D-Arg; *18*, taurine; *19*, β-Ala; *20*, D-Ala; *21*, L-Ala; *22*, γ-aminobutyrate (GABA); *23*, L-Tyr; *24*, D-Tyr; *25*, L-Val; *26*, D-Met; *27*, L-Met; *28*, L-Trp; *29*, D-Val; *30*, D-Trp; *31*, D-Phe; *32*, L-Phe; *33*, L-ornithine; *34*, L-Ile; *35*, D-Ile; *36*, DL-Lys; *37*, D-Leu; and *38*, L-Leu.

With the exception of glycine, all amino acids are present in two symmetrical forms, D- or L- form or -enantiomers. The L-enantiomer is enzymatically converted to the D-form and vice versa, which is called racemization or isomerization (Figure 1). The chemical properties of D- and L- amino acids are identical and can mainly be distinguished by their optical behavior because pure solutions rotate the plane of polarized light into opposite directions. Interestingly, D- and L- amino acids may also differ in taste; D-amino acids are generally sweet and those of the L-form range from being bitter to neutral [6]. Amino acid compositions can be determined by reversed-phase high-performance liquid chromatography (HPLC). HPLC analysis after derivatization with *o*-phthalaldehyde and *N*-acetyl-L-cysteine (chiral derivatization reagent) is very convenient and accurate [7]. Chromatography was performed using a modified method of Nimura et al. [8], enabling the simultaneous determination of D- and L-amino acids [9].

Although L-amino acids were previously thought to only exist in eukaryotes, a bacterial peptidoglycan layer of the cell wall contains D-amino acids, such as D-alanine, D-aspartic acid, and D-glutamic acid, which presumably provide stability against proteolytic degradation. In addition, D-amino acids have also been found in bacterial products, such as the peptide antibiotics bacitracin and mycobacillin [10]. Based on these findings, D-amino acids were not thought to occur naturally in eukaryotes. However, over the past four decades, various D-amino acids have been found in animals. D-Alanine exists in the milkweed bug [11], and both D-serine and D-glutamic acid exist in the silkworm [11,12]. D-Aspartate has also been identified in a variety of organisms, including rat [13] and human [14]. D-Aspartate acts as a negative regulator of melatonin synthesis in the rat pineal gland [15] and it is also involved in the synthesis of testosterone in rat testes [16]. Notably, large quantities of D-alanine have been detected in several species of bivalve mollusks [17,18] and crustaceans [19], and large quantities of D-aspartate have been detected in cephalopods [20,21] and in the bivalve mollusk *Scapharca broughtonii* [22]. In bivalve mollusks and crustaceans, D-alanine is thought to function as an osmolyte in isosmotic regulation [23]. In addition, D-aspartate in cephalopods is hypothesized to be involved in neural transmission because high concentrations of D-aspartate are present in the brain, optic lobes, and axoplasm but not in other non-nervous organs [20,21]. Furthermore, D-aspartate is also involved in anaerobic energy metabolism in *S. broughtonii* [24,25].

The occurrence of D-amino acids in terrestrial plants has been discussed in some previous reviews [26,27]. Ogawa et al. [28] suggested that free D-aspartate and D-alanine might participate in the development of pea seedlings. However, the technique used to identify D-amino acids in animals and terrestrial plants detected D-amino acids in various species of seaweeds, which led to an extensive amino acid analysis of seaweeds for comparison with terrestrial plants [29-31]. An example of D-aspartate detection in the brown alga *Sargassum fusiforme* (formerly *Hizikia fusiformis*) is shown in Figure 2. In general, brown algae contain high levels of D-aspartate, and D-alanine is observed in red, green, and brown algae (see below) [29,30].

Artificial regulation of the composition of D- and L-amino acids is important to produce commercially valuable seaweeds. Thus, the differences between free amino acid compositions under various environmental stresses could enhance knowledge related to the appropriate amino acids for producing algae with good flavor and nutritional value. Herein, current knowledge about the distribution of free D-amino acids and the influence of environmental stresses on their composition in seaweeds is summarized.

Table 1. Comparative account of free amino acids in different species of seaweeds

	Brown Algae						Red Algae									Green Algae	
compound	Analipus japonicus	Sargassum confusum	Sargassum fusiforme	Sargassum horneri	Scytosiphon lomentaria	Undaria Pinnatifida	Ahnfeltiopsis paradoxa	Chondria crassicaulis	Chondrus ocellatus	Chondrus yendoi	Gloiopeltis furcata	Lomentaria hakodatensis	Polysiphonia senticulosa	Pyropia yezoensis	Bangia fuscopurpurea	Cladophora opaca	Ulva pertusa
Asp	1.82	3.03	0.76	1.95	0.57	1.11	0.84	2.91	1.30	0.25	0.16	0.52	1.70	1.15	3.91	3.52	0.17
Asn	0.26	0.24	0.24	1.80	0.10	0.83	N.D.	0.18	2.55	0.17	1.31	0.34	0.12	3.34	9.27	5.66	0.05
Ser	0.12	0.58	0.04	0.07	0.13	0.42	0.16	0.11	0.38	0.06	0.13	0.00	0.09	0.42	0.36	0.26	0.14
Glu	2.91	1.36	0.89	1.63	1.33	2.49	4.69	1.26	3.27	1.83	0.43	1.25	1.50	8.94	26.61	0.50	0.63
Gln	1.10	2.28	0.67	0.81	0.37	0.87	0.39	1.70	0.11	0.08	0.11	0.21	2.31	3.48	3.27	1.28	1.33
His	N.D.	N.D.	N.D.	N.D.	N.D.	N.D.	N.D.	N.D.	N.D.	N.D.	N.D.	N.D.	N.D.	N.D.	N.D.	N.D.	N.D.
Gly	0.27	0.13	0.03	0.33	0.14	4.23	0.62	0.62	0.44	0.19	0.53	0.32	0.90	1.46	1.31	0.43	0.77
Thr	0.62	0.62	0.15	0.34	0.39	N.D.	1.82	0.56	2.18	0.50	0.55	1.29	1.18	2.36	0.92	0.54	0.12
Cit	N.D.	N.D.	N.D.	N.D.	N.D.	N.D.	15.87	0.13	0.02	3.56	0.70	0.13	0.16	10.57	7.20	N.D.	N.D.
Arg	0.08	0.06	N.D.	0.03	0.06	0.24	N.D.	N.D.	N.D.	N.D.	N.D.	N.D.	0.11	N.D.	N.D.	0.87	0.03
Tau	0.19	0.03	0.01	0.10	0.10	N.D.	1.79	0.06	7.05	2.83	5.31	2.43	0.32	14.99	23.09	1.68	0.06
β-Ala	0.07	0.06	0.03	0.07	0.02	N.D.	0.13	0.03	0.32	N.D.	N.D.	0.02	0.61	0.20	0.11	0.34	0.10
Ala	4.29	7.05	1.65	2.29	6.10	55.62	2.62	5.75	1.17	1.09	3.04	1.12	3.90	4.76	19.41	3.11	0.18
GABA	0.01	N.D.	N.D.	0.01	N.D.	N.D.	N.D.	N.D.	N.D.	N.D.	N.D.	N.D.	0.03	N.D.	N.D.	0.02	N.D.
Tyr	0.17	0.09	0.06	0.07	0.10	0.02	0.14	0.05	0.04	0.05	0.06	0.11	N.D.	0.09	N.D.	0.09	0.02
Val	0.29	0.17	0.09	0.18	0.15	0.08	0.49	0.31	0.50	0.13	0.21	0.21	0.25	1.40	0.34	0.28	0.12
Met	N.D.	N.D.	N.D.	N.D.	N.D.	N.D.	N.D.	N.D.	N.D.	N.D.	N.D.	N.D.	N.D.	N.D.	N.D.	N.D.	N.D.
Trp	0.18	0.22	0.08	N.D.	0.08	N.D.	N.D.	N.D.	0.07	N.D.	N.D.	N.D.	N.D.	0.23	N.D.	0.06	N.D.
Phe	0.16	0.27	0.08	0.15	0.07	0.05	2.01	0.05	0.25	0.16	0.24	0.07	0.13	0.23	0.06	0.10	0.02
Orn	N.D.	N.D.	N.D.	N.D.	N.D.	N.D.	N.D.	N.D.	N.D.	N.D.	N.D.	N.D.	N.D.	N.D.	N.D.	0.11	N.D.
Ile	0.85	0.59	0.56	0.61	0.44	0.47	1.66	1.59	2.45	0.56	0.60	0.62	0.32	0.52	0.10	0.84	0.45
Lys	0.17	0.23	0.06	0.11	0.12	0.13	0.04	0.08	0.02	N.D.	0.02	0.04	0.07	0.97	0.67	0.23	0.20
Leu	0.17	0.11	0.07	0.14	0.08	0.05	0.33	0.19	0.37	0.12	0.13	0.10	N.D.	0.66	0.20	0.20	0.02

The levels are given in μmol/g wet. N.D., not detected. Samples were collected from Okirai Bay in Iwate Prefecture, Japan, in June 2013.
Cit, Citrulline; Tau, Taurine; GABA, γ-Aminobutyric acid; Orn, Ornithine

2. D-AMINO ACIDS IN SEAWEEDS

2.1. D-Aspartate

With advances in analytical techniques over the past four decades, various D-amino acids have been detected in plants [32], especially in seaweeds [29]. For example, *S. fusiforme*, a popular edible alga in Japan, contains high levels of free D-aspartate in fresh blade samples [31], whereas various terrestrial plants have relatively low levels of D-aspartate [32]. Recently, we detected D-aspartate in 4 out of 17 (23.5%) species examined and its concentration was relatively high in the brown algae *S. fusiforme* and *Sargassum horneri* compared to that of green and red algae, in which D-aspartate was either not detected or present at very low levels (Table 2). Nagahisa et al. [29] reported that the brown algae *Costaria costata* and *Sargassum yezoensis* contained D-aspartate at levels approximately equal to algae that contained high levels of L-aspartate. Thus, the percentage [D / (D + L) × 100 (%)] of D-aspartate in *C. costata* was high (64.1%). Similarly, the percentage of D-aspartate in *S. fusiforme* was also high (50.0%) (Table 2). By contrast, as shown in Table 2, the percentages of D-aspartate in the red algae *Chondria crassicaulis* and *Polysiphonia senticulosa* were 1.0% and 1.8%, respectively, indicating that a high level of the D-enantiomer was unique to aspartate in brown algae.

Although D-aspartate has been identified in its free form in seaweeds, its functions and origin have not been elucidated. The marine brown alga *S. fusiforme* contains a substantial amount of free D-aspartate in its blade (Table 2), as demonstrated by HPLC analysis. However, its intracellular localization, which provides information about its functions and origin, has not yet been determined. Therefore, we prepared a polyclonal antibody against D-aspartate and investigated its cellular localization in *S. fusiforme* using immunohistochemistry [33]. A glutaraldehyde conjugate of D-aspartate and bovine serum albumin (BSA) was prepared using the method of Storm-Mathisen et al. [34], which utilizes anti-γ-aminobutyrate (GABA) and anti-glutamate antiserum in immunohistochemical studies.

Unlike vascular plants such as Spermatophyta or Pteridophyta, seaweeds do not differentiate into roots, stems, and leaves. However, some brown algae, such as *Saccharina* (formerly *Laminaria*), consist of cortical and medullary layers, which are probably required for long-distance transport similar to sieve tubes in vascular plants. In addition, *Saccharina* absorbs nutrients from the sea water through all parts of the body, and the organic matter synthesized from the absorbed nutrients is concentrated in the medullary layer and subsequently translocated to the growth point. The medullary layer of *Saccharina* plays a role in the translocation of the synthesized organic matter [35]. Immunohistochemical approaches have indicated specific localization in the medullary layer rather than the cortical layer of the blade in *S. fusiforme* (Figure 3), suggesting that D-aspartate in *S. fusiforme* may be concentrated in the medullary layer in order to be translocated to a particular region in this alga to perform a physiological function.

Because the blade of *Sargassum* also consists of cortical and medullary layers, D-aspartate may accumulate in the medullary layer and be translocated to specific regions via a long-distance transport system in *S. horneri* that contains high levels of D-aspartate (Table 2).

Table 2. Distributions of free D- and L-aspartate in seaweeds

Species	Asp (μmol/g wet) D	L	% D/(D+L)×100
Brown Algae			
Analipus japonicus	N.D.	1.82	—
Sargassum confusum	N.D.	3.03	—
Sargassum fusiforme	0.38	0.38	50.0
Sargassum horneri	0.54	1.41	27.7
Scytosiphon lomentaria	N.D.	0.57	—
Undaria Pinnatifida	N.D.	1.43	—
Red Algae			
Ahnfeltiopsis paradoxa	N.D.	0.84	—
Chondria crassicaulis	0.03	2.88	1.0
Chondrus ocellatus	N.D.	1.31	—
Chondrus yendoi	N.D.	0.25	—
Gloiopeltis furcata	N.D.	0.16	—
Lomentaria hakodatensis	N.D.	0.52	—
Polysiphonia senticulosa	0.03	1.68	1.8
Pyropia yezoensis	N.D.	1.15	—
Bangia fuscopurpurea	N.D.	3.91	—
Green Algae			
Cladophora opaca	N.D.	3.52	—
Ulva pertusa	N.D.	0.17	—

The levels are given in μmol/g wet. N.D., not detected. Samples were collected from Okirai Bay in Iwate Prefecture, japan, in June 2013.

The results of immunohistochemical staining also revealed that D-aspartate was concentrated in the cytosol of *S. fusiforme* cells and was not present in the cell wall, chloroplast, or vacuole (Figure 3c). The cytosolic localization of D-aspartate in *S. fusiforme* excludes the possibility that it is present in attached or symbiotic materials, such as marine bacteria. These results suggest that D-aspartate is in fact synthesized in algae, but its biosynthesis and metabolism remains to be fully elucidated.

2.2. D-Alanine

D-Alanine is frequently found in all classes of seaweeds (Table 3). Nagahisa et al. [30] reported the distribution of free D-alanine in 15 out of the 24 (63%) species of algae examined. For example, quantitative analysis indicated that the brown algae *Analipus japonicus* (formerly *Heterochordaria abietina*) and *Sargassum nigrifolium* contained 83 nmol/g and 78 nmol/g wet weight of D-alanine, respectively. We also surveyed seaweeds containing free D-alanine and found that 14 out of the total of 17 (82.4%) species examined contained the amino acid (Table 3). These results show a wider distribution of D-alanine in seaweeds compared to that of D-aspartate. However, the levels of D-alanine were very low compared to those of L-alanine, and the percentage [D / (D + L) ×100 (%)] of D-alanine was below 2.6%. Thus, there was no correlation between D-alanine content and its percentage (see Table 3).

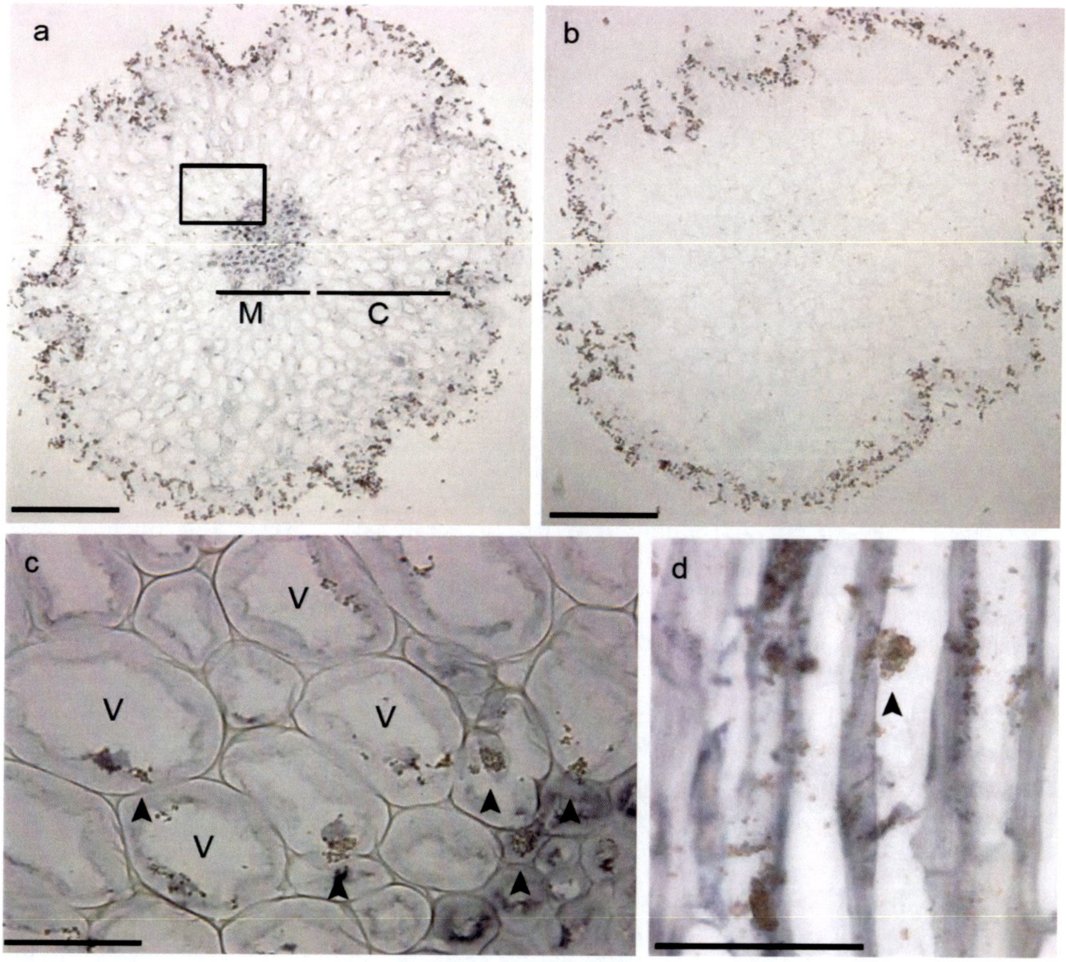

Figure 3. Immunohistochemical staining of blades of *Sargassum fusiforme* with anti-D-aspartate antibody.
(a) Horizontal section of the *S. fusiforme* blade incubated with anti-D-aspartate antibody at a final dilution of 1:250. M, medullary layer; and C, cortical layer. (b) The same section as shown in (a) stained with anti-D-aspartate antibody and 20 mM D-aspartate. (c) A horizontal section of an *S. fusiforme* blade. The boxed area in (a) is presented at a higher magnification. The cytosol is stained violet and arrowheads indicate chloroplasts. V, vacuole. (d) Vertical section of the *S. fusiforme* blade. Arrowheads indicate chloroplasts that were originally brown. The bars in (a) and (b) indicate 500 μm and those in (c) and (d) indicate 25 μm.

Table 3. Distributions of free D- and L-alanine in seaweeds

Species	Ala (μmol/g wet) D	L	% D/(D+L)×100
Brown Algae			
Analipus japonicus	0.04	4.25	1.0
Sargassum confusum	0.01	7.03	0.2
Sargassum fusiforme	N.D.	1.65	—
Sargassum horneri	N.D.	2.29	—
Scytosiphon lomentaria	0.04	6.07	0.6
Undaria Pinnatifida	0.01	26.39	0.0
Red Algae			
Ahnfeltiopsis paradoxa	0.04	2.59	1.4
Chondria crassicaulis	0.02	5.73	0.3
Chondrus ocellatus	0.01	1.16	0.7
Chondrus yendoi	0.01	1.09	0.5
Gloiopeltis furcata	0.03	3.01	1.0
Lomentaria hakodatensis	0.01	1.11	0.5
Polysiphonia senticulosa	0.07	3.83	1.9
Pyropia yezoensis	0.09	4.67	1.9
Bangia fuscopurpurea	0.29	19.12	1.5
Green Algae			
Cladophora opaca	0.08	3.03	2.6
Ulva pertusa	N.D.	0.18	—

The levels are given in μmol/g wet. N.D., not detected. Samples were collected from Okirai Bay in Iwate Prefecture, japan, in June 2013.

Ogawa et al. [28] suggested that free D-alanine might play a role in the development of pea seedlings. However, its function is still unclear because the physiological role of free D-alanine in seaweeds has not been extensively studied.

3. Physiological Function of D-Aspartate in *Sargassum fusiforme*

The role of D-amino acids remains uncertain in seaweeds. However, their high concentrations and cytosolic localization (Figure 3 and Table 2) led us to speculate about the involvement of D-aspartate in physiological regulations in *S. fusiforme*. Despite the lack of direct evidence, the data supporting this possibility are discussed below.

3.1. Involvement in Developmental Regulation

S. fusiforme from the sea near Iwate Prefecture, Japan germinated in late summer and grew from autumn to spring, with rapid growth in early spring. In fact, the content of D-aspartate increased during the early growth period, especially in February (Figure 4). These observations suggest that D-aspartate plays an important role in the early growth stages of *S. fusiforme* [31].

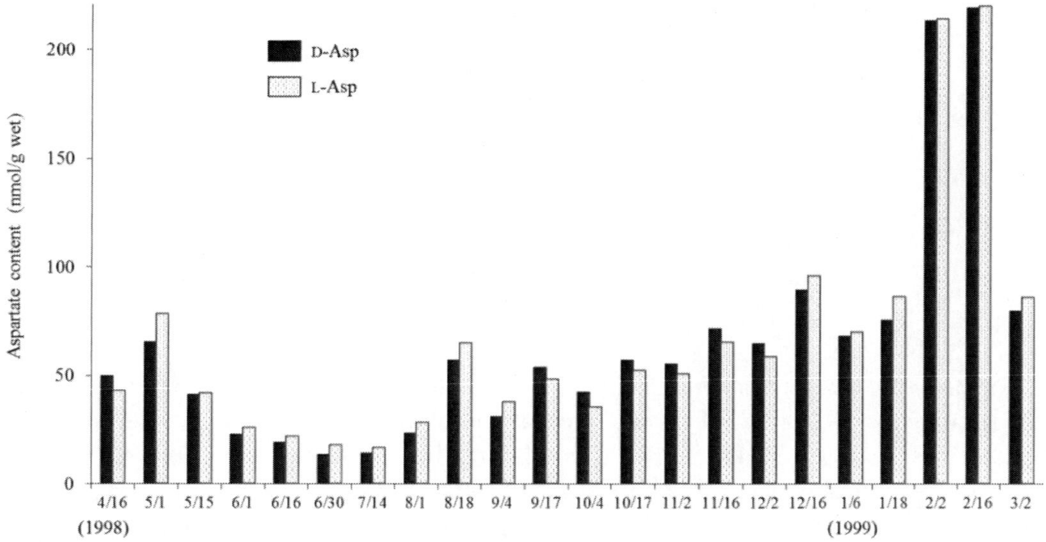

Figure 4. Seasonal variation of D-aspartate content in *Sargassum fusiforme* blade.
Samples were collected from Okirai Bay in Iwate Prefecture, Japan between April 1998 and March 1999.

3.2. Involvement in the Hypoxic Stress Response

Aspartate functions as a key substrate in anaerobic energy metabolism during the initial stages of anoxia in many mollusks, including bivalves with a high tolerance for anoxia [36,37]. Watanabe et al. [25] reported the effects of hypoxic stress on free aspartate levels in the muscle of the blood shell *S. broughtnii*. When the shell was reared in hypoxic seawater, the aspartate levels in the foot muscle rapidly decreased to one-fourth its original value within 24 hours. By contrast, the levels of many other amino acids (alanine, glutamic acid, and glycine) increased under hypoxic conditions. In particular, L-alanine, but not D-alanine, levels significantly increased during the initial stages of hypoxia and became approximately 5 times higher than that of the aerobic control level after 24 hours, suggesting that the end products under hypoxic stress have functional significance. In these experiments, D-aspartate and L-aspartate were not distinguished; thus, the function of D-aspartate under hypoxic stress is still unknown. To address this problem, we examined changes in the contents of several amino acids, including D-aspartate, in the brown alga *S. fusiforme* under hypoxic stress as well as their subsequent recovery from the stress (Figure 5).

Of the amino acids examined, only D- and L-aspartate showed changes in concentrations during hypoxia in a parallel manner. Both of them rapidly decreased over 12 hours and then maintained constant aspartate levels until 24 hours. As shown in Figure 5, D- and L-aspartate levels decreased by 42.43% and 42.35%, respectively, over a 12-hour period of hypoxia. However, the L-aspartate levels decreased more rapidly during the hypoxia period (0–6 hours) than those of D-aspartate and increased more rapidly in the recovery period (24–36 hours), suggesting that L-aspartate is first converted to other metabolites and then D-aspartate is converted to L-aspartate by aspartate racemase [EC 5.1.1.13], an enzyme that catalyzes the interconversion of D- and L-aspartate. These observations suggest that aspartate in *S. fusiforme* may be involved in anaerobic energy metabolism during the initial stages of anoxia as well as in bivalves, because the rates of photosynthesis and respiration under water loss or desiccation at the intertidal zone where *S. fusiforme* inhabits decreased [38], thereby preventing photosynthesis, which is required for survival.

4. Difficulty in the Biochemical Identification of the Aspartate Racemase in Seaweeds That Produces D-Aspartate

Research on the biosynthetic pathway of D-amino acids in marine algae is limited. Although traces of D-amino acids are present in seawater [39], the concentration of D-aspartate in several species of brown algae is significantly higher. In addition, we have confirmed the cytosolic localization of D-aspartate (Figure 3), which rules out the amino acid being produced by attached or symbiotic bacteria. Thus, D-aspartate in seaweeds could be synthesized *in vivo* by aspartate racemase.

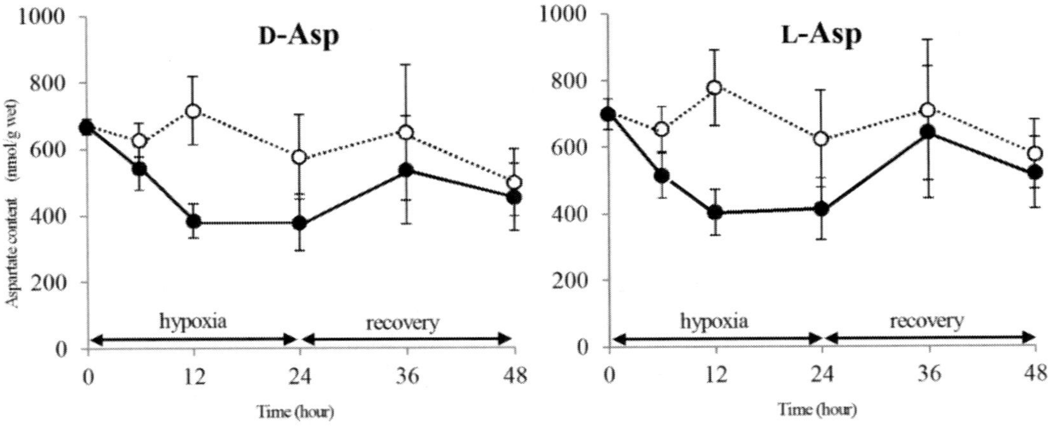

Figure 5. The effects of hypoxia and recovery on the contents of free aspartate in *Sargassum fusiforme*. Algae were placed individually in a bottle containing 500 mL of artificial seawater at a temperature of 10 °C. The bottle was flushed with nitrogen until the dissolved oxygen became less than 0.6 mg/L. The bottles were tightly sealed off, and algae were sampled after 0 hours, 6 hours, 12 hours, and 24 hours. For the recovery experiments performed after 24 hours, algae were transferred to artificial seawater, which was well-aerated to contain dissolved oxygen above 8.0 mg/L. The seawater was sampled after 36 hours and 48 hours. The experiment was performed in the dark, and algae were acclimated in well-aerated artificial seawater under an illumination intensity of 75 $\mu mol \cdot m^{-2} \cdot s^{-1}$ using cool white fluorescent lamps before the experiment began (for 24 hours). (○), control (aerobic condition); and (●), hypoxic stress followed by recovery from the stress. Vertical bars represent the standard deviation (S. D.).

Aspartate racemase has recently been found in the bivalve mollusk *S. broughtonii* [22] and subsequently in mouse [40]. In addition, Funakoshi et al.[41] reported that D-amino acid transaminase [EC 2.6.1.21], which catalyzes the transamination of D-aspartate and other D-amino acids, was found in the terrestrial plant *Arabidopsis thaliana*. However, there is no information about aspartate racemases in seaweeds to date. Thus, it is necessary to identify and characterize the various enzymes involved in the biosynthesis of D-aspartate in *S. fusiforme* and *S. horneri*. However, in our preliminary experiments, the purification of the enzyme from marine brown algae was difficult, and we have not been able detect the activity of aspartate racemase. The concentration of protein in these algae is extremely low and there are highly viscous polysaccharides, such as alginic acid, that prevent protein extraction, and currently hinders the study of the biosynthesis of D-amino acids in seaweeds.

5. Putative Racemase Genes in Seaweeds

The molecular biological identification of genes encoding racemases can sometimes help to overcome challenges faced when using biochemical approaches. Thus, we searched for genomic and expressed sequence tag (EST) data for seaweeds, such as *Ectocarpus siliculosus* [42], *Porphyra umbilicalis* [43,44], *Porphyra purpurea* [43,44], *Pyropia yezoensis* [45], and *Chondrus crispus* [46], and compared those sequences to those from unicellular algae, such as *Cyanidioschyzon merolae* [47] and *Chlamydomonas reinhardtii* [48]. Genes encoding putative serine racemases [EC 5.1.1.18] were identified, although there was no homologue for the known alanine and aspartate racemases. Because little is known about genes encoding racemases in seaweeds, this section represents the first study of putative racemase genes in seaweeds as well as the structural characterization of their products.

5.1. No Alanine Racemase Gene Homolog

The presence of alanine racemase [EC 5.1.1.1] in eukaryotes was first confirmed in the fungus *Tolypocladium inflatum* [49], whereas the first prokaryotic alanine racemase was reported in *Enterococcus faecalis* (formerly *Streptococcus faecalis*) [50]. Subsequently, alanine racemase activity detected in the aquatic shell [18] and alfalfa [51], which were the first reports of this type of enzymatic activity in an animal and plant, respectively. For algae, the activity of alanine racemase was characterized in microalgae, such as the diatom *Thalassiosira* sp. [52] and the unicellular green alga *C. reinhardtii* [53]. However, there is no report of alanine racemase in seaweeds, although the occurrence of D-alanine was confirmed in various kinds of macroalgal species [30].

Despite detecting alanine racemase activity in eukaryotes, nucleotide and amino acid sequences of this enzyme have been reported only for the fission yeast *Schizosaccharomyces pombe* [54] and the marine kuruma prawn *Marsupenaeus japonicus* [55]. However, recent progress in genomic sequencing has provided an abundance of information, indicating the presence of the alanine racemase genes in terrestrial plants. Predicted amino acid sequences from cDNA demonstrated that, in addition to bacterial enzymes, plant and other eukaryotic alanine racemases belong to the fold-type III of pyridoxal 5′-phosphate (PLP)-dependent enzymes. PLP, a coenzyme of phosphorylated and oxidized vitamin B_6, is known to bind to an active site lysine residue in alanine racemase to regulate amino acid metabolism [56,57].

Figure 6 shows that seaweeds have the fold-type III of PLP-dependent enzyme as a proline synthetase co-transcribed bacterial homolog protein [58]. Even though the lysine residue for PLP-binding is conserved at the N-terminal region in the aligned protein sequences, these seaweed proteins lack a region corresponding to the C-terminal domain of the fission yeast alanine racemase (Figure 6). Since this domain contains the tyrosine residue known as an important catalytic residue, these proteins are not alanine racemases. To date, little is known about the physiological significance of D-alanine in seaweeds. However, the differential distribution of this amino acid in foldfasts and blades [30] suggests physiological roles in a tissue-dependent manner. Since the presence of D-alanine strongly suggests that seaweeds have alanine racemase genes, molecular cloning of putative alanine racemase genes and the functional analysis of them could help to resolve the physiological roles of D-alanine and alanine racemization in seaweeds.

5.2. The Presence of Serine Racemase/Dehydratase Genes

Eukaryotic serine racemase activity producing D-serine was initially detected in the silkworm *Bombyx mori* [59] and subsequently in the rat brain [60]. To date, many genes encoding eukaryotic serine racemases have been cloned from the rat [61], human *Homo sapiens* [62], *A. thaliana* [63], barley *Hordeum vulgare* [64], fission yeast *S. pombe* [65], rice *Oryza sativa* [66], and slime mold [67]. The primary structures of eukaryotic serine racemases are different from those of bacterial racemases. Eukaryotic enzymes belong to fold-type II PLP-dependent enzymes, whereas bacterial enzymes are fold-type III [56,57]. In addition, eukaryotic enzymes catalyze the dehydration of D-serine and L-serine to produce pyruvate and ammonia, indicating that the eukaryotic serine racemase is a bifunctional enzyme, which is sometimes called a serine racemase/dehydratase. Moreover, Mg^{2+} and Ca^{2+} affect the serine racemase and dehydratase activities of eukaryotic serine racemases [63,64,66,67]. Figure 7 shows that the PLP- and Mg^{2+}-binding sites comprise highly conserved amino acid residues [66].

The genomic sequence and EST data clearly indicate the presence of putative serine racemase genes in seaweeds, such as *P. umbilicalis*, *P. purpurea*, *C. crispus*, and *E. siliculosus*. The alignment in Figure 7 indicates highly conserved primary structures and amino acid residues involved in PLP-and Mg^{2+}-binding in serine racemases from algae, terrestrial plants, and mammals, although *Porphyra* enzymes contain a long insertion after the PLP-binding lysine residues. These findings suggest a functional significance of seaweed enzymes. However, amino acid substitutions are found at the Mg^{2+}-binding aspartate in red algal enzymes and at the PLP-associating glycine and serine in algae (Figure 7). The enzymatic characterization of serine racemases from rice, barley, and *A. thaliana* indicated similar bifunctional activities with racemization and hydration [63,64,66] to those of mammalian enzymes [68,69]. These results are consistent with the highly conserved amino acid residues involved in PLP- and Mg^{2+}-binding in all of these enzymes (Figure 7). Thus, the amino acid substitutions mentioned above suggest differences in enzymatic activity and/or catalytic efficiency in algae compared to those in terrestrial plants and mammals. However, the serine racemase from *E. siliculosus* is an exception, as it has conserved PLP- and Mg^{2+}-binding residues (Figure 7).

Despite the presence of putative serine racemase genes in seaweeds, D-serine has not been detected in the examined seaweeds (Yokoyama, unpublished). Thus, it is unclear why homologues of the serine racemase gene are present in seaweed genomes. To this point, D-serine regulates glutamate receptor-like Ca^{2+} channels to control pollen tube growth in *Arabidopsis* pistils, in which the serine racemase gene is involved in *de novo* D-serine synthesis [70]. This finding clearly demonstrates the functional significance of D-serine in plants. Therefore, the enzymatic and functional characterization of seaweed enzymes may provide new insight into the enzymatic activity and physiological function of the eukaryotic serine racemase.

```
P. yezoensis    MSRLIHRVYLGFTLPCSLASRVSRLPGIHSSCAGRRLIQGHPPILFANRWTSAAAAASPLVTMATPDRPPTSTASPA   77
P. purpurea     -------------------------------------------------------------------RPPASTASPD   10
P. umbilicalis  --------------------------------------------------------PTPLPPAPPP-PPPPTPSWRR   20
C. crispus      MS-------------------------------------------------------------------SETTSVTN   10
C. merolae      MS-----------------------------VRWKRL-QAQLGFL-----------SSGLRMQHTISESKAQGERAQ   36
E. siliculosus  MS-------------------------------------EAELK-----------------------SEPKATG-AS   16
C. reinhardtii  MA-------------------------------------------------------------------------DK    4
S. pombe        MR-------------------------------GAR-----------------------------------------    5

                                                   @
P. yezoensis    AI-ATALDTIRSRIEAAVEAASPPPTAPVRLIAVSKTRPPEAIAAAFDAGQVDFGENYVQ---ELVGKAGACPPGVR   149
P. purpurea     AI-AAALESIRGRIDAAVGSASPPPTAPVKLIAVSKTRPPEAIAAAFDAGQVDFGENYVQ---ELVGKAGDCPPGVR   83
P. umbilicalis  WT-PSAAASTR-----RWGRRPPPPTAPVTLIAVSKTRPPDAIAAAFDAGQVDFGENYVQ---ELVGKVADCPPGVR   88
C. crispus      GV-AEALRRVQERITAAASGCGRPP---PELVAVSKTKPVELLLEAYEAGQRSFGENYVQ---EIVAKSPQLPSDIK   80
C. merolae      TI-AERLCKVRERIQVAALAAKRDP-ARVRLVAVSKTHAAGTVREAYDAGQRHFGENYVQ---ELVEKAEQLPTDIN   108
E. siliculosus  PI-ADNLEAVRKRVEEVTGEGEAGK-PVPRLVAVSKTKPLENLQDAYDAGQRIFGENYAQ---ELIDKSPQMPDDVV   88
C. reinhardtii  PI-AALALQDVSRMKQATERANRT--HPVRLVAVSKTKPAEALQEAYDAGQRVFGENYVQARCEMLDKAPALPGDVQ   78
S. pombe        --AVIDLRAIGYNYNLIRELLDEKN-PDAHILCIIKANA-------YGHGAVEV--------AQFLAKYCSAEGFGV   64

P. yezoensis    WHYIGALQSNKAKVLVGQVPG-LACVQSVDRVKTARALNTAVV-NRSEKAAA-------AGSDNGDSRASGDDGRAL   218
P. purpurea     WHYIGALQSNKAKVLVGEVSG-LACVQSVDRSKTARALNNAVV-NRTQKAAAGDGDGE-EAGEGAGADDGGDGGDTL   157
P. umbilicalis  WHYIGALQSNKAKVLVGEVPG-LACVQSVDRSKTARALNNAVV-NRQQKAAVTDGDGQAAAGEGAGAADGGDGGDSL   163
C. crispus      WHFIGGLQSNKAKVL-AGVPN-LYMVESVDRQKTATALNKAFA-AVGR-------------------------AEKL   129
C. merolae      WHFIGHLQSNKVRTLLS-VPN-LWCVETVDRPKLANTLNRLMD-ELRPD------------------------GGPI   158
E. siliculosus  WHFIGHLQSNKAKALVAGVPN-LAVLETLD---TVKLANKLQS-ACVSSER----------------------KRPL   138
C. reinhardtii  WHFIGHLQSNKVKAVLEGVPN-LAMVETVDSAKLADKLNKTLETVSGR-------------------------TAPL   129
S. pombe        ASIEE-ALELRLSGIQNRILLLEGFFTDEDELSLIDKYNFSITIHCEEQLKSFMNYPF---------------KKPV   124

P. yezoensis    DVMVQVNTSGEDSKSGCAPD---EAVAVARYVTDECASLRLVGLMTIGAPDTQSERPPAFDVLLRVRGEVEAALGLG   292
P. purpurea     DVMVQVNTSGEESKSGCAPD---EAVELARYVTDECRALRLIGLMTIGAPDTESDRPPAFEVLLRVRGEVEDALDRG   231
P. umbilicalis  DVMVQVNTSGEESKSGCAPD---EAVELARYVSDECRALRLVGLMTIGAPDTESDRPPAFEVLLRVRGEVEAALDRG   237
C. crispus      KVMVQVNTSGEESKSGCQPG---ETAQLAKFVDTECDNLELVGLMTIGAFDT-SDEPEAFKILATERDAVAEVVERA   202
C. merolae      PVMVQVNVSGEASKAGIEPA---AAPELVEHILQACPRLKLLGLMTIGSPDP-SPEPVAFQRLSHLRDQIQDRFRFQ   231
E. siliculosus  GVYLQIDTSGEDSKAGIYHSDLDACLSLARHLKDNCPALELKGLMTIGAPGD----MECFDRLNACRDAVAGGLGME   211
C. reinhardtii  AVMVQVNTSGEESKYGVEPT---ECVSLAKHIAQNCPKLRLAGLMTIGMPDY-SSRPECFICLSGCRDAVATELGLR   202
S. pombe        EVHLKLDSG--MNRLGFTPA---EYADKYRLLKN---HKNVSGIVKATHFAF-ADIPEKSEY-TLKQWRIFEKAAGC   192

P. yezoensis    GD-GGSRLGLSMGMSGDFEAA--VRCGATT------VRVGSAIFGARDYSA   334
P. purpurea     EGEGGGRASASPWA                                       245
P. umbilicalis  EG-GGRGGRRPLQWA                                      252
C. crispus      AS----TLTLSMGMSADFEAA--IRMGSDS------VRVGSTIFGAREYPPK   242
C. merolae      E-----PLELSMGMSDDFEAA--VRMGSTN------VRIGSTIFGPRVYPQ   269
E. siliculosus  AQ----ALELSMGMSGDYEEA--IRRGSTN------VRVGSTIFGARFYPNKT   252
C. reinhardtii  PE----QLELSMGMSGDFEQA--IEMGSTN------IRVGSTIFGAREYKPKSPPASPKKE   251
S. pombe        LPD----PLSAGGGV---------IVVGWLNTIHMDWLRTGSMLYGLDPYDLDAKAPELPKPLIPAMKLMSTIVCVK   256

                         *
S. pombe        HVEKDQPIGYGGAYVTTRDSLIGVVAIGYGDGFPQVRNGCPVIINGKRVPTVGKVCMDMLAIDVTDVPDVKRGDDVV   333

S. pombe        LWGNPELTIEEVSTFSNENPFEIITGLTRRVPLQYTF   370
```

Figure 6. Multiple amino acid sequence alignment of seaweed proline synthetase co-transcribed bacterial homologus compared to that of fission yeast.
The putative PLP-binding lysine residue is shown in a bold letter indicated by @, whereas * represents the catalytic tyrosine residue found in the *S. pombe* protein. The following are the GenBank accession numbers or EST clone identities of the proteins examined: *P. yezoensis*, contig_21487_g5299; *P. purpurea*, esContig5734; *P. umbilicalis,* esContig05694; *C. crispus*, XP_005710953; *C. merolae*, XP_00553996; *E. siliculosus*, Esi0516_0006; *C. reinhardtii*, XP_001689869; and *S. pombe*, NP_595052 (Alr2).

```
P. purpurea       M-------------------------------ADPAPIVTLTDIQAAAARLDGVAHVTPLLTCASLDVALSGGT---   43
P. umbilicalis    M-----RSRRCSSGSRGCHRRR----------------------RRRRRRRRWYRHHLPIPSWTAAATAAAE-----   45
C. crispus        M-------------------------PSPLEDACASTGVSPATIREASERIRPHAHVTPILTSSLIDAWLCGSSPAL   52
C. merolae        M------------------------------------GVNRLHI------------ATPCIESREFSRFIGS-----   24
E. siliculosus    MHEVQRRSRRCL--SVGPEARDLGAMSNDASDGSSPAVLSLAGIREAADRISPHTQITPVLTCGALNEMS-G-----   69
C. reinhardtii    M-------------------------------TGDNYAASLEDIQAAQKRIAPFAHVTSVQTCSALNELAGG-----   41
A. thaliana       M-----------------------------EANREKYAADILSIKEAHDRIKPYIHRTPVLTSESLNSIS-G-----   42
O. sativa         MG-----SR---------------GGSGGDGAESHGYAADIHSIREAQARIAPYVHKTPVLSSTSIDAIV-G-----   51
H. sapiens        M--------------------------------CAQYCISFADVEKAHINIRDSIHLTPVLTSSILNQLT-G-----   39

                                 *@
P. purpurea       TRFYFKAEALQRTGSFKFRG---AYNAVAALVERLEGVPSPPPLPPPALVLAPLADSVADRGDNCGGRHSGLPGDGE  117
P. umbilicalis    -----------------------TAGCLATA----------------------------------------------   53
C. crispus        TNFFFKAECFQKTGSFKFRG---ALNAVGQL----------------------------------------------   80
C. merolae        QHVWLKLENLQRSGSFKIRGIGLLCSESLRL----------------------------------------------   55
E. siliculosus    REVFFKCELFQKTGSFKARG---ACNAVLLT----------------------------------------------   97
C. reinhardtii    RQLFFKCEIFQRTGAFKFRG---ACNSVQLL----------------------------------------------   69
A. thaliana       RSLFFKCECLQKGGAFKFRG---ACNAVLSL----------------------------------------------   70
O. sativa         KQLFFKCEC-QKAGAFKIRG---ASNSIFAL----------------------------------------------   79
H. sapiens        RNLFFKCELFQKTGSFKIRG---ALNAVRSLV------------------------------------------PD-   70

                                                                  *
P. purpurea       DGRDEERAAAVGTTPR-------PPCWSAAALPGARGH-LTLLTHSSGNHGQAVAAAAAVVGARAHIVLPAASAAVK  186
P. umbilicalis    ------RPAVTRSAPRQRSPPRPPPRWSAAAAPGARGY-LTFLSHSSGNHGQAVAAAAAAVGARAHIVVPAASAAVK  123
C. crispus        -------------------------------QQSTSTRGKPVVTHSSGNHAQALSAAAQVYGVAAHVVIPQD-AIPK  126
C. merolae        -----------------------------------QHC-TEIICSSGGNAGLAAAAAASALGLPVQVYVPETTPAYM   96
E. siliculosus    --------------------------------PPE--C-PGVVTHSSGNHAQALAWAASVRGIPAHIVMPRSCMAVK  139
C. reinhardtii    --------------------------------SDEEAK-KGVVTHSSGNHAAALALAAKLRGIPAHIVVPHTTPQCK  113
A. thaliana       --------------------------------DAEQAA-KGVVTHSSGNHAAALSLAAKIQGIPAYIVVPKGAPKCK  114
O. sativa         --------------------------------DDDEAS-KGVVTHSSGNHAAAVALAAKLRGIPAYIVIPRNAPACK  123
H. sapiens        --------------------------------ALERKP-KAVVTHSSGNHGQALTYAAKLEGIPAYIVVPQTAPDCK  114

P. purpurea       RTAIKAYGGTITTCAAGMASRRATVERLLRSPAVAAGTVAAAAPVALSPAVVAGAA-GAFPLPSPPELRVLVDPHNS  262
P. umbilicalis    RAAIKAYGGTITTCAAGMVARRAAVARLLHPPAVAAAASAVSSADAL-PIAVAAAA-GS-PSPPLPELRVLVDPHNS  197
C. crispus        SEAARAYGGTIHECAPGMDSRRSAAEKVLH------------------------DT-GG----------ILVHPFLD  168
C. merolae        RARLRSEGATVHVHGKVWDETNRKALEVVE-----------------------RSS-GA----------LYVHPFDH  139
E. siliculosus    VDAVRGYGASVTLCENSKSSRESTAADIVS------------------------KLPGS----------SLIHSSND  182
C. reinhardtii    VDAVRAYGGTLYFCEPTVDAREAKCAEVAA------------------------TT-GA----------TLVPPYNY  155
A. thaliana       VDNVIRYGGKVIWSEATMSSREEIASKVLQ------------------------ET-GS----------VLIHPYND  156
O. sativa         VDNVKRYGGHIIWSDVSIESRESVAKRVQE------------------------ET-GA----------ILVHPFNN  165
H. sapiens        KLAIQAYGASIVYCEPSDESRENVAKRVTE------------------------ET-EG----------IMVHPNQK  156

                                                      * *                          #   #    #
P. purpurea       ADVIAGHGTLALELLAQS------PTPLDALIVPVGSGSMLAGVAAAVRSVAIAPGGRPLRLFGAQPAVAA---AAA  330
P. umbilicalis    ADVIAGHGTLALELLAQA------PTPLDALIVPVGSGSMLAGVAAVVRSAAATPAGRPVRLFGAQPAVAA---AAA  265
C. crispus        PNVVSGQGTIGIELIQQV------KD-LNAIVVPVGGGGLISGITIAVKSACP-----HVKIIGAEAALSN---AAS  230
C. merolae        PLIWTGHSTLVDE-LDQR------PA---AIVLSVGGGGLLLGVVQGLQRRGWT----DVPVVAAETVGAA---SFA  199
E. siliculosus    ARVQCGIGTIGLELVSQVKDLTGGAD-LDAVIVTIGGGGLISGVTTAVKALCP-----NARIIGAQPEKAA---DAF  250
C. reinhardtii    GPVIAGQGTIALEFLEQV------PD-LDALVVPVSGGGMISGIALAAKALKP-----GIKIIAAEPGGTNGGADVA  220
A. thaliana       GRIISGQGTIALELLEQI------QE-IDAIVVPISGGGLISGVALAAKSIKP-----SIRIIAAEPKGAD---DAA  218
O. sativa         KNTISGQGTVSLELLEEV------PE-IDTIIVPISGGGLISGVALAAKAINP-----SIRILAAEPKGAD---DSA  227
H. sapiens        PAVIAGQGTIALEVLNQV------PL-VDALVVPVGGGGMLAGIAITVKALKP-----SVKVYAAEPSNAD---DCY  218

P. purpurea       ASLAAGRRVP-NNPTGA-AATVADGLRA-DLCAAGWAALARGDVEEV-LLVGEADIGGWMRVVWERLKVTIEPSAAV  403
P. umbilicalis    ASLAAGRRVA-NDPVGA-ATTVADGLRA-DLCAAGWAALARGDVEEV-LLVGEPDIGGWMRVVWERLKVMIEPSAAV  338
C. crispus        RSLAAGKRIE-FYN----PSTIADGIRA-GIGELGW-EVLGRMVEEV-VSVQEPEIEDTMKTFMERMKVVIEPSAAV  299
C. merolae        ATFNANGVLQRIPT----IDTVAKSLGALQVSEACA-ELIRSRAHPLYSYVCTDQEAVEAMLAFATLHRMLVEPACG  271
E. siliculosus    KSKQSGTLCS-PAEGEALPDTIADGLRT-SLGNKAW-PVVRDSVDEI-VTVSEDDIRVWMGVVYKRMKLVIEPSAAV  323
C. reinhardtii    LSKMAGWVVP-CAK----PNTIADG-QG-RLGDLTW-PVVRDLVDCV-VVVDEQEIVGAMRLIMERMKLVVEPSGAV  289
A. thaliana       QSKVAGKIIT-LPV----TNTIADGLRA-SLGDLTW-PVVRDLVDDV-VTLEECEIIEAMKMCYEILKVSVEPSGAI  287
O. sativa         QSKAAGKIIT-LPS----TNTIADGLRA-FLGDLTW-PVVRDLVDDI-IVVDDNAIVDAMKMCYEMLKVAVEPSGAI  296
H. sapiens        QSKLKGKIMPNLYP----PETIADGVKS-SIGLNTW-PIIRDLVDDI-FTVTEDEIKCATQLVWERMKLLIEPTAGV  288

                                               *
P. purpurea       GVAAAAGERFGRL-------GLRHVGVVLSGGNVDLGAPLPWQRGGV  443
P. umbilicalis    GVAAAAGERFGRL-------GMRHVGVVLCGGNVDLGAPLPWQRGGI  378
C. crispus        AVAACKTKQFRER-------GFRRVAVVLCGGNVDLQA-LPWNCRR  337
C. merolae        AALTLVTSPRGRAFLHEHIGQTGNIVVVVCGGNMITPS-MLMELAKVTGASGDSWSA  328
E. siliculosus    GVAAIMSPQVRAMPS-----EMRKVGVVLCGGNVDVTL-LQDLLSVAGPSSD  369
C. reinhardtii    GLAAVLSPAFGASAQTR---GCKRVGVVLCGGNVDLGARGLWDCWIPKSARRED  340
A. thaliana       GLAAVLSNSFRNNPSCR---DCKNIGIVLSGGNVDLGS-LWDSFKSSK  331
O. sativa         GLAAALSDEFKQSSAWH---ESSKIGIIVSGGNVDLGV-LWESLYKR  339
H. sapiens        GVAAVLSQHFQTVSP-----EVKNICIVLSGGNVDLTSSITWVKQAERPASYQSVSV  340
```

Figure 7. Comparison of amino acid sequences of serine racemase/dehydratase from seaweeds, microalgae, terrestrial plants, and human.

The PLP-binding lysine and conserved PLP-interacting five residues are marked by @ and *, respectively. Putative magnesium binding residues are indicated by #, and amino acid residues are shown in bold letters. The following are GenBank accession numbers or EST clone identities of the proteins examined: *P. purpurea*, esContig8644; *P. umbilicalis,* esContig04983; *C. crispus*, CFD33870; *C. merolae*, CMR455C; *E. siliculosus*, Esi0109_0058; *C. reinhardtii*, EDO99697; *A. thaliana*, AEE83032; *O. sativa*, Q7XSN8; and *H. sapiens*, AAK58495.

CONCLUSION

The presence of D-amino acids and genes encoding putative serine racemase has been confirmed in seaweeds. However, little is known about the enzymatic activity of these racemases, and thus confirmation of the enzymatic activity of seaweed racemases using molecular biological approaches is necessary to understand the biosynthesis and biological function of D-serine in seaweeds. However, there is no information about aspartate racemase in brown seaweeds containing high levels of D-aspartate, although an aspartate racemase has recently been found in the bivalve mollusk *S. broughtonii* [22]. Furthermore, Funakoshi et al. [41] reported that a D-amino acid transaminase, which catalyzes the transamination of D-aspartate and other D-amino acids, was found in the terrestrial plant *A. thaliana*. Because the purification of the corresponding enzymes from marine brown algae is difficult due to their low concentrations, the identification and characterization of genes encoding these enzymes is one of our short-term goals.

D-Amino acids have been recognized to be tastier compounds compared to their L-enantiomer counterparts. For example, various D-amino acids, such as D-alanine, D-glutamate, D-leucine, D-aspartic acid, and D-isoleucine, have been found in fermented drinks, such as wines, berries, and vinegar [71-73]. Among these D-amino acids, D-alanine is known to be sweet [74,75]. Thus, artificial alteration of the D-alanine content can improve the taste of fermented drinks. Similarly, the modulation of D-alanine content in commercially valuable seaweeds could control their taste and commercial values because low levels of free D-alanine are present in many seaweeds. D-Alanine content can also increase or decrease under appropriate environmental stress conditions because the D-aspartate content is influenced by hypoxic stress in *S. fusiforme*. Therefore, the identification of culture conditions to increase D-alanine content could contribute to producing high quality seaweeds.

In summary, the biochemical identification and characterization of D-amino acid racemases and molecular biological analysis of the regulation of racemase gene expression could provide detailed understanding of its biological function, which could be applied to both the basic and applied research of seaweeds.

ACKNOWLEDGMENTS

This work was supported in part by JSPS KAKENHI Grant Numbers 24580307, 2566016003.

REFERENCES

[1] Lewis, EJ; Gonzalves, EA. The protein, peptide, and free amino-acid contents of some species of marine algae from Bombay. *Annals of Botany*, 1962, 26, 301-316.

[2] Lewis, EJ; Gonzalves, EA. Periodic studies of the proteins, peptides, and free amino-acids in *Enteromorpha prolifera* f. *capillaris* and *Ulva lactuca* var. *rigida*. *Annals of Botany*, 1962, 26, 317-327.

[3] Fleurence, J. Seaweed proteins: biochemical, nutritional aspects and potential uses. *Trends in Food Science & Technology*, 1999, 10, 25-28.

[4] Ito, K. Free amino acids and peptides in marine algae. *Nippon Suisan Gakkaishi* (in Japanese), 1969, 35, 116-129.

[5] Impellizzeri, G; Mangiafico, S; Oriente, G; Piattelli, M; Sciuto, S. Amino acids and low-molecular-weight carbohydrates of some marine red algae. *Phytochemistry*, 1975, 14, 1549-1557.

[6] Zagon, J; Dehne, LI; Bögl, KW. D-Amino acids in organisms and food. *Nutrition Research*, 1994, 14, 445-463.

[7] Aswad, DW. Determination of D- and L-aspartate in amino acid mixtures by high-performance liquid chromatography after derivatization with a chiral adduct of *o*-phthaldialdehyde. *Analytical Biochemistry*, 1984, 137, 405-409.

[8] Nimura, N; Kinoshita, T. *o*-Phthalaldehyde-*N*-acetyl-L-cysteine as a chiral derivatization reagent for liquid chromatographic optical resolution of amino acid enantiomers and its application to conventional amino acid analysis. *Journal of Chromatography A*, 1986, 352, 169-177.

[9] Yokoyama, T; Kan-no, N; Ogata, T; Kotaki, Y; Sato, M; Nagahisa, E. Presence of free D-amino acids in microalgae. *Bioscience, Biotechnology, and Biochemistry*, 2003, 67, 388-392.

[10] Friedman, M. Chemistry, nutrition, and microbiology of D-amino acids. *Journal of Agricultural and Food Chemistry*, 1999, 47, 3457-3479.

[11] Corrigan, JJ. D-Amino acids in animals. *Science*, 1969, 164, 142-149.

[12] Sekimizu, K; Larranaga, J; Hamamoto, H; Sekine, M; Furuchi, T; Katane, M; Homma, H; Matsuki, N. D-Glutamic acid-induced muscle contraction in the silkworm, *Bombyx mori*. *Journal of Biochemistry*, 2005, 137, 199-203.

[13] Hashimoto, A; Nishikawa, T; Oka, T; Hayashi, T; Takahashi, K. Widespread distribution of free D-aspartate in rat periphery. *FEBS Letters*, 1993, 331, 4-8.

[14] Man, EH; Sandhouse, ME; Burg, J; Fisher, GH. Accumulation of D-aspartic acid with age in the human brain. *Science*, 1983, 220, 1407-1408.

[15] Ishio, S; Yamada, H; Hayashi, M; Yatsushiro, S; Noumi, T; Yamaguchi, A; Moriyama, Y. D-Aspartate modulates melatonin synthesis in rat pinealocytes. *Neuroscience Letters*, 1998, 249, 143-146.

[16] D'Aniello, A; Cosmo, AD; Cristo, CD; Annunziato, L; Petrucelli, L; Fisher, G. Involvement of D-aspartic acid in the synthesis of testosterone in rat testes. *Life Sciences*, 1996, 59, 97-104.

[17] Felbeck, H; Wiley, S. Free D-amino acids in the tissues of marine bivalves. *Biological Bulletin*, 1987, 173, 252-259.

[18] Matsushima, O; Katayama, H; Yamada, K; Kado, Y. Occurrence of free D-alanine and alanine racemase activity in bivalve molluscs with special reference to intracellular osmoregulation. *Marine Biology Letters*, 1984, 5, 217-225.

[19] D'Aniello, A; Giuditta, A. Presence of D-alanine in crustacean muscle and hepatopancreas. *Comparative Biochemistry and Physiology Part B: Biochemistry and Molecular Biology*, 1980, 66, 319-322.

[20] D'Aniello, A; Giuditta, A. Identification of D-aspartic acid in the brain of *Octopus vulgaris* Lam. *Journal of Neurochemistry*, 1977, 29, 1053-1057.

[21] D'Aniello, A; Giuditta, A. Presence of D-aspartate in squid axoplasm and in other regions of the cephalopod nervous system. *Journal of Neurochemistry*, 1978, 31, 1107-1108.
[22] Watanabe, T; Shibata, K; Kera, Y; Yamada, R. Occurrence of free D-aspartate and aspartate racemase in the blood shell *Scapharca broughtonii*. *Amino Acids*, 1998, 14, 353-360.
[23] Abe, H; Yoshikawa, N; Sarower, MG; Okada, S. Physiological function and metabolism of free D-alanine in aquatic animals. *Biological & Pharmaceutical Bulletin*, 2005, 28, 1571-1577.
[24] Shibata, K; Watanabe, T; Yoshikawa, H; Abe, K; Takahashi, S; Kera, Y; Yamada, R. Nucleotides modulate the activity of aspartate racemase of *Scapharca broughtonii*. *Comparative Biochemistry and Physiology Part B: Biochemistry and Molecular Biology*, 2003, 134, 713-719.
[25] Watanabe, T; Shibata, K; Kera, Y; Takahashi, S; Yamada, R. Effects of hypoxic and osmotic stress on the free D-aspartate level in the muscle of blood shell *Scapharca broughtonii*. *Amino Acids*, 2005, 28, 291–296.
[26] Robinson, T. D-Amino acids in higher plants. *Life Sciences*. 1976, 19, 1097-1102.
[27] Gamburg, KZ; Rekoslavskaya, NI. Formation and functions of D-amino acids in plants. *Soviet plant physiology*, 1992, 38, 904-912.
[28] Ogawa, T; Kimoto, M; Sasaoka, K. Identification of D-aspartic acid and D-glutamic acid in pea seedlings. *Agricultural and Biological Chemistry*, 1977, 41, 1811-1812.
[29] Nagahisa, E; Kanno, N; Sato, M; Sato, Y. Occurrence of free D-aspartic acid in marine macroalgae. *Biochemistry international*, 1992, 28, 11-19.
[30] Nagahisa, E; Kanno, N; Sato, M; Sato, Y. Occurrence of free D-alanine in marine macroalgae. *Bioscience, Biotechnology, and Biochemistry*, 1995, 59, 2176-2177.
[31] Nagahisa, E; Kanno, N; Sato, M; Sato, Y. Variations in D-aspartate content with season and part of *Hizikia fusiformis*. *Fisheries Science*, 1994, 60, 777-779.
[32] Brückner, H; Westhauser, T. Chromatographic determination of L- and D-amino acids in plants. *Amino Acids*, 2003, 24, 43-55.
[33] Yokoyama, T; Amano, M; Sekine, M; Homma, H; Tokuda, M; Sato, M. Immunohistochemical localization of endogenous D-aspartate in the marine brown alga *Sargassum fusiforme*. *Bioscience, Biotechnology, and Biochemistry*, 2011, 75, 1481-1484.
[34] Storm-Mathisen, J; Leknes, AK; Bore, AT; Vaaland, JL; Edminson, P; Haug, F-MS; Ottersen, OP. First visualization of glutamate and GABA in neurones by immunocytochemistry. *Nature*, 1983, 301, 517-520.
[35] Schmitz, K. Translocation. In: Lobban CS and Wynne MJ (eds). *The Biology of Seaweeds*, Oxford: Blackwell Scientific Publications, 1981, 534-558.
[36] Collicutt, JM; Hochachka, PW. The anaerobic oyster heart: Coupling of glucose and aspartate fermentation. *Journal of Comparative Physiology B*, 1977, 115, 147-157.
[37] Gäde, G; Meinardus, G. Anaerobic metabolism of the common cockle *Cardium edule*. V. Changes in the level of metabolites in the foot during aerobic recovery after anoxia. *Marine Biology*, 1981, 65, 113-116.
[38] Ji, Y; Tanaka, J. Effect of desiccation on the photosynthesis of seaweeds from the intertidal zone in Honshu, Japan. *Phycological Research*, 2002, 50, 145-153.

[39] Bada, JL; Lee, C. Decomposition and alteration of organic compounds dissolved in seawater. *Marine Chemistry*, 1977, 5, 523-534.

[40] Kim, PM; Duan, X; Huang, AS; Liu, CY; Ming, G; Song, H; Snyder, SH. Aspartate racemase, generating neuronal D-aspartate, regulates adult neurogenesis. *Proceedings of the National Academy of Sciences of the United States of America*, 2010, 107, 3175-3179.

[41] Funakoshi, M; Sekine, M; Katane, M; Furuchi, T; Yohda, M; Yoshikawa, T; Homma, H. Cloning and functional characterization of *Arabidopsis thaliana* D-amino acid aminotransferase– D-aspartate behavior during germination. *FEBS Journal*, 2008, 275, 1188-1200.

[42] Cock, JM; Sterck, L; Rouzé, P; Scornet, D; Allen, AE; Amoutzias, G; Anthouard, V; Artiguenave, F; Aury, JM; Badger, JH; Beszteri, B; Billiau, K; Bonnet, E; Bothwell, JH; Bowler, C; Boyen, C; Brownlee, C; Carrano, CJ; Charrier, B; Cho, GY; Coelho, SM; Collén, J; Corre, E; Da Silva, C; Delage, L; Delaroque, N; Dittami, SM; Doulbeau, S; Elias, M; Farnham, G; Gachon, CM; Gschloessl, B; Heesch, S; Jabbari, K; Jubin, C; Kawai, H; Kimura, K; Kloareg, B; Küpper, FC; Lang, D; Le Bail, A; Leblanc, C; Lerouge, P; Lohr, M; Lopez, PJ; Martens, C; Maumus, F; Michel, G; Miranda-Saavedra, D; Morales, J; Moreau, H; Motomura, T; Nagasato, C; Napoli, CA; Nelson, DR; Nyvall-Collén, P; Peters, AF; Pommier, C; Potin, P; Poulain, J; Quesneville, H; Read, B; Rensing, SA; Ritter, A; Rousvoal, S; Samanta, M; Samson, G; Schroeder, DC; Ségurens, B; Strittmatter, M; Tonon, T; Tregear, JW; Valentin, K; von Dassow, P; Yamagishi, T; Van de Peer, Y; Wincker, P. The *Ectocarpus* genome and the independent evolution of multicellularity in brown algae. Nature, 2010, 465, 617-621.

[43] Chan, CX; Zäuner, S; Wheeler, G; Grossman, AR; Prochnik, SE; Blouin, NA; Zhuang, Y; Benning, C; Berg, GM; Yarish, C; Eriksen, RL; Klein, AS; Lin, S; Levine, I; Brawley, SH; Bhattacharya, D. Analysis of *Porphyra* membrane transporters demonstrates gene transfer among photosynthetic eukaryotes and numerous sodium-coupled transport systems. *Plant Physiology*, 2012, 158, 2001-2012.

[44] Stiller, JW; Perry, J; Rymarquis, LA; Accerbi, M; Green, PJ; Prochnik, S; Lindquist, E; Chan, CX; Yarish, C; Lin, S; Zhuang, Y; Blouin, NA; Brawley, SH. Major developmental regulators and their expression in two closely related species of *Porphyra* (Rhodophyta). *Journal of Phycolology*, 2012, 48, 883-896.

[45] Nakamura, Y; Sasaki, N; Kobayashi, M; Ojima, N; Yasuike, M; Shigenobu, Y; Satomi, M; Fukuma, Y; Shiwaku, K; Tsujimoto, A; Kobayashi, T; Nakayama, I; Ito, F; Nakajima, K; Sano, M; Wada, T; Kuhara, S; Inouye, K; Gojobori, T; Ikeo, K. The first symbiont-free genome sequence of marine red alga, Susabi-nori (*Pyropia yezoensis*). *PLoS One*, 2013, 8, e57122.

[46] Collén, J; Porcel, B; Carré, W; Ball, SG; Chaparro, C; Tonon, T; Barbeyron, T; Michel, G; Noel, B; Valentin, K; Elias, M; Artiguenave, F; Arun, A; Aury, JM; Barbosa-Neto, JF; Bothwell, JH; Bouget, FY; Brillet, L; Cabello-Hurtado, F; Capella-Gutiérrez, S; Charrier, B; Cladière, L; Cock, JM; Coelho, SM; Colleoni, C; Czjzek, M; Da Silva, C; Delage, L; Denoeud, F; Deschamps, P; Dittami, SM; Gabaldón, T; Gachon, CM; Groisillier, A; Hervé, C; Jabbari, K; Katinka, M; Kloareg, B; Kowalczyk, N; Labadie, K; Leblanc, C; Lopez, PJ; McLachlan, DH; Meslet-Cladiere, L; Moustafa, A; Nehr, Z; Nyvall Collén, P; Panaud, O; Partensky, F; Poulain, J; Rensing, SA; Rousvoal, S; Samson, G; Symeonidi, A; Weissenbach, J; Zambounis, A; Wincker, P; Boyen, C.

Genome structure and metabolic features in the red seaweed *Chondrus crispus* shed light on evolution of the Archaeplastida. *Proceedings of the National Academy of Sciences of the United States of America*, 2013, 110, 5247-5252.

[47] Matsuzaki, M; Misumi, O; Shin-i, T; Maruyama, S; Takahara, M; Miyagishima, S; Mori, T; Nishida, K; Yagisawa, F; Nishida, K; Yoshida, Y; Nishimura, Y; Nakao, S; Kobayashi, T; Momoyama, Y; Higashiyama, T; Minoda, A; Sano, M; Nomoto, H; Oishi, K; Hayashi, H; Ohta, F; Nishizaka, S; Haga, S; Miura, S; Morishita, T; Kabeya, Y; Terasawa, K; Suzuki, Y; Ishii, Y; Asakawa, S; Takano, H; Ohta, N; Kuroiwa, H; Tanaka, K; Shimizu, N; Sugano, S; Sato, N; Nozaki, H; Ogasawara, N; Kohara, Y; Kuroiwa, T. Genome sequence of the ultrasmall unicellular red alga *Cyanidioschyzon merolae* 10D. *Nature*, 2004, 428, 653-657.

[48] Merchant, SS; Prochnik, SE; Vallon, O; Harris, EH; Karpowicz, SJ; Witman, GB; Terry, A; Salamov, A; Fritz-Laylin, LK; Maréchal-Drouard, L; Marshall, WF; Qu, LH; Nelson, DR; Sanderfoot, AA; Spalding, MH; Kapitonov, VV; Ren, Q; Ferris, P; Lindquist, E; Shapiro, H; Lucas, SM; Grimwood, J; Schmutz, J; Cardol, P; Cerutti, H; Chanfreau, G; Chen, CL; Cognat, V; Croft, MT; Dent, R; Dutcher, S; Fernández, E; Fukuzawa, H; González-Ballester, D; González-Halphen, D; Hallmann, A; Hanikenne, M; Hippler, M; Inwood, W; Jabbari, K; Kalanon, M; Kuras, R; Lefebvre, PA; Lemaire, SD; Lobanov, AV; Lohr, M; Manuell, A; Meier, I; Mets, L; Mittag, M; Mittelmeier, T; Moroney, JV; Moseley, J; Napoli, C; Nedelcu, AM; Niyogi, K; Novoselov, SV; Paulsen, IT; Pazour, G; Purton, S; Ral, JP; Riaño-Pachón, DM; Riekhof, W; Rymarquis, L; Schroda, M; Stern, D; Umen, J; Willows, R; Wilson, N; Zimmer, SL; Allmer, J; Balk, J; Bisova, K; Chen, CJ; Elias, M; Gendler, K; Hauser, C; Lamb, MR; Ledford, H; Long, JC; Minagawa, J; Page, MD; Pan, J; Pootakham, W; Roje, S; Rose, A; Stahlberg, E; Terauchi, AM; Yang, P; Ball, S; Bowler, C; Dieckmann, CL; Gladyshev, VN; Green, P; Jorgensen, R; Mayfield, S; Mueller-Roeber, B; Rajamani, S; Sayre, RT; Brokstein, P; Dubchak, I; Goodstein, D; Hornick, L; Huang, YW; Jhaveri, J; Luo, Y; Martínez, D; Ngau, WC; Otillar, B; Poliakov, A; Porter, A; Szajkowski, L; Werner, G; Zhou, K; Grigoriev, IV; Rokhsar, DS; Grossman, AR. The *Chlamydomonas* genome reveals the evolution of key animal and plant functions. *Science*, 2007, 318, 245-250.

[49] Hoffmann, K; Schneider-Scherzer, E; Kleinkauf, H; Zocher, R. Purification and characterization of eucaryotic alanine racemase acting as key enzyme in cyclosporin biosynthesis. *Journal of Biological Chemistry*, 1994, 269, 12710-12714.

[50] Wood, WA; Gunsalus, IC. D-Alanine formation: a racemase in Streptococcus *faecalis*. *Journal of Biological Chemistry*, 1951, 190, 403-416.

[51] Ono, K; Yanagida, K; Oikawa, T; Ogawa, T; Soda, K. Alanine racemase of alfalfa seedlings (*Medicago sativa* L.): First evidence for the presence of an amino acid racemase in plants. *Phytochemistry*, 2006, 67, 856-860.

[52] Yokoyama, T; Tanaka, Y; Sato, M; Kan-no, N; Nakano, T; Yamaguchi, T. Nagahisa E. Alanine racemase activity in the microalga *Thalassiosira* sp. *Fisheries Science*, 2005, 71, 924-930.

[53] Nishimura, K; Tomada, Y; Nakamoto, Y; Kawada, T; Ishii, Y; Nagata, Y. Alanine racemase from the green alga *Chlamydomonas reinhardtii*. *Amino Acids*, 2007, 32, 59-62.

[54] Uo, T; Yoshimura, T; Tanaka, N; Takegawa, K; Esaki, N. Functional characterization of alanine racemase from *Schizosaccharomyces pombe*: a eucaryotic counterpart to bacterial alanine racemase. *Journal of Bacteriology*, 2001, 183, 2226-2233.

[55] Yoshikawa, N; Okada, S; Abe, H. Molecular characterization of alanine racemase in the Kuruma prawn *Marsupenaeus japonicus*. *Journal of Biochemistry*, 2009, 145, 249-258.

[56] Jansonius, JN. Structure, evolution and action of vitamin B_6-dependent enzymes. *Current Opinion in Structural Biology*, 1998, 8, 759-769.

[57] Schneider, G; Käck, H; Lindqvist, Y. The manifold of vitamin B_6 dependent enzymes. *Structure*, 2000, 8, R1-6.

[58] Ikegawa, S; Isomura, M; Koshizuka, Y; Nakamura, Y. Cloning and characterization of human and mouse *PROSC* (proline synthetase co-transcribed) genes. *Journal of Human Genetics*, 1999, 44, 337-342.

[59] Uo, T; Yoshimura, T; Shimizu, S; Esaki, N. Occurrence of pyridoxal 5'-phosphate-dependent serine racemase in silkworm, *Bombyx mori*. *Biochemical and Biophysical Research Communications*, 1998, 246, 31-34.

[60] Wolosker, H; Sheth, KN; Takahashi, M; Mothet, JP; Brady, RO Jr; Ferris, CD; Snyder, SH. Purification of serine racemase: biosynthesis of the neuromodulator D-serine. *Proceedings of the National Academy of Sciences of the United States of America*, 1999, 96, 721-725.

[61] Wolosker, H; Blackshaw, S; Snyder, SH. Serine racemase: a glial enzyme synthesizing D-serine to regulate glutamate-*N*-methyl-D-aspartate neurotransmission. *Proceedings of the National Academy of Sciences of the United States of America*, 1999, 96, 13409-13414.

[62] De Miranda, J; Santoro, A; Engelender, S; Wolosker, H. Human serine racemase: moleular cloning, genomic organization and functional analysis. *Gene*, 2000, 256, 183-188.

[63] Fujitani, Y; Nakajima, N; Ishihara, K; Oikawa, T; Ito, K; Sugimoto, M. Molecular and biochemical characterization of a serine racemase from *Arabidopsis thaliana*. *Phytochemistry*, 2006, 67, 668-674.

[64] Fujitani, Y; Horiuchi, T; Ito, K; Sugimoto, M. Serine racemases from barley, *Hordeum vulgare* L., and other plant species represent a distinct eukaryotic group: gene cloning and recombinant protein characterization. *Phytochemistry*, 2007, 68, 1530-1536.

[65] Yamauchi, T; Goto, M; Wu, HY; Uo, T; Yoshimura, T; Mihara, H; Kurihara, T; Miyahara, I; Hirotsu, K; Esaki, N. Serine racemase with catalytically active lysinoalanyl residue. *Journal of Biochemistry*, 2009, 145, 421-424.

[66] Gogami, Y; Ito, K; Kamitani, Y; Matsushima, Y; Oikawa, T. Occurrence of D-serine in rice and characterization of rice serine racemase. *Phytochemistry*, 2009, 70, 380-387.

[67] Ito, T; Murase, H; Maekawa, M; Goto, M; Hayashi, S; Saito, H; Maki, M; Hemmi, H; Yoshimura, T. Metal ion dependency of serine racemase from *Dictyostelium discoideum*. *Amino Acids*, 2012, 43, 1567-1576.

[68] Foltyn, VN; Bendikov, I; De Miranda, J; Panizzutti, R; Dumin, E; Shleper, M; Li, P; Toney, MD; Kartvelishvily, E; Wolosker, H. Serine racemase modulates intracellular D-serine levels through an α,β-elimination activity. *Journal of Biological Chemistry*, 2005, 280, 1754-1763.

[69] Cook, SP; Galve-Roperh, I; Martínez del Pozo, A; Rodríguez-Crespo, I. Direct calcium binding results in activation of brain serine racemase. *Journal of Biological Chemistry*, 2002, 277, 27782-27792.

[70] Michard, E; Lima, PT; Borges, F; Silva, AC; Portes, MT; Carvalho, JE; Gilliham, M; Liu, LH; Obermeyer, G; Feijó, JA. Glutamate receptor-like genes form Ca^{2+} channels in pollen tubes and are regulated by pistil D-serine. *Science*, 2011, 332, 434-437.

[71] Erbe, T; Brückner, H. Chiral amino acid analysis of vinegars using gas chromatography – selected ion monitoring mass spectrometry. *Zeitschrift für Lebensmittel-Untersuchung und -Forschung A*, 1998, 207, 400-409.

[72] Erbe, T; Brückner, H. Chromatographic determination of amino acid enantiomers in beers and raw materials used for their manufacture. *Journal of Chromatography A*, 2000, 881, 81-91.

[73] Kato, S; Ishihara, T; Hemmi, H; Kobayashi, H; Yoshimura, T. Alterations in D-amino acid concentrations and microbial community structures during the fermentation of red and white wines. *Journal of Bioscience and Bioengineering*, 2011, 111, 104-108.

[74] Solms, J; Vuataz, L; Egli, RH. The taste of L- and D-amino acids. *Experientia*, 1965, 21, 692-694.

[75] Kawai, M; Hayakawa, Y. Complex taste – Taste of D-amino acids. *Chemical Senses*, 2005, 30 (Suppl 1), i240-i241.

In: Seaweeds
Editor: Vitor H. Pomin

ISBN: 978-1-63117-571-8

Chapter 8

SEAWEED POLYPHENOLS AS POTENTIAL ANTI-OXIDANT AGENTS

Rashmi Chandrakant Vinayak* and Vasanth L. Kallapur
Department of Microbiology and Biotechnology,
Karnatak University, Dharwad, Karnataka State, India

ABSTRACT

Recent developments in the medicinal field have reported a number of diseases linked to free radicals that have been allied to an unhealthy lifestyle, exposure of chemicals, pollution, cigarette smoking, drugs, illness, and stress etc. In the past, coastal populations across the globe consumed dietary foods such as edible seaweeds as a part of their habitual diet including many several Asian countries, that protected them from the free radical induced damages. Their anti-oxidant composition alone presented a useful alternative to non-natural substances, while at the same time providing worthwhile nutritional benefits as foods, food supplements, nutraceuticals and medicines. Previous studies have shown that phenolic compounds are the main contributors to the anti-oxidant activity of various seaweeds. In fact, a positive correlation has been well documented between total phenol content (TPC) and anti-oxidant activities. Many types of seaweed are already in use as alternative and complementary medicine, and obtainable in numerous ways, such as, teas, syrups, oils, liquid extracts, tinctures, and dry extracts (pills or capsules) for therapeutic purposes. However, economic utilization of seaweeds from the perspective of anti-oxidant benefits to human health still remains in its infancy. This chapter will highlight current lack of data in this area and help outline seaweed extracts and seaweed derived isolates with major health benefits that merit consequent analytical studies and possible utilization as natural anti-oxidants.

Keywords*:* Anti-oxidants, seaweeds, polyphenols, free radical scavengers, nutraceuticals, dietary supplements

---

* Corresponding author: Dr. Rashmi C. Vinayak, Faculty at Department of Microbiology and Biotechnology, Karnatak University,Dharwad, Karnataka, India E-mail: rashmi.vinayak@gmail.com Tel: +91 832 2450354, Fax: +91 832 2450603

I. INTRODUCTION

The marine environment provides a large collection of distinctive and exceptional biodiversity of novel bioactive molecules that provide health benefits above and beyond traditional nutrients. Algae form one of the most important prospective marine nutraceutical resources that are distinctively divided in two main groups, namely macro-algae (seaweed) and micro-algae. Seaweeds are primitive non-vascular, photosynthesizing plants found in the intertidal zone that accumulate nutrients and trace elements by direct absorption from surrounding sea waters. Since seaweeds lack many of the distinct organs found in terrestrial plants, entire seaweed plants are accessible as a biomass supply. There are about 6000 species of seaweeds divided into three main classes based on habitat and color. The green seaweeds (Chlorophyceae) are mostly from the upper tidal zone; brown seaweeds (Phaeophyceae) are mostly in the mid tidal zone; and red seaweeds (Rhodophyceae) are mostly from the lower tidal zone. Seaweeds have been utilized for centuries in many countries; particularly in Asia, for different uses. Traditionally, they have been used since ancient times as food, fodder, fertilizer, and as other sources of medicinal drugs, with earliest records dating back to 2700 B.C. In the west, seaweeds are chiefly used as raw materials for industrial production of agar, carrageenan, and alginates, while they continue to be widely consumed as food in some East and South East Asian countries. In addition, seaweeds have been utilized in the preparations of salad, soups, pasta, and jellies and are nutritionally valuable, in both fresh as well as dried forms.

Of late, algae have been explored for a number of anti-oxidants such as polyphenols and sulfated polysaccharides. In comparison with their terrestrial counterparts *i.e.,* plants, algae are potentially good sources of anti-oxidants and have an advantage over many other organisms in that they can appropriately produce a large amount of desirable and specific bioactive compounds. Although polyphenols have been reported in general, there has been relatively little information on those from marine origin. Hence an effort has been made in the present discourse to review the literature as well as provide recent developments with respect to occurrence of seaweed polyphenols and their anti-oxidant properties.

II. SEAWEEDS AS A RICH SOURCE OF ANTI-OXIDANTS

Reactive oxygen species (ROS) and reactive nitrogen species (RNS) are essential for energy supply, detoxification, chemical signaling, and immune functions. They help maintain homeostasis in normal healthy cells when produced in the human body in a well regulated manner. However, if this balance is lost, which can happen in response to xenobiotics or environmental stress, the level of ROS and RNS can increase and become a major contributing factor to the development of several medical conditions, such as cardiovascular diseases, including atherosclerosis, various types of cancers, diabetes and neurodegenerative diseases, like Parkinson's and Alzheimer's disease (Devasagayam et al., 2004). The harm caused by ROS, such as Superoxide anion (O^{-2}), hydroxyl ($HO^{\cdot}$), hydro peroxy ($HOO^{\cdot}$) radicals and RNS, such as nitric oxide (NO), peroxynitrite ($ONOO^{-}$), and nitrogen dioxide (NO_2) that are formed naturally during many metabolic processes are alleviated by the presence of anti-oxidants. Interestingly, the body possesses several defense mechanisms against such radical-induced oxidative stress, which involve preventative mechanisms, repair mechanisms, physical defenses and anti-oxidant defenses including, both enzymatic and non-enzymatic systems. Enzymatic anti-oxidant defenses include superoxide dismutase (SOD), glutathione peroxidase (GPX), catalase (CAT), etc., whereas non-enzymatic anti-oxidants constitute of ascorbic acid (vitamin C), α-tocopherol (vitamin E), glutathione (GSH), carotenoids, NADPH, phenolic acids, tannins, flavonoids, etc (Mallick and Mohn, 2000; Wu and Cederbaum, 2003; Devasagayam et al., 2004). These various agents act by one or more mechanisms like reducing activity, free radical-scavenging, potential complexing of pro-oxidant metals and quenching of singlet oxygen. They work together to thwart, interrupt and restore the injury caused by the free radicals (Devasagayam et al., 2004). It is, therefore, possible to reduce the risks of chronic diseases and prevent disease progression by either enhancing the body's natural anti-oxidant defenses or by supplementing with proven dietary anti-oxidants (Stanner et al., 2004). This is one of the reasons why discovery and synthesis of novel anti-oxidants is a major active research area in recent years.

Butylated hydroxyl anisol (BHA), butylated hydroxyl toluene (BHT), and tertiarybutyl hydroquinone (TBHQ) are some of the synthetic anti-oxidants used in food and pharmaceutical industries. But these anti-oxidants have been suspected to cause toxicity. Therefore, interest is more focused towards finding natural anti-oxidants as additives in food systems. Being derived from natural resources, these anti-oxidant molecules are easily accepted by the consumers and may not warrant any safety measures from the regulatory authorities. The interest in natural anti-oxidants is, perhaps, more in the pharmaceutical industry due to urgent need towards anti-ageing pharmacopeia.

In recent times, the seaweed extracts and fractions have been shown to be a rich source of anti-oxidants and diverse varieties of anti-oxidants have been isolated from various species of seaweeds (Fujimoto and Kaneda, 1984; Cahyana et al., 1992; Nagai and Yukimoto, 2003; Huang and Wang, 2004; Wang et al., 2009; Mancini-Filho et al., 2009; Rioux et al., 2010; Hu et al., 2010; Vinayak et al., 2011a, 2011b). The prospective anti-oxidant compounds were recognized as a few pigments (fucoxanthin, astaxanthin, carotenoid e.g.) and polyphenols (phenolic acid, flavonoid, tannins etc.,), which are extensively dispersed in seaweeds and are known to display superior antioxidative activities. Some of the active components identified from the marine seaweeds were; phylopheophytin in *Eisenia bicyclis* (Cahyana et al., 1992), phlorotannins in *Sargassum kjellamanianum* (Yan et al., 1996) and fucoxanthin in *Hijikia fusiformis* (Yan et al., 1999), a low molecular weight sulphated polysaccharide from *Laminaria japonica* (Wang et al., 2008), and mycosporin like amino acids (MAA's) from red seaweeds (Nakayama et al., 1999). Their activities have been ascribed to their ROS scavenging ability, quenching singlet and triplet oxygen, chelating ability, reducing power and the inhibition of lipid peroxidation (Karawita et al., 2005).

Different species of seaweed display varying degrees of anti-oxidant activities. Brown seaweeds generally show better anti-oxidant properties than that of green and red seaweeds. In particular, some species such as *Ecklonia cava, Ecklonia kurome, , Fucus vesiculosus, F. Serratus, F. Spiralis, Hijikia fusiformis* and *S. Ringgoldianum* have been reported to possess remarkably high anti-oxidant potential *in vitro*, which is linked to their total polyphenolic (TPC) content (Karawita et al., 2005; Cerantola et al., 2006; Nakai et al., 2006; Shibata et al., 2008). A positive correlation has been well documented between TPC and anti-oxidant activities of different seaweed extracts (Vinayak et al., 2011a and b). High degree of correlation between TPC and 1,1-Diphenyl-2- picryl hydrazine (DPPH) radical scavenging activity has been reported for different solvent extracts (R^2 = 0.971) (Siriwardhana et al., 2003) as well as different organic and aqueous fractions of *H. Fusiformis* (R^2 =0.921) (Karawaita et al., 2005). However, it should be made known that a number of other studies have also accounted reduced or low correlation coefficients between TPC and anti-oxidant activities of some seaweed extracts indiacting that some other co-extracted active compounds, such as fucoxanthin and sterols in solvent extracts and water soluble sulphated polysaccharides, proteins or peptides, ascorbic acid, and glutathione (GSH) in water and enzymatic extracts, may also contribute to the overall anti-oxidant properties.

In vitro studies of seaweed polyphenols, especially phlorotannins isolated from several brown seaweeds have been reported to possess extraordinary anti-oxidant properties; however, because of lack of reliable *in vivo* data their possible anti-oxidant effects in biological systems are not very persuasive. A few recent studies have shed some light on anti-oxidant properties of seaweed polyphenols *in vivo* and their potential application in pharmaceutical and neutraceutical fields. The anti-oxidant and antidiabetic effects of methanol extract (MEE) from brown alga *Ecklonia stolonifera* were recently studied in KK-A^y diabetic mice model (Iwai et al., 2008). Continuous administration of diets containing MEE for 4 weeks inhibited lipid peroxidation in plasma, red blood cells, liver and kindney of unfasted KK-Ay mice in a dose dependent manner. The formation of plasma thiobarbituric acid reactive species (TBARS) was almost completely inhibited in both low and high level MEE diet groups and the levels were significantly lower than those in control group. The potent anti-oxidant activity of MEE appeared to have beneficial effects for the prevention of diabetic oxidative stress and related complications. The effective components in MEE were identified as phlorotannin compounds.

III. Seaweed Polyphenols

Polyphenols are a large, diverse group of secondary metabolites belonging to the tetraterpene family possessing phenolic functionalities. No less than 8,000 different bioactive compounds considered to be polyphenolic compounds have been derived from diverse terrestrial plants such as, algae, photosynthetic bacteria etc., till date (Bravo, 1998). They are largely derived from the phenylpropanoid-acetate pathway, and related biochemical pathways (Figure 1), with a wide array of physiological roles in plants such as, defending plants, determining durability, establishing flower color and contributing substantially to certain taste and odors. However, marine seaweeds are only just now gaining interest as a potential source of polyphenols. These phenolic compounds have been known to vary from comparatively simple chemical structures such as, phenolic acids and other polyphenolic compounds to the more complex structures like phlorotannins. Seaweeds are known to possess a wide variety of anti-oxidant molecules, such as ascorbate and glutathione (GSH) when fresh, as well as secondary metabolites, including carotenoids (α- and β-carotene, fucoxanthin, astaxanthin etc.,), mycosporine-like amino acids (mycosporine-glycine), catechins (e.g., catechin, epigallocatechin, epigallocatechin), gallate, phlorotannins (e.g., phloroglucinol), eckol and tocopherols (α-, χ-, δ-tocopherols) (Yuan et al., 2005). Among polyphenols, phlorotannins, phenolic acids, flavonoids, isoflavones can be mentioned (Dai and Mumper, 2010).

The anti-oxidant capacity of phenols has been linked to the impediment of several diseases including cancer, coronary heart diseases, inflammatory disorders, neurological degeneration, and aging (Madhavi et al., 1996; Wollaston and Anklam, 2000). Seaweeds generally produce these compounds in abundance for defending themselves against stress and herbivores (Li et al., 2011). ROS are exceedingly reactive and can cause cellular dysfunction and cytotoxicity when produced in organisms during metabolic reactions (Alviano and Alviano, 2009). Polyphenols generate non-reactive end products by supplying hydrogen to these ROS (Gupta and Abu-Ghannam, 2011).

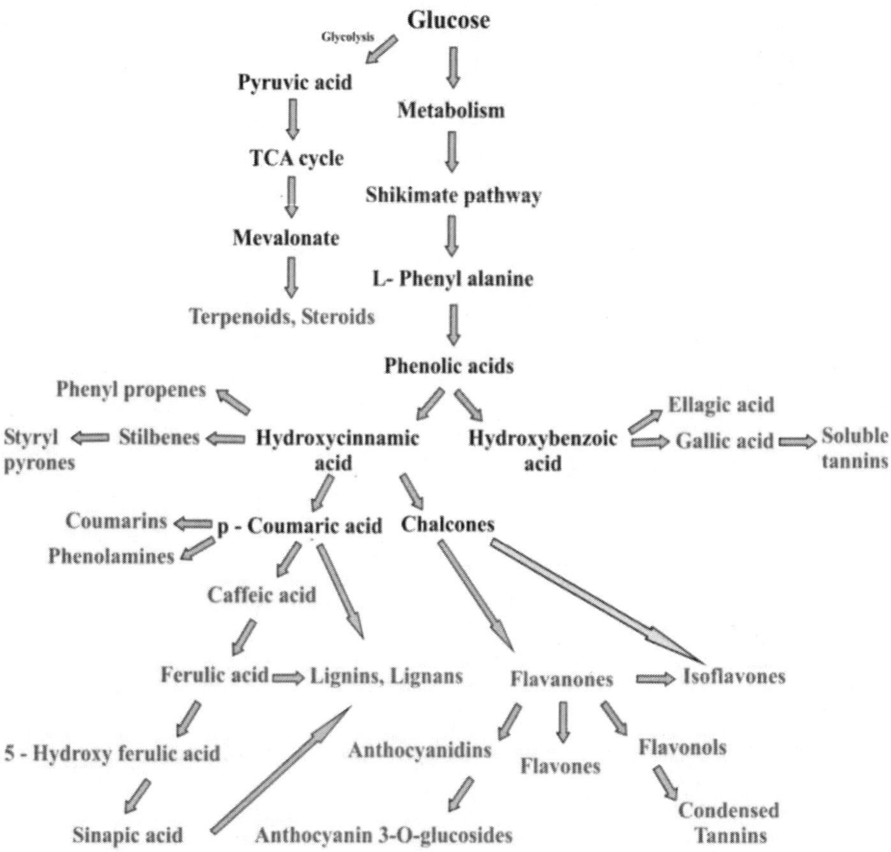

Figure 1. Outilne of the biosynthetic pathways leading to the synthesis of specific phenolic subclasses in higher plants.

Seaweed extracts contain appreciable amounts of polyphenols and its variation appear to be correlated with plant size, age, tissue type as well as environmental factors such as salinity, nutrients, light availability, seasonal variations and water depth (Pavia and Toth, 2000). *Ascophyllum spp* comprises of noticeably more polyphenols than other seaweeds, while *Ulva* spp have the lowest content of these compounds (Craige, 2011; Keyrous et al., 2011). Phlorotannin levels reported for *F. Vesiculosus* was known to seasonally vary, with higher TPC levels usually occurring in summer and early autumn where as lower amounts were noted throughout winter (Connan et al., 2004). Besides, *F. vesiculosus* and *A. nodosum* when exposed to sun were found to have largely higher TPC than their shaded counterparts.

Sub classes	Structure	Seaweed species	Biological activities	References
Fucols	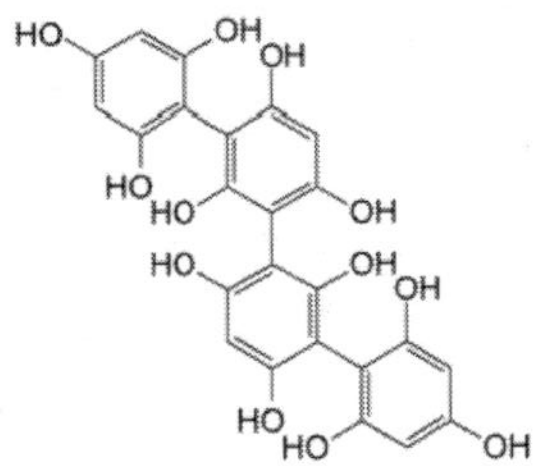	*Ecklonia cava, Ecklonia kurome,Ecklonia stolonifera, Eisenia bicyclis, Eisenia arboreae, Fucus vesiculosus, Fucus spiralis, Fucus serratus, Ascophyllum nodosum, Sargassum kjellmanianum, Sargassum pallidum, Sargassum ringgoldianum, Sargassum siliquastrum, Ishige okamurae*	Anti-oxidant, anti-inflammatory, anti-allergic, anti-tumor, anti-diabetic, anti-bacterial, HIV-1 reverse transcriptase and protease inhibitor and chemoprevention in many vascular diseases.	Kang et al., 2003; Kang et al., 2004; Fukuyama et al., 1989; Fukuyama et al., 1990; Lee et al., 1998; Han et al., 2000; Nagayama et al., 2002; Ahn et al., 2004; Ahn et al., 2007
Fuhalols	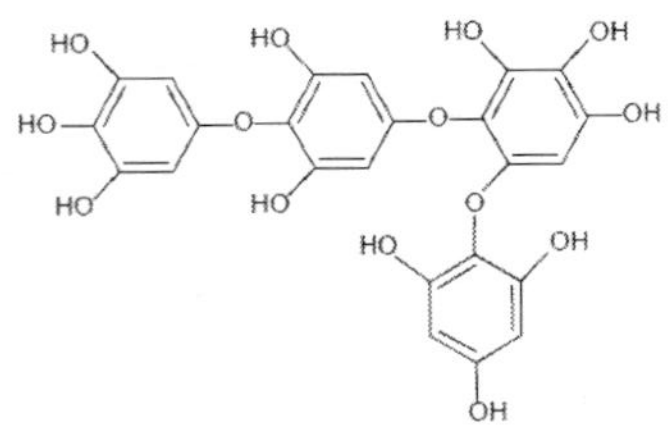			

Eckols	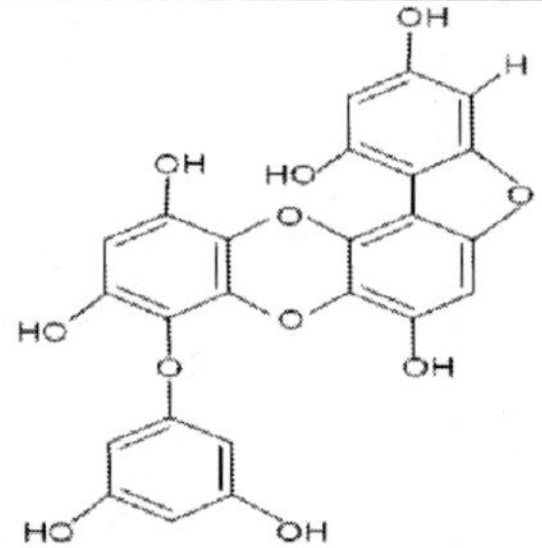			
Fuco - phloroethols	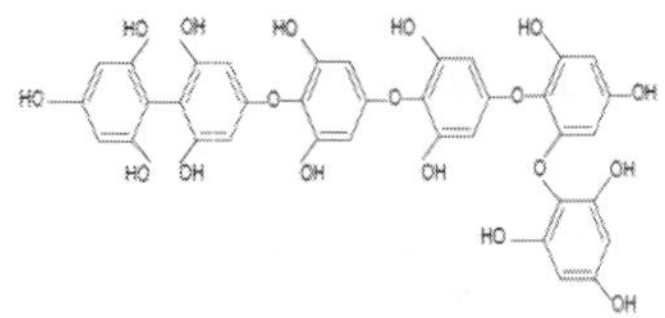			
Phlorethols	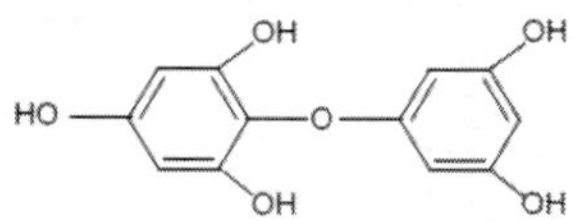			
Isofuhalols	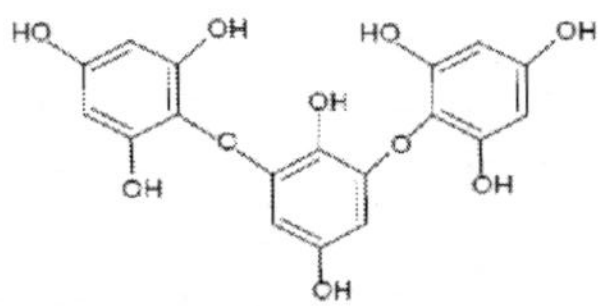			

Figure 2. Phlorotannins and their therapeutic properties.

Studies have revealed that seaweed polyphenols possess many biological activities, including anti-inflammatory (Shin et al., 2006), hepatoprotective (Zhao et al., 2004; Kim et al., 2005), antitumor (Kim et al., 2006; Yuan and Walsh, 2006), antihypertensive (Jung et al., 2006), and HIV-1 reverse transcriptase (Ahn et al., 2004) activities in addition to antidiabetic activity (Iwai, 2008). Their varied physiological activities, as a result, may offer many advantages for potential applications in the nutraceutical, pharmaceutical, and cosmetic industries. However, the assessment of the effects wielded by seaweed polyphenols can cause several setbacks when extending from simple experimental systems to the complex human system. The main hurdle is the bioavailability and the complexity in understanding the intricate mechanism of absorption and metabolism. So far, seaweed polyphenols have been studied for their biological activities *in vitro* using cultivated cell lines at the laboratory scale, for the most part using water or organic extracts. These conservative procedures have quite a few disadvantages for instance low selectivity, low extraction competency, solvent residue and environmental pollution. A critical analysis will be required to delineate the absorption time of seaweed polyphenols, classification of their constituents, and the actual molecular actions that emphasize the biological effects of seaweed polyphenols. Moreover, several pioneering technologies for extraction, separation, and purification of bioactive compounds from marine seaweeds require to be set up.

a. Phlorotannins

Phlorotannins are the largest group of polyphenols that are produced in brown seaweeds biosynthetically via acetate-malonate pathway, also known as polyketide pathway (Arnold and Targett, 2002).They are basically tannin derivatives composed of several phloroglucinol units (1,3,5-trihydroxybenzene) linked to each other through aryl-aryl bonds and diaryl ether bonds to form different subgroups of phlorotannins ranging between molecular sizes of 126 kda to 650 kda (Ragan and Glombitza, 1986; Boettcher and Targett, 1993; mcclintock and Baker, 2001; Singh and Bharate, 2006). They are known to represent up to 15% of the dry weight of brown seaweeds (Ragan and Glombitza, 1986; Targett and Arnold, 1998), varying considerably among different taxonomic groups as well as among different geographical areas. The highest levels are found in Fucales and Dictyotales (Ragan and Glombitza 1986; Van Alstyne et al., 1999a).

Phlorotannins can be methodically classified into six main subclasses (Figure 2.) on the type of structural linkages between the phloroglucinol sub-units (diphenylethers or biphenyls) and on the number of additional hydroxyl groups: Fucols, fuhalols, eckol, fucophlorethols, phloroethols and isofuhalols (Ragan and Glombitza, 1986; Targett and Arnold, 1998; Jung et al., 2006; Parys et al., 2007). Fucols are phlorotannin polymers in which phloroglucinol units are purely connected only by C-C (aryl-aryl) bonds. Fuhalols are formed when phloroglucinol units are connected with para- and ortho-arranged ether bridges containing one additional OH-group in every third ring connected exclusively via ether bonds. Eckols are distinguished by the presence of at least one three ring moiety with a dibenzo-1, 4-dioxin element substituted by a phenoxy group at the C-4 position isolated only from some specific genera of brown algae including *Ecklonia*, *Eisenia* and *Alarieae*. Fucophloroethols contain both biaryl and aryl-ether linkages. Carmalols are further derivatives of phlorethols containing a dibenzodioxin moiety. Phlorethols and isofuhalols are small, distinct, specialized groups isolated from specific seaweed genera such as *Chorda filum.*

The members of brown seaweed are reported to be the richest in phlorotannins amongst all other classes of marine seaweeds (Okada et al., 2004). The most well known phlorotannins from brown seaweeds such as, *Ecklonia bicyclis*, *Ecklonia cava* and *Ecklonia kurome* are phloroglucinol and its phloroglucinol tetramer, eckol, phlorofucofuroeckol A, dieckol, 8,8′-bieckol and dioxinodehydroeckol (Li et al., 2009; Shibata et al., 2004; Jung et al., 2010). In addition, a few more novel phlorotannins such as phlorofucofuroeckol A, triphloroethol B, 2-phloroeckol, 7-phloroeckol, diphloroethol, fucofuroeckol A from *Ecklonia stolinifera* and phlorofucofuroeckol A from *Ecklonia kurome* (Wijesekara et al., 2010) have been reported from other edible seaweeds. Shibata et al., (2008) evaluated the anti-oxidant properties of *E. bicyclis*, *E. cava* and *E. kurome* by DPPH, superoxide anion scavenging activities as well as inhibition of phospholipid peroxidation in the liposome system and found them to be 2 to 10 times superior to those of catechin, α-tocopherol and ascorbic acid. These phlorotannin oligomers also demonstrated significant protective effects against peroxidation of phospholipid liposomes.

A partially purified phlorotannin-rich fraction of *S. ringgoldianum* was found to have the highest superoxide radical scavenging activity, in a survey of 25 common seaweed species collected from Japanese coast (Nakai et al., 2006). The activity exhibited by *S. ringgoldianum* fraction was around five times higher than that of catechin.

Phlorotannins mainly derived from brown algae, can function as herbivore deterrents, digestive inhibitors, and antibacterial agents; thus, they are putative defensive agents against the natural enemies of these algae (Glombitza et al., 1997; Targett and Arnold, 1998; Luder and Clayton, 2004). Nagayama et al., (2002) found the antibacterial or bactericidal activity of seaweed phlorotannins against food-borne pathogenic bacteria as being more effective than that of catechins. Antibacterial and anti-oxidant activities of phlorotannins could be useful as a food preservative.

b. Flavanoids

The term "flavanoid" is used to describe a broad collection of phenolic compound that occur universally in plant foods having a wide spectrum of chemical and biological activities (Kahkonen *et al* 1999). They are largely associated with their ability to modulate cell-signaling pathways.They perform important functions in plants including, acting as cell wall sustaining materials (Wallace and Fry, 1994) and as colourful attractants for birds and insects helping seed dispersal and cross pollination (Harborne, 1994). They are also known to involve in the defence mechanisms of plants under different environmental stress conditions such as wounding, infection, and excessive light or UV irradiation (Bennet and Wallsgrove 1994, Dixon and Paiva 1995).

In recent years, they have aroused considerable interest because of their potential beneficial effects on human health such as anti-oxidant, anti-inflammatory, anti hepatotoxic, anti-tumoral, anti-allergic, anti-viral, anti-bacterial, anti-platelet and anti-ulcer actions, as well as a direct cytoprotective effect on coronary and vascular systems. They have also exhibited anti-coagulant, anti-hyperlipidemia, anti-nephritic, vasodilative effects and integrase inhibition of human immuno deficiency virus (HIV) type 1 (Tapas et al., 2008). These characteristic properties place them among the most attractive natural substances available to enrich the current therapy options.

The basic flavonoid structure is characterized by a flavan nucleus (Figure 3.), containing 15 carbon atoms arranged in three rings (C6-C3-C6), which are labeled as A, B and C. Flavonoid are themselves divided into six subgroups: flavones, flavonols, flavanols, flavanones, isoflavones, and anthocyanins, according to the oxidation state of the central C ring (Figure 4.). Their structural variation in each subgroup is partly due to the degree and pattern of hydroxylation, methoxylation, prenylation, or glycosylation. In excess of 4500 different falvonoids have been described till date. Some of the most common flavonoids like quercetin, a flavonol abundant in onion, broccoli, and apple; catechin, a flavanol found in tea and several fruits; naringenin, the main flavanone in grapefruit; cyanidin-glycoside, an anthocyanin abundant in berry fruits (black currant, raspberry, blackberry, *etc.*); and daidzein, genistein and glycitein, the major isoflavones in soybean have been reported (D'Archivio et al., 2007).

The best demonstrated properties of almost every group of flavonoids are their protective effects in biological systems owing to their ability to transmit free radical electrons. For example, chelate metal catalysts (Ferrali et al., 1997), activate anti-oxidant enzymes (Elliott et al., 1992), reduce alpha-tocopherol radicals (Hirano et al., 2001), and inhibit oxidases (Cos et al., 1998). They are estimated to express radical scavenging effect due to the several hydroxyl (OH) groups present in the flavanoid structure (Qian et al., 1999; Yan et al., 1999) or occasionally pro-oxidant effect as a source of reactive oxygen species (Oshima et al., 1998; Chan et al., 1999). It is has been argued that if the amount of hydroxyl groups in flavonoids are elevated, the anti-oxidant capacity is also enhanced (Cao et al., 1997), but it appears that their position in flavonoid skeleton is the critical structural element. It is apparent from the above findings, the flavones and catechins have emerged to be the most powerful flavonoids for defending the body against ROS.

Yoshie et al., (2000) and Santoso et al., (2002) identified a number of flavonoid compounds such as catechin (e.g gallocathecin, epicathecin, catechin gallate), flavonols and flavonol glycosides in red, brown and green algae. Flavanoids such as catechin and epicatechin (up to 4.77 mg g^{-1} dry weight) or the flavonol glycoside rutin (up to 11.4 mg g^{-1}dry weight) have been found to be abundant in *Porphyra* species. The flavone glycoside, hespiridin is commonly present in green, brown and red seaweeds and is particularly high in certain green algae (117 mg g^{-1} dry weight; Holdt and Kraan, 2011).

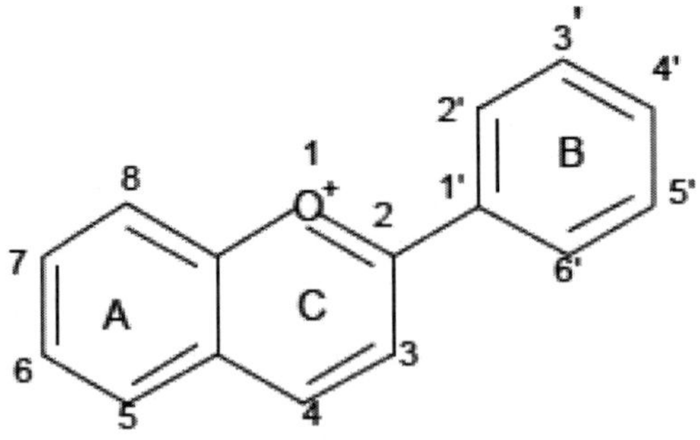

Figure 3. Structure of flavan nucleus.

Sub classes	Structure	Seaweed species	Biological activities	References
Flavones (Apigenin, Chrysin, Luteolin)	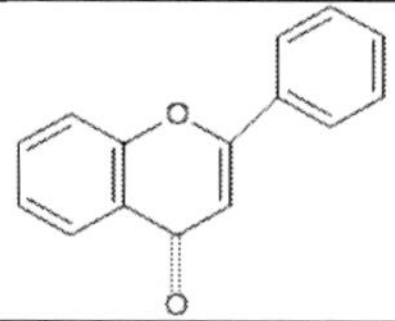	*Acanthophora spicifera, Caulerpa serrulata*	Anti-oxidant, anti-inflammatory, anti-hapatotoxic, antitumoral, anti-allergic, anti-viral, anti-bacterial, anti-platelet, anti-ulcer effects, cryoprotective effect on coronary and vascular systems, pancreas and liver, anti-coagulant, anti-hyperlipidemia, anti-nephritic,vasodilative effects and human immuno deficiency virus (HIV) type 1 integrase inhibition	Devi et al., 2008, Tapas et al., 2008
Flavanols (Catechin, Gallocatechin)	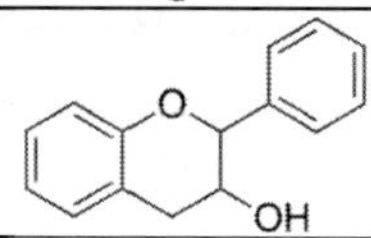			
Flavonols (Kampherol, Myrecetin, Quercetin)	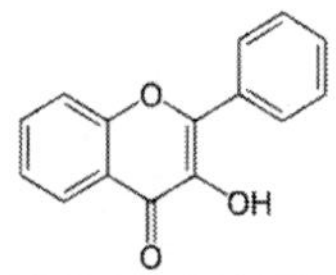			
Flavanones (Eriodictyol,Hesperitin,Narin genin)	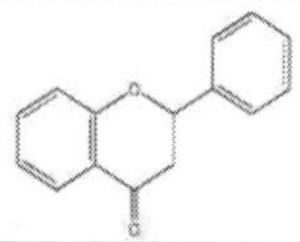			
Isoflavones (Daidzein, Genistein, Glycetein, Formononetin)	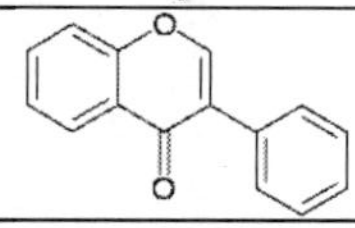			
Anthocyanins (Cyanidin, Delphinidin, Malvidin, Pelargonidin, Peonidin, Petunidin)	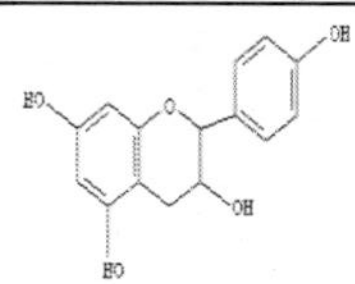			

Figure 4. Flavanoids and their therapeutic properties.

Sub classes	Structure	Seaweed species	Biological activities	References
Hydrobenzoic acid (Gallic acid, Ellagic acid, Protocatechulic acid, *p*-Hydroxybenzoic acid)	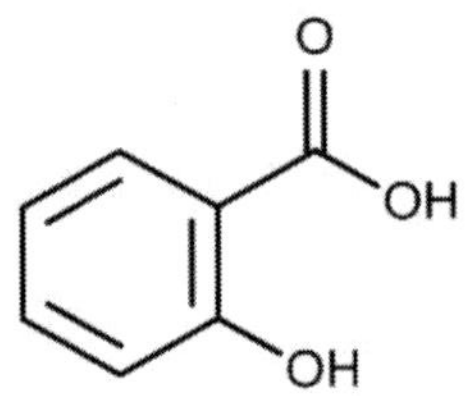	*Halimeda monile, H. Incrassata*	Anti-oxidant, anti-HIV, antitumorogenic activity immunostimulant activities	Mancini-Filho et al., 2009, Nair et al., 1991; Varnavas et al., 1991; Liu et al., 1995; Cardenas et al., 2006; Martínez-Toméet al., 2004; Sousa et al., 2004; Sousa et al., 2004; Anselmi et al.,2004; Mansouri et al., 2005
Hydrocinnamic acid (Coumaric acid, Caffeic acid, Ferulic acid, Chlorogenic acid and Sinapic acid)	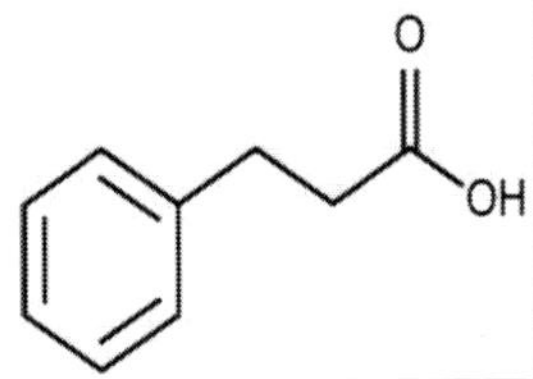			

Figure 5. Phenolic acids and their therapeutic properties.

Sub classes	Structure	Seaweed species	Biological activities	References
Bromocatechol (3-BC, 4-BC)		*Codium fragile, Odonthalia dentata, Polysiphonia morrowii,P. Nigra, P. Elongata, P. Fastigiata, P. Lanosa, P. Morrowii, P. Nigrescens, and P. Urceolata, Pterosiphonia complanata, Pterocladiella capillaceae,Rhodomelia confervoides, Symphyocladia latiuscula, Tichocarpus crinitus, Vertebrata kanosa and Vidalia volubilis,*	Anti-oxidant, anti-bacterial, anti-fungal,anti-viral, anti-inflammatory, anti-proliferative, anti-proliferative, anti-fouling, anti-feedant, cytotoxic, ichthyotocix and insecticidal activity	Li et al., 2011
Bromophenol (2-BP, 4-BP, 2,4-BP, 2,6-BP, 2,4,6-TBP)	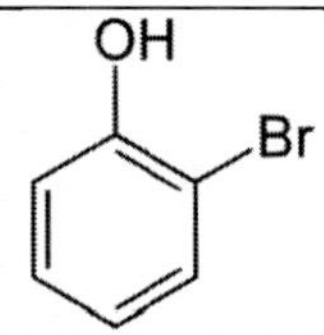			

Figure 6. Halogenated Phenols and their therapeutic properties.

There are reports to demonstrate a positive relationship between total flavonoid content and anti-oxidant activity in different seaweeds extracts (Jimenez et al., 2001; Athukorala et al., 2006). A quantitative data revealed the amount of flavanoids in 27 Japanese seaweeds like rutin, quercitrin, hesperidin, myricetin, morin, luteolin, quercetin, apigenin, kaempferol, baicalein, caffeic acid and catechol (Yoshie-Stark et al., 2003). Further it was found, that quercetin, apigenin, kaempferol, and baicalein were not universally distributed in the seaweeds examined. In general, red seaweed had larger amount of these compounds than that of brown and green seaweeds.

The total flavonoid content of *Kappaphycus alvarezii* using various organic and aqueous solvent extracts and their anti-oxidant activities were reported using DPPH method (Lalopua et al., 2011). The study revealed hexane extract to have highest total flavonoid content (15.93 %) than that of the water extracts (0.187 %). The radical scavenging activity of the hexane solvent extract increased with increasing concentrations and the highest activity could be obtained at 5 μg/ml hexane extract. The results indicated that hexane extract of *Kappaphycus alvarezii* could be a rich potential source of natural anti-oxidants.

Studies have shown high scavenging activities for *Ulva* species like, *Ulva clathrata* and *Ulva prolifera*. Between the two, the former displayed a high phenolic and flavonoid content and consequently showed a high DPPH radical scavenging activity with a low IC_{50} (Ramazan et al., 2013). Results showed that the phenolic and flavonoid contents of the extracts had a positive correlation with the DPPH radical scavenging activity ((r^2=0.93).), suggesting that they could be the main contributors of anti-oxidant activities in these *Ulva* species.

Quercetin, kaempferol, morin, myricetin and rutin, by acting as anti-oxidants, have been previously demonstrated to have anti-inflammatory, antiallergic, antiviral, as well as anticancer activity. They have also been suggested to play a protective role in liver diseases, cataracts, and cardiovascular diseases (Tapas et al., 2008). In addition to acting as free radical scavengers, quercetin and silybin, were shown to exert a protective effect in liver reperfusion ischemic tissue damage (Hillwell, 1994; Fraga et al., 1987). The scavenging activity of flavonoids has been reported to be in the order: Myrcetin > quercetin > rhamnetin > morin > diosmetin > naringenin > apigenin > catechin > 5, 7-dihydroxy-3', 4', and 5'-trimethoxyflavone > robinin > kaempferol > flavones (Ratty, 1988).

The role of flavonoids in the protection of health and prevention of diseases appear encouraging, in spite of the fact that the data is still inadequate and inconsistent. Moreover, reports on bioavailability of flavonoids from a range of foods, their level and mechanism of absorption in the human body and long-term consequences of chronic flavonoid ingestion are especially scarce. Hence, there is a need to improve analytic techniques to allow collection of more data on their absorption and excretion. Eventually, natural, hemisynthetic and synthetic flavonoids alone or in combination with other defensive and/or therapeutic strategies can develop into effective future drugs against some widespread degenerative diseases such as cancer, diabetes and cardiovascular complications.

c. Phenolic Acids

Phenolic acids or phenolcarboxylic acids are a type of aromatic acids compounds containing a phenolic ring and an organic carboxylic acid function (C6-C1 skeleton). Phenolic acids are also known to be the secondary metabolites extensively spread throughout the plant kingdom. They confer unique taste, flavour, and health-promoting properties found in vegetables and fruits (Tomas-Barberan and Espin, 2001) and are crucial for plants growth and reproduction. They are produced in response to environmental factors such as, light, chilling, pollution etc., and to defend injured plants (Valentine et al., 2003).

There are many different phenolic acids found in nature and they can be divided into two categories: derivatives of benzoic acid and derivatives of cinnamic acid (Figure 5.). Caffeic acid is the most abundant phenolic acid in many fruits and vegetables, most often esterified with quinic acid as in chlorogenic acid, which is the major phenolic compound in coffee. Another common phenolic acid is ferulic acid, which is present in cereals and is esterified to hemicelluloses in the cell wall (D'Archivio et al., 2007). However, not much data is available on the phenolic acids in seaweeds.

Phenolic acids have been the subject of a great number of chemical, biological, agricultural, and medical studies. Among the phenolic acids analyzed, chicoric acid and caffeic acid have previously been tested for anti-HIV and immunostimulant activities (King et al., 1999; Kim et al., 2000; Beale Robinson, 2000; Lee et al., 2003; Charvat et al., 2006). On the other hand, cinnamic acid and ferulic acid have been tested for antitumorogenic activity (Nair et al., 1991; Varnavas et al., 1991; Liu et al., 1995; Cardenas et al., 2006) with promising results. Again, obtained from different sources, gentisic acid (Martínez-Toméet al., 2004), caffeic acid, p-coumaric acid (Sousa et al., 2004), ferulic acid (Anselmi et al.,2004), and cinnamic acid (Mansouri et al., 2005) have shown to possess anti-oxidant activity. Since, phenolic acids are easily absorbed through the walls of intestinal tract; they may be beneficial to human health by virtue of their anti-oxidants that prevent cellular damage due to free-radical oxidation reactions. They may also promote anti-inflammatory conditions in the body when consumed regularly.

Water soluble simple phenolic acids are present in all three classes of seaweeds and include caffeic acid and cinnamic acid (Yuan, 2008). Recently many studies have focused on simple water and ethanol based extraction from seaweeds for obtaining simple phenolic acids Yan et al., 1998.

Previous works have shown the ability of *Halimeda incrassata* to capture free radicals in neuronic cells cultivation and its anti-oxidant effect was verified in rats exposed to oxidative stress induced by intoxication with methyl mercury (Fallarero et al., 2003). Hydrophilic and lipophilic fractions of *Halimeda incrassata* have further displayed potent anti-oxidant activity when studied using the ²-carotene-linoleate assay system (Novoa et al., 2011). The highest anti-oxidant activities values were found in the polar fractions containing phenolic acids. The strong activity was partly attributed to the presence of salicylic and ferulic acids, however further research is required to demonstrate the presence of other phenolics if any and their role as anti-oxidants.

Free phenolic acids (FPA) fractions from the seaweed *Halimeda monile* were evaluated for their anti-oxidant activity by investigating their ability to protect the expression of hepatic enzymes in rats, under experimental ccl_4 injury. The results revealed that phenolic compounds from the seaweed *Halimeda monile* with anti-oxidant activity were able to strengthen the mechanism of liver repair regulated by hepatic enzymes and by the scavenging action on free radicals *in vivo*. The total free phenolic acids (FPA) content was found to be 27.76 mg/g dry seaweed, with salicylic, cinnamic, gallic, and caffeic acids being the major components (Hsu, 2008; Kang et al., 2005). The different polar fractions of *H. Monile*, such as FPA, soluble phenolic acid esters (PASE) and insoluble phenolic acid esters (PAIE) showed almost similar activity profiles.

Phenolic acids exert anti-oxidant and anti-carcinogenic properties *in vitro* as well as *in vivo*. However, further epidemiologic studies are needed to show whether a high phenolic acid intake can induce a reduced risk for coronary heart disease or cancer in humans when consumed in considerable amounts with food.

d. Halogenated Phenols

Halogenated compounds are basically mono- and dihydroxy C_6-C_1, C_6-C_2 and C_6-C_3 phenols containing one or two bromine atoms (Fenical 1975; Fenical and Norris; Fenical 1982) and are mainly dispersed in several different classes of primary and secondary metabolites, including indoles, terpenes, acetogenins, phenols, fatty acids and volatile halogenated hydrocarbons (such as, bromoform, chloroform, dibromomethane) (Holdt and Kraan, 2011). Halogenations often provides phenols with interesting key features and have been researched for the past decades for their biological and pharmacological properties, with outcomes showing antibacterial, antifungal, antiviral, anti-inflammatory, antiproliferative, antifouling, anti-feedant, cytotoxic, ichthyotoxic, and insecticidal activities (Holdt and Kraan, 2011).

Marine seaweeds synthesize diverse halogenated compounds in large amounts naturally (e.g: bromophenols; Figure 6.) by diverse and unique biosynthetic pathways (Holdt and Kraan, 2011). The earliest marine bromophenol was isolated from the red alga *Polysiphonia morrowii* in 1955 (Saito and Ando, 1955) and consequently quite a few bromophenols have been isolated and identified from red, brown and green algae since then (Li et al., 2007; Liu et al., 2011) with 2, 4, 6-tri-bromophenol being the most predominant. Total bromophenols content in seaweeds range from 8 to 180 μg/kg, varying substantially depending on season; although concentration of 2, 4, 6-tri-bromophenol remains rather high all year round. Seasonal studies on *U. Lactuca* have also illustrated that the levels of bromophenol content increased noticeably at the end of the summer, whereas they decreased during the rest of the year. The bromophenols possibly shield the seaweed from strong light intensities experienced throughout the summer (Pedersen et al., 1996; Flodin et al., 1999). More than 30 monoaryl and diaryl bromophenols have been discovered in the *Rhodomelaceae* family (Zhao *et al* 2004). Seaweeds are not the only source of these metabolites, as they are also found in ascidians and sponges (Liu et al., 2011).

Many species of red seaweeds have been demonstrated to generate an extensive array of halogenated compounds (Dembitsky and Toltikov, 2003). For instance, bromocatechol was found in red seaweeds *Halopitys incurvus, Polysiphonia nigra, P. Elongata, P. Fastigiata, P. Lanosa*, *P. Morrowii, P. Nigrescens, P. Urceolata, Pterosiphonia complanata and Vidalia volubilis* and *Odonthalia dentata*, , (Dembitsky and Toltikov, 2003; Lee et al., 2007). In addition, bromophenol was reported to be present in green seaweed *Codium fragile* and the red seaweed *Pterocladiella capillacea* (Whitfield et al., 1999).

Bromophenols have been revealed to possess strong antioxidative properties (Dummermuth et al., 2003). Recently, bromophenols isolated from the red algae *Symphyocladia latiuscula*, were reported to possess DPPH radical scavenging activities (Choi *et al* 2000; Duan et al., 2007). All these bromophenols were fully substituted by different groups and highly brominated, and many of them possessed a 3, 4-dihydroxy-2, 5, 6-tribromobenzyloxy unit in the molecule. All of them showed DPPH radical scavenging activities varying between IC_{50} values of 7.5 and 24.7 μm, which was more potent than the positive control butylated hydroxytoluene (IC_{50} = 81.8 μm), (Choi et al., 2000; Duan et al., 2007). It appears that the number of hydroxyl groups in the molecule and their common 3, 4-dihydroxy- 2, 5, 6-tribromobenzyloxy unit may be an important contributing factor for their anti-oxidant activity (Duan et al., 2007).

Lim *et al.,* (2006) established that methanolic extracts of *Rhodomela confervoides*, competently scavenged DPPH at IC_{50} 90 μg/ml and totally inhibited H_2O_2 induced lipid peroxidation in rat brain homogenate at a concentration of 20 μg/ ml. Subsequent study by Li et al., (2011) reported 19 naturally occurring bromophenols, six of which were new, from the seaweed *R. Confervoides*, possessing anti-oxidant activities. Majority of the isolated compounds showed noteworthy anti-oxidant activities against DPPH and 2, 2′-Azino-bis (3-ethylbenzothiazoline-6-sulfonic acid) (ABTS) radicals *in vitro* and were stronger than or comparable to the positive controls butylated hydroxytoluene (BHT) and ascorbic acid. . To this day, approximately 30 bromophenols with anti-oxidant activity have been reported from this marine seaweed indicating it to be an excellent source of natural anti-oxidants.

Three new and three known bromophenols were isolated and identified from the red seaweed *Polysiphonia urceolata* from Quindao (China) on the basis of extensive analysis of spectroscopic data and the structures of these compounds were determined to be 7-bromo-9,10-dihydrophenanthrene-2,3,5,6-tetraol, 4,7-dibromo-9,10-dihydrophenanthrene-2,3,5,6-tetraol, 1,8-dibromo-5,7-dihydrodibenzo(c,e)oxepine-2,3,9,10-tetraol, urceolatol, 3-bromo-4,5-dihydroxybenzaldehyde, and 3,5-dibromo-4-hydroxybenzaldehyde (Li et al., 2008). Each of the isolated compounds were evaluated for DPPH radical scavenging activities and all were found to be potent anti-oxidants with IC_{50} values ranging from 6.1 to 35.8 microm, which was very high compared to the positive control, butylated hydroxytoluene (BHT) (IC_{50} of 83.8 microm), making these metabolites potential leads as anti-oxidant agents.

Crude extracts of *Vertebrata lanosa* have demonstrated high anti-oxidant activities (biochemical assays) besides having a high phenolic content (Zubia et al., 2009). This was further supported by studies of Olsen et al., (2013), wherein the anti-oxidant activities of four bromophenols isolated from *V. Lanosa* namely 2,2′,3-tribromo-3′,4,4′, 5-tetrahydroxy-6′-hydroxymethyldiphenylmethane, bis (2, 3-dibromo-4,5-dihydroxylbenzyl) ether, and 5,5″-oxybis (methylene)bis (3 - bromo - 4 - (2′,3′-dibromo-4′,5′-dihydroxylbenzyl) benzene -1, 2-diol) were evaluated using oxygen radical absorbance capacity (ORAC), cellular anti-oxidant activity (CAA) and cellular lipid peroxidation anti-oxidant activity (CLPAA) assays. 2, 2′, 3-tribromo-3′, 4, 4′, 5-tetrahydroxy-6′-hydroxymethyldiphenyl methane illustrated its superiority against luteolin in both the CAA and CLPAA assays and quercetin in the CLPAA assay.

A series of bromophenols and corresponding debrominated phenol compounds isolated from red alga *Tichocarpus crinitus* demonstrated their abilities to inhibit free radicals using DPPH assay. The most active free radical scavengers were the debrominated phenolic derivatives 3, 4-dihydroxybenzyl alcohol and 3, 3’, 4, 4’-tetrahydroxydiphenylmethane when compared with 2, 6-di-tert-butyl-4-methylphenol (BHT) at 100μg/ml (Lee et al., 2007).

In addition to anti-oxidant activities, different biological activities have been reported for the bromophenols including, antimicrobial, anticancer, anti-diabetic and anti-thrombotic effects, which make them interesting compounds in the development of new pharmaceutical agents (Li et al., 2011).

IV. Anti-Oxidant Mechanisms of Polyphenols

Plant and seaweed derived phenolic compounds have been shown to act as potent anti-oxidants in a variety of systems. However, neither the anti-oxidant mechanism nor the structure-activity relationship of seaweed polyphenols has been fully elucidated up to date. The generally considered mechanisms have been allied to their free radical scavenging, singlet oxygen quenching, transition metal ion chelating, and reducing power capacities.

Many studies have documented and extensively reviewed the potent free radical scavenging activities of phenolic compounds derived from both plant and algal origin (Nakai et al., 2006, Rice-Evans et al., 1997; Shahidi and Naczk, 2004a). Polyphenols have been found to possess the ability to capture free radicals, particularly peroxy radicals, which are the key reactive intermediates/chain propagators, thereby terminating the chain reaction (Shahidi and Naczk, 2004a).

Phenolic compounds can effortlessly transfer a hydrogen atom to lipid peroxyl radicals, ensuing in the formation of a stable phenoxyl radical and less reactive cis-, trans-lipid hydroperoxide. Alternatively, it has also been suggested that phenolic compounds can neutralize lipid peroxyl radicals by single electron transfer (Ou et al., 2002).

However, there are conflicting reports in the literature as regards to the metal chelating abilities of polyphenols. Some studies have shown that polyphenols derived from brown algae are potent ferrous ion chelators (Chew et al., 2008; Senevirathne et al., 2006), with their metal chelating ability being dependent upon their unique phenolic structure and the number and location of hydroxyl groups (Santoso et al., 2004).In contrast, other authors have reported that overall metal chelation played a trivial role in the anti-oxidant activities of several plant derived phenolic compounds (Rice-Evans et al., 1996). In agreement with this, wang et al., (2009) reported that the ferrous ion-chelating ability of ten species of Icelandic seaweed extracts correlated neither with TPC nor with DPPH radical scavenging activity or reducing power assay or metal chelation assay. Therefore, the major role of seaweed polyphenols appeared to be as potent radical scavengers and primary, chain-breaking anti-oxidants (Siriwardhana *et al* 2003). Other components such as polysaccharides, proteins, or peptides may be more significant for the chelating effects detected.

Reducing power has been suggested as an important indicator of the overall anti-oxidant potential of phenolic anti-oxidants. In general, the reducing properties are associated with the presence of reductones, which exhibit their anti-oxidant activities through their action of breaking the free radical chain by the donation of a hydrogen atom (Gordon, 1990). Reductones are thought not only to react directly with peroxides but also to avert peroxide formation by way of reacting with certain precursors of peroxide. Seaweed polyphenols may act in a comparable way as reductones given that they possess inherent ability to donate electrons and react with free radicals, changing them into more stable end products and blocking free radical chain reaction from proceeding further.

Due to a universal fundamental mechanism (electron/hydrogen donation), there is a reasonably good linear relationship between TPC, the reducing activity and DPPH radical scavenging activity (Duan et al., 2006; Wang et al., 2009).

V. STRUCTURE-ANTI-OXIDANT ACTIVITY RELATIONSHIP OF POLYPHENOLS

Several studies have reported that high molecular weight phlorotannins exhibit potent anti-oxidant activities than the monomer phloroglucinol due to their unique molecular skeleton (Ahn et al., 2007).

The anti-oxidant power of polyphenols is greatly linked to the phenol rings that act as electron traps for scavenging free radicals such as peroxy, superoxide-anions and hydroxyl radicals. The molecular skeleton of phlorotannins from brown algae consists of 8 interconnected phenol rings (O'Sullivan et al., 2011), while their terrestrial counterparts consist of only 3 to 4 rings, as in the case of green tea catechins (Amitosomy and Raj, 2011; Hemat, 2007). Therefore, phlorotannins have very strong anitioxidant properties (Amitosomy and Raj, 2011). For instance, phlorotannins isolated from *Eisenia bicyclis* have shown even 10 times higher anti-oxidant activity in contrast to ascorbic acid and α-tocopherol radicals (Gupta and Abu-Ghannam, 2011).

Oligomerization of phloroglucinol is believed to be another factor critical for the augmentation of the radical scavenging activity (Nakamura et al., 1996). Nakamura et al. (1996) observed that the fractions containing phlorotannin oligomers of higher molecular weight such as dieckol, phlorofuroeckol and 8-8' bieckol were more effective to extent the induction time of auto-oxidation of methy α-linolenate than low molecular weight fravtions containing phloroglucinol and eckol. Alternatevely, Cerantola et al., (2006) isolated two types of structurally distinct phlorotannins from *Fucus spiralis* extracts and compared their DPPH radical scavenging activity. The first polymer had relatively simple structure and belonged to the fucol subclass whereas the second one was more complex and linked though aryl-aryl and aryl-ether bonds, thus belonging to the fucophlorethol subclass. Both polymeric phlorotannins showed equally high scavenging activity in comparison to ascorbic acid and phloroglucinol monomer, indicating that there was no clear correlation between the types of structural links and anti-oxidant activity of phlorotannins.

Other studies have also showed that presence of O-ridged linkages (ether linkages) and the number and arrangement of phenolic hydroxyl groups in phlorotannin skeletons have profound influence on the anti-oxidant property. Phlorotannin compounds extracted from *Ecklonia* species such as *E. Cava* and *E. Stolonifera*, harvested from Jeju island of Korea, have been reported to possess extraordinary anti-oxidant potential than the other species collected from the same area (Shin et al., 2006). The phlorotannins of *E. Cava* and *E stolonifera* belong to a unique polyphenolic category, characterized by a dibenzo-1, 4-dioxin unit in the molecular skeleton. The unique structure appears to be restricted to some specific brown algae such as *Eisenia* and *Ecklonia* species and may be responsible for their potent anti-oxidant activities compared with other brown algae. Moreover, several phlorotannin oligomers isolated from these species, including eckol, phlorofucofuroeckol A, dieckol, and

8-8'-bieckol were found to possess varying degrees of scavenging activities against DPPH and superoxide anion radicals (Shibata et al., 2008).

Still further, Yan *et al* (1996) reported that high molecular weight phlorotannins from *S. kjellmanianum* were more effective than those of their low molecular weight counterparts to prevent rancidity in fish oil. The anti-oxidant activity of high molecular weight phlorotannins at a concentration of 1% was about 2.6 times higher than that of 0.02% butylated hydroxytoluene (BHT). The diverse radical scavenging properties of these phlorotannin polymers were assumed to be contributed due to the number of hydroxyl groups attached to the eckol skeleton.

VI. Functional Value of Seaweeds As Dietary Supplements

Seaweeds are readily available for consumption as a whole foods source in coastal communities of most countries in South-East Asia (Yuan and Walsh 2006). The Japanese consume seaweeds on an average of 1.6 kg dry weight per person, per year and form the largest group of consumers (Chandini et al., 2010). Their consumption has been epidemiologically linked to have a protective effect against a range of modern day diseases. Even scientists have advocated several health benefits derived from eating seaweeds in Asian countries (Nisizawa 2002). The official Japanese Food Guide (Rhatigan 2009) also encourages seaweed to be included as a part of nutritional diet.

Recent interests in seaweed as food additives and supplements have been recognized due to their nutritional and functional potential. They encompass a variety of distinctive phytochemicals such as, lipids, minerals, polysaccharides, vitamins and polyphenols not found in terrestrial flora (Darcy-Vrillon, 1993). A wide range of studies have described the high anti-oxidant capacity of a number of polyphenols present abundantly in edible seaweeds that were far superior to conventional anti-oxidants (Confrades et al., 2010; Yuan et al., 2005; Wang et al., 2009; Gupta et al., 2011) Certain brown seaweeds such as, *Ascophyllum nodosum* have been reported to possess very high concentrations of polyphenols (14 % of its dry weight) as compared to in terrestrial plants (2-3%). As such, edible seaweeds form the only appropriate nutritional supply of some of these factors.

In recent years, seaweeds have proved to be a boon to the pharmaceutical and agri-food processing industries owing to their extensive range of bioactive constituents that vary from anti-oxidants to minerals. Therefore, seaweeds rich in marine phytochemicals suggest exciting possibilities as constituents in the development of new food products and must be indisputably included as a part of well balanced diet.

All over the world commercial manufacturing of microalgae for human nourishment is already a reality. Various combinations of microalgae or mixtures with added health foods can be seen in the market in the form of tablets, powders, capsules, pastilles and liquids, as dietary additives. In fact, more studies are progressing towards using marine macroalgae for manufacturing of new foods, such as health beverages, processed meat products, pastas, biscuits, bread, snack foods, candies, yoghurts, soft drinks, with the purpose of exploiting their naturally occurring anti-oxidant compounds and other nutritive constituents (Nagai and Yukimoto 2003, López-López et al., 2009).

The addition of seaweed and seaweed isolates to foods has previously revealed its possible use in the development of anti-obesity foods due to its ability to boost satiety and lessen the after meal absorption rates of glucose and lipids in acute human feeding studies. (Hall et al., 2012). They can benefit health as well as improve food acceptability

Previous studies in animal models and cell culture have also exemplified that seaweed phytochemicals have the capability to slow down the progression of carcinogenesis (Ohigashi et al., 1992; Yuan *et al*, 2005b). This is also supported by a number of studies conducted on Japanese and Chinese populations that have shown including seaweeds in diet as a food ingredient is an important factor for low frequency of breast and prostate cancer incidences as compared to North America and Europe (Lacey et al., 2002).

Polyphenols extracted from whole tissue extracts from certain brown algae (e.g., *Ecklonia* spp.) provided highly protective effects against UVB induced skin carcinogenesis in mice not considering whether the polyphenols were applied topically, or consumed in diet, thus demonstrating that the seaweed derived anti-oxidants are unchanged by digestive process (Hwang et al.,2006).

In vitro studies also suggest a potential for phenolic compounds from seaweeds to have the potential to influence macronutrient uptake and perform as a therapeutic agent to help fight metabolic diseases by inhibiting the action of digestive enzymes (Apostolidis and Lee, 2010; Ben Rebah et al., 2008) In addition, a recent study recommends the use of a polyphenol-rich extract of *Ecklonia stolonifera* improved glycaemic control in a non-insulin dependent diabetic mouse model (Iwai, 2008). A similar effect was also noted in a chemically induced diabetic mouse model (Zhang et al., 2007). Anti-inflammatory properties of a phlorotannin-rich extract from *Ecklonia cava* have also been demonstrated *in vitro* (Shin *et al*, 2006).

The above experimental data highlights a number of remarkable ways in which these anti-oxidant compounds isolated from seaweed could benefit health. And may well lessen our reliability on traditional foods and, may be, add a new flavor to the range of exotic spices that already exist.

CONCLUSION

Marine seaweeds are a valuable natural source of compounds with incredible anti-oxidant and added bioactive properties. The major objective of scientists nowadays is to harness these natural anti-oxidants that will substitute for the synthetic ones in food, pharmaceutical, and cosmetic industries. However, research is required prior to utilization of this enormous unexploited resource to be used as such. Utilising natural anti-oxidants from aquatic source offers a resourceful option to improve oxidative stability, flavour quality and nutritional value of food merchandise, as well as for consumption in functional foods or nutritional supplement. Studies pertaining to seaweed derived anti-oxidants are still in their early years and sustained research endeavors are required in turn to expand this underexplored supply into a dietary anti-oxidant or food supplement. An improved understanding of the precise mechanisms implicated in the anti-oxidant defense and the structure-anti-oxidant activity relationship of polyphenolic compounds needs to be undertaken to enable researchers to recognize the fragile kinks in the shield of emerging diseases of modernity, devise new ways to battle free radicals, and estimate the protective influence of latest methodologies. To aid this, further purification, characterization and structural studies of polyphenols must be performed. While extraction of bioactive natural compounds from seaweeds is preferred, very little has ensued in this area to thoroughly study their prospects. Innovative animal and human clinical studies need to be undertaken to methodically assess their health implications and underlying hazards. In this regard, considerable research efforts regarding the toxicity features ought to be carried out prior to their actual use in clinical settings. Furthermore, the impact of processing conditions on the stability and bioavailability of seaweed polyphenols needs to be standardized to avoid its unfavorable interactions with added food ingredients such as proteins, carbohydrates and lipids, and develop into toxic, allergenic or carcinogenic by-products.

The demand for anti-oxidants is immense and the requirement for natural anti-oxidants is greater than ever. The prospect for focused commodities like seaweed extracts and purified polyphenols from seaweed with valuable health outcomes is thus enormous and requires further work to truly understand their further potential.

References

Ahn, G. N., Kim, K. N., Cha, S. H., Song, C. B., Lee, J., Heo, M. S., Yeo, I. K., Lee, N. H., Jee, Y. H., Kim, J . S., Heu, M. S., Jeon, Y. J. (2007) Anti-oxidant activities of phlorotannins purified from *Ecklonia cava* on free radical scavenging using ESR and H_2O_2 - mediated DNA damage. *Eur. Food Res. Technol.* 226, 71-79.

Ahn, M. J., Yoon, K. D., Min, S. Y., Lee, J. S., Kim, J. H., Kim, T. G., et al., (2004). Inhibition of HIV-1 reverse transcriptase and protease by phlorotannins from the brown alga *Ecklonia cava. Biol Pharmaceut Bull.* 27, 544-547.

Alviano, D. S, Alviano, C. S. (2009) Plant extracts: search for new alternatives to treat microbial diseases. *Curr. Pharm. Biotechnol.* 10, 106-121.

Anselmi, C., Centini ,M., Andreassi, M., Buonocore, A., Rosa, C., Facino, R.M., Sega, A., Tsuno, F. (2004) Conformational analysis: A tool for the elucidation of the anti-oxidant properties of ferulic acid derivatives in membrane models. *J. Pharmaceut. Biomed.* 35, 1241-1249.

Antonisamy, M. J., Raj, E .D. S. (2011) UV–VIS and HPLC studies on *Amphiroa anceps* (Lamarck) Decaisne. *Arabian J. Chem.* [E Pub ahead of Print].

Apostolidis, E., Lee, C. M. (2010) *In vitro* potential of *Ascophyllum nodosum* phenolic anti-oxidant-mediated alpha-glucosidase and alpha-amylase inhibition. *J. Food Sci.* 75, 97-102.

Arnold, T. M., Targett, N. M. (2002) Marine tannins: The importance of a mechanistic framework for predicting ecological roles. *J Chem Ecol* 28, 1919-1934.

Arnold, T. M., Targett, N. M. (2003). To grow and defend: lack of trade-offs for brown algal phlorotannins. *Oikos* 100, 406-408.

Athukorala, Y., Kim, K. N. Jeon, Y. J. (2006) Antiproliferative and anti-oxidant properties of an enzymatic hydrolysate from brown alga, *Ecklonia cava. Food Chem. Toxicol.* 44, 1065-1074.

Beale, K., Robinson, E.W. (2000) Combinations of reverse transcriptase, protease, and integrase inhibitors can be synergistic *in vitro* against drug-sensitive and RT inhibitor-resistant molecular clones of HIV-1. *Antivir. Res.* 46, 223-232.

Ben Rebah, F., Smaoui, S., Frikha, F., Gargouri, Y., Miled N. (2008) *Inhibitory effects* of *Tunisian marine algal extracts* on *digestive lipases*. *Appl. Biochem. Biotechnol.* 151, 71-79.

Bennet, R.. C., Wallsgrove, R. M. (1994) Tansley Review No. 72: Secondary metabolites in plant defence mechanisms, *New Phytol.* 127, 617–633.

Boettcher, A. A., Targett, N. M. (1993) Role of polyphenolic molecular size in reduction of assimilation efficiency in *Xiphister mucosus*. *Ecology.* 74, 891-903.

Bravo, L. (1998) Polyphenols: chemistry, dietary sources, metabolism, and nutritional significance. *Nutr. Rev.* 56, 317-333.

Cahyana, A. H., Shuto, Y., Kinoshita, Y. (1992) Pyropheophytin a as an antioxidative substance from the marine alga, *Arame* (*Eisenia bicyclis*). *Biochem. Biophys. Res. Commun.* 56, 1533-1535.

Cao, G., Sofic, E., Prior, R.L. (1997) Anti-oxidant and pro-oxidant behavior of flavonoids; structure-activity relationships. *Free Radical Biol Med.* 22, 749-760.

Cardenas, M., Marder, M., Blank, V. C., Roguin, L. P. (2006) Antitumor activity of some natural flavonoids and synthetic derivatives on various human and murine cancer cell lines. *Bioorgan. Med. Chem.* 14, 2966-2971.

Cerantola, S., Breton, F., Ar Gall, E. and Deslandes, E. (2006) Co-occurrence and anti-oxidant activities of fucol and fucophlorethol classes of polymeric phenols in *Fucus spiralis*. *Bot. Mar.* 49, 347-351.

Chan, T., Galati, G., O'Brien, P.J. (1999) Oxygen activation during peroxidase catalyzed metabolism of flavones or flavanones. *Chem. Biol. Interact.* 122, 15-25

Chandini, S. K., Ganesan, P., Bhaskar, N. (2008). *In vitro* anti-oxidant activities of three selected brown seaweeds of India. *Food Chem.* 107,707-713.

Charvat, T. T., Lee, D. J., Robinson, E. W., Chamberlin, A. R. (2006) Design, synthesis, and biological evaluation of chicoric acid analogs as inhibitors of HIV-1 integrase. *Bioorgan. Med. Chem.* 14, 4552-4567.

Chew, Y.L., Lim, Y.Y., Omar, M., Khoo, K. S. (2008) Anti-oxidant activity of three edible seaweeds from two areas in South East Asia. *LWT-Food Sci. Technol.* 41, 1067-1072.

Choi, J. S., Park, H. J., Jung, H. A., Chung, H.Y., Jung, J. H., Choi, W. C. (2000) A cyclohexanonyl bromophenol from the red alga *Symphyocladia latiuscula*. *J. Nat. Prod.* 63, 1705-1706.

Cofrades S, López-López I, Solas MT, Bravo L. (2008) Jiménez-Colmenero F: Influence of different types and proportions of added edible seaweeds on characteristics of low-salt gel/emulsion meat systems. *Meat Sci.* 79,767-776.

Connan, S., Goulard, F., Stiger, V., Deslandes, E., Ae Gall, E. (2004) Interspecific and temporal variation in phlorotannin levels in an assemblage of brown algae. *Bot. Mar.*, 47, 410-416.

Cos, P., Ying, L., Calomme, M., Hu, J. P., Cimanga, K., Van Poel, B., Pieters, L., Vlietinck A.J., Vanden Berghe, D. (1998) Structure-activity relationship and classification of flavonoids as inhibitors of xanthine oxidase and superoxide scavengers. *J Nat. Prod.* 61, 71-76.

Craige, J. S. (2011) Seaweed extract stimuli in plant science and agriculture. *J. Appl. Phycol.* 23, 371-393.

Dai, J., Mumper, R.J. (2010) Plant Phenolics: Extraction, analysis and their anti-oxidant and anticancer properties. *Molecules.* 15, 7313-7352.

D'Archivio, M., Filesi, C., Di Benedetto, R., Gargiulo, R., Giovannini, C., Masella, R. (2007) Polyphenols, dietary sources and bioavailability. *Ann. Ist. Super. Sanita.* 43, 348-361.

Darcy-Vrillon, B. (1993) Nutritional aspects of the developing use of marine macroalgae for the human food industry. *Int. J. Food Sci. Nutr.* 44, 523-535.

Dembitsky, V. M., Tolstikov, G. A. (2003) *Natural halogenated mononuclear phenol compounds and their derivatives. Chem.Sustain. Develop.* 11, 567-575.

Devasagayam, T., Tilak, J., Boloor, K., Sane, K.S. (2004) Free radicals and anti-oxidants in human health: Current status and future prospects. *J. Assoc. Physicians India 52*, 794-804.

Dixon, R. A., Paiva, N. L. (1995) Stress-induced phenylpropanoid metabolism. *Plant Cell.* 7, 1085–1097.

Duan, X. J., Li, X. M., Wang, B. G. Highly brominated mono- and bis-phenols from the marine red alga *Symphyocladia latiuscula* with radical-scavenging activity. *J. Nat. Prod.* 70, 1210-1213.

Duan, X. J., Zhang, W. W., Li, X. M. and Wang, B. G. (2006) Evaluation of anti-oxidant property of extract and fractions obtained from a red alga, *Polysiphonia urceolata. Food Chem.* 95, 37-43.

Dummermuth, A. L., Karsten, U., Fisch, K. M., Konig, G.M., Wiencke, C. (2003) Responses of marine macroalgae to hydrogen-peroxide stress. *J. Exp. Mar. Biol. Ecol.* 289, 103-121.

Elliott, A. J., Scheiber, S. A., Thomas, C., and Pardini, R. S (1992) Inhibition of glutathione reductase by flavonoids. A structure-activity study. *Biochem. Pharmacol.* 44, 1603-1608.

Fallarero, A., Loikkanen, J. J., Mannisto, P. T., Castañeda, O., Vidal, A. (2003), Effects of aqueous extracts of *Halimeda incrassata* (Ellis) Lamouroux and *Bryothamnion triquetrum* (S. G. Gmelin) Howe on hydrogen peroxide and methyl mercury-induced oxidative stress in GT1 – 7 mouse hypothalamic immortalized cells. *Phytomed.* 10, 39-47.

Farasat, M., Khavari-*Nejad* R. A., Nabavi, S. M. B., *Namjooyan,* F. (2013) anti-oxidant properties of two edible green seaweeds from northern coasts of the Persian Gulf. *Jundishapur J. Nat. Pharmaceu. Prod.* 8, 47-52.

Fenical, W. (1982) Natural products chemistry in the marine environment. *Science* 215: 923-928.

Fenical, W., Norris, J.N.(1975) Chemotaxonomy in marine algae: chemical separation of some *Laurencia* species (Rhodophyta) from the Gulf of California). *J. Phycol.* 11, 104-108.

Ferrali, M., Signorini, C., Caciotti, B., Sugherini, L., Ciccoli, L., Giachetti, D., Comproti, M. (1997) Protection against oxidative damage of erythrocyte membrane by the flavonoid quercetin and its relation to iron chelating activity. *FEBS Lett.* 416, 123-129.

Flodin, C., Helidoniotis, F., Whitfield, F. B. (1999) Seasonal variation in bromophenol content and bromoperoxidase activity in *Ulva lactuca. Phytochem.* 51,135-138.

Fraga, C. G., Mactino, U. S., Ferraro, G. E., Coussio, J. F., Boveris, A. (1987) Flavonoids as anti-oxidants evaluated by *in vitro* and *in situ* liver chemiluminescence. *Biochem. Med. Metabol. Biol.* 36, 717-720.

Fujimoto, K., Kaneda, T. (1984) Separation of antioxygenic (anti-oxidant) compounds from marine algae. *Hydrobiol.* 116/117, 111-113.

Fukuyama, Y., Kodama, M., Miura, I., Kinzyo, Z., Kido, M., Mori, H., Nakayama, Y., Takahashi, H. (1989) Structure of an anti-plasmin inhibitor, Eckol, isolated from the brown alga Ecklonia kurome OKAMURA and inhibitory activities of its derivativeson plasmin inhibitors *Chem. Pharm. Bull.* 37, 349-353.

Fukuyama, Y., Kodama, M., Miura, I., Kinzyo, Z., Mori, H., Nakayama, Y., Takahashi, H. Anti-plasmin inhibitor. VI. Structure of phlorofucofuroeckol A, a novel phlorotannin with both dibenzo-1,4-dioxin and dibenzofuran elements, from Ecklonia kurome Okamura. (1990) *Chem. Pharm. Bull.* 38, 133-135.

Glombitza, K. W., Hauperich, S., and Keusgen, M. (1997) Phlorotannins from the brown algae *Cystophora torulosa* and *Sargassum spinuligerum. Nat. Roxins.* 5, 58-63.

Gordon, M. H. *(1990)* The mechanism of *anti-oxidants* action *in-vitro*. In: *Food Anti-oxidants. Hudson, B.J.F (Ed.)*, Elsevier Applied Science, pp. *1–18*.

Gupta, S., Abu-Ghannam, N. (2011) Recent developments in the application of seaweeds or seaweed extracts as a means for enhancing the safety and quality attributes of foods. *Innov. Food Sci. Emerg.* Technol. 12, 600-609.

Hall, A.C., Fairclough A. C., Mahadevan, K., Paxman, J.R. (2012) Ascophyllum nodosum enriched bread reduces subsequent energy intake withbno effect on post-prandial glucose and cholesterol in healthy, overweight males. A pilot study. *Appetite.* 58, 379-386.

Han, E. S., Kim, J. W., Eom, M. O., Kang, I. H., Kang, H. J., Choi, J. S., Ha, K. W., Oh, H. Y. (2000) Inhibitory effect of *Eckloia stolonifera* on gene mutation on mouse lymphoma tk +/+ locus in L5178Y-3.2.C cell and bone marrow micro-nuclei formation in ddY mice. *.Environ. Mutagen Carcinogen* 20, 104-111

Harborne, J. B. (1994) The Flavonoids: Advances in research since 1986. London, UK: Chapman and Hall, pp. 619-652.

Hemat, R. A. S. (2007) Fat and muscle dysfunction. In: Andropathy. Hemat, R. S. S (Ed.), Urotext, pp. 83-85.

Hillwell, B. (1994) Free radicals, anti-oxidants and human disease: Curiosity, cause or constipation? Lancet. 344, 721-724.

Hirano, R., Sasamoto, W., Matsumoto, A., Itakura, H., Igarashi, O., and Kondo, K. (2001) Anti-oxidant ability of various flavonoids against DPPH radicals and LDL oxidation. *J Nutr Sci Vitaminol.* 47, 357-362.

Holdt, S., Kraan, S. (2011) Bioactive compounds in Seaweed: functional food applications and legislation.*J Appl Phycol.* 23, 543–597.

Hsu, C. Y. (2008), Anti-oxidant activity of *Pyrrosia petiolosa. Fitoterapia.* 79, 64-66.

Hu T, Liu, D., Chen, Y., Wu, J., Wang, S. (2010) Anti-oxidant activity of sulfated polysaccharide fractions extracted from *Undaria pinnitafida in vitro. Int. J. Biol. Macromol.* 46, 193-198.

Huang. H. L., Wang, B. G. (2004) Anti-oxidant capacity and lipophilic content of Seaweeds collected from the Qingdao coastline. *J. Agric. Food Chem.* 52, 4993-4997.

Hwang, H., Chen, T., Nines, R. G., Shin, H. C., Stoner, G. D. (2006) Photochemoprevention of UVB-induced skin carcinogenesis in SKH-1 mice by brown algae polyphenols. *Int. J. Cancer.* 119, 2742-2749.

Iwai K. (2008) Antidiabetic and anti-oxidant effects of polyphenols in brown alg*a Ecklonia stolonifera* in genetically diabetic KK-A(y) mice. *Plant Foods Hum. Nutr.* 63,163-169.

Jimeńez-Escrig, A., Jimeńez-Jimeńez, I., Pulido, R. Saura-Calixto, F. (2001) Anti-oxidant activity of fresh and processed edible seaweeds. *J. Sci. Food Agr.* 81, 530-534.

Jung, H. A., Oh, S. H., Choi, J. S. (2010) Molecular docking studies of phlorotannins from *Eisenia bicyclis* with BACE1 inhibitory activity. *Bioorg. Med. Chem. Lett.* 20, 3211-3215.

Jung, H.A., Hyun, S.K., Kim, H.R., Choi, J.S. (2006) Angiotensin converting enzyme I inhibitory activity of phlorotannins from *Ecklonia stolonifera*. *Fish. Sci.* 72, 1292-1299.

Kahkonen, M. P., Hopia, A. I., Vuorela, H. J., Rauha, J. P., Pihlaja, K., Kujala, T. S., Heinonen M. (1999) Anti-oxidant activity of plant extracts containing phenolic compounds. *J. Agr. Food Chem.* 47, 3954-3962.

Kang, H. S., Chung, H. Y., Kim, J. Y., Son, B. W., Jung, H. A., Choi, J. S. (2004) *Inhibitory phlorotannins from the edible brown alga ecklonia stolonifera on* total reactive oxygen species (ROS) generation. *Arch. Pharm. Res.* 27, 194-198.

Kang, K., A., Bu, H. D., Park D. S., Go G. M., Jee Y., Shin T., Hyun J. W. (2005) Anti-oxidant activity of ethanol extract of *Callophyllis japonica*. *Phytother. Res.* 19, 506 -510.

Kang, K., Park, Y., Hwang, H. J., Kim, S. H., Lee, J. G., Shin, H. C. (2003) *Antioxidative properties of brown algae polyphenolics and their perspectives as chemopreventive agents* against vascular risk factors. *Arch. Pharm. Res.* 26, 286-293.

Karawita, R., Siriwardhana, N., Lee, K., Heo, M., Yeo, I., Lee, Y. and Jeon, Y. (2005) Reactive oxygen species scavenging, metal chelation, reducing power and lipid peroxidation inhibition properties of different solvent fractions from *Hizikia fusiformis*. *Eur. Food Res. Tech.* 220, 363-371.

Keyrouz, R., Abasq, M.L., Le Bourvellec, C. (2011) Total phenolic contents, radical scavenging and cyclic voltammetry of seaweeds from Brittany. *Food Chem.* 126, 831-836.

Kim, M. M., Ta, Q. V., Mendis, e., Rajapakse, N., Jung, W. K., Byun, H. G., Jeon, Y. J., Kim, S. K. (2006). Phlorotannins in *Ecklonia cava* extract inhibit matrix metalloproteinase activity. *Life sci.* 779, 1436-1433.

Kim, S. N., Lee, J. Y., Kim, H. J., Shin, C. G., Park, H., Lee, Y.S. (2000) Synthesis and HIV-1 integrase inhibitory activities of caffeoyl glucosides. *Bioorg. Med. Chem. Lett.* 10, 1879-1882.

Kim, Y. C., An, R. B., Yoon, N. Y., Nam, T. J., Choi, J. S. (2005) Hepatoprotective constituents of the edible brown algae *Ecklonia stolonifera* on tacrine-induced cytotoxicity in Hep G2 cells. *Arch. Pharm. Res.* 28, 1376-1380.

King, P. J., Ma, G., Miao, W., Jia, Q., McDougall, B. R., Reinecke, M. G., Cornell, C., Kuan, J., Kim, T. R., Robinson, W. E. (1999) Structure-activity relationships: Analogues of the dicaffeoylquinic and dicaffeoyltartaric acids as potent inhibitors of human immunodeficiency virus type 1 integrase and replication. *J. Med. Chem.* 42, 497-509.

Lacey Jr,J. V., Devesa, S. S., Brinton, L. A. (2002) Recent trends in breast cancer incidence and mortality. *Environ. Mol. Mut.*39, 82-88.

Lalopua, V., Purnomo, H., Sukoso., Aulani'am. (2011) Red seaweed (*Kappaphycus alvarezii* DOTY) from Mollucas island water as potential flavonoid resource of natural anti-oxidant. *Livestock Res. Rural Dev.* 23, *Article #254.*

Lee, J. H., Kim, N. D., Choi, J. S., Kim, Y. J., Moon, Y., H, Lim, S. Y., Park, K. Y. (1998) Inhibitory effects of the methanolic extract of an edible brown alga, Ecklonia stolonifera and its component, phloroglucinol on aflatoxin B1 mutagencity *in vitro* (Ames test) and on benzo (a) pyrene or N- methyl N- nitrosourea clastogenicity *in vivo* (mouse micronucleus test). *Nat. Prod. Sci.* 4, 105-114.

Lee, J. H., Lee, T. K., Kang, R. S., Shin, H. J., Lee, H. S. (2007) The *in vitro* anti-oxidant activities of the bromophenols from the red alga *Tichocarpus crinitus* and phenolic derivatives. *J. Korean Magn. Reson. Soc.* 11, 56-63.

Lee, J. Y., Yoon, K. J., Lee, Y.S. (2003) Catechol-substituted L - chicoric acid analogues as HIV integrase inhibitors. *Bioorgan. Med. Chem.* 13, 4331-4334.

Leopoldini, M., Russo, N., Chiodo, S., Toscano, M. (2006) Iron chelation by the powerful anti-oxidant flavonoid quercetin. *J. Agric. Food Chem.* 54, 6343-6351.

Li, K., Li, X. M., Gloer, J.B., Wang, B. G. (2011) Isolation, characterization, and anti-oxidant activity of bromophenols of the marine red alga *Rhodomela confervoides. J. Agric. Food Chem. 59*, 9916-9921.

Li, K., Li, X. M., Ji, N. Y., Wang, B. G. (2007) Natural bromophenols from the marine red alga *Polysiphonia urceolata* (Rhodomelaceae): structural elucidation and DPPH radical-scavenging activity. *Bioorg. Med. Chem.* 15, 6627-6631.

Li, K., Li, X. M., Ji, N.Y., Wang, B. G. (2008) Bromophenols from the marine red alga *Polysiphonia urceolata* with DPPH radical scavenging activity. *J. Nat. Prod.* 71, 28-30.

Li, Y.X., Wijesekaraa, I., Li, Y.,Kima, S.K (2011) Phlorotannins as bioactive agents from brown algae. *Process Biochem*. 46, 2219-2224.

Lim, C., Jin, D., Sung, J., Lee, J., Choi, H., Ha, I., Han, J. (2006) Anti-oxidant and anti-inflammatory activities of the methanolic extract of *Neorhodomela aculeate* in hippocampal and microglial cells. *Biol. Pharm. Bull.* 29, 1212-1216.

Liu, L., Hudgins, W. R., Shack, S., Yin, M. Q., Samid, D. (1995) Cinnamic acid: A natural product with potential use in cancer intervention. *Int. J. Cancer.* 62, 345-350.

Liu, M., Hansen, P.E., Lin, X. (2011) Bromophenols in marine algae and their bioactivities. *Mar. Drugs*, 9, 1273-1292.

López-López, I., Bastida, S., Ruiz-Capillas, C., Bravo, L., Larrea, M. T., Sánchez-Muniz, F., Cofrades, S., Jimnez-Colmenero, F. (2009) Composition and anti-oxidant capacity of low-salt meat emulsion model systems containing edible seaweeds. *Meat Sci.* 83, 492-498.

Lüder, U. H., Clayton, M. N. (2004) Induction of phlorotannins in the brown macroalga Ecklonia radiata (Laminariales, Phaeophyta) in response to simulated herbivory-the first microscopic study. *Planta.* 218, 928-937

Mallick ,N., Mohn, F. H. (2000) Reactive oxygen species: Response of algal cells. *J. Plant Physiol.* 157, 183-193.

Mancini-Filho, J., Novoa, A. V., González, A. E., de Andrade –Wartha, E. R., de Oe Silva, A. M., Pinto, J. R., Mancini, D. A. (2009) Free phenolic acids from the seaweed *Halimeda monile* with anti-oxidant effect protecting against liver injury. *Z. Naturforsch. C.* 64, 657-663.

Mancini-Filho, J., Novoa, A. V., González, A. E., de Andrade-Wartha, E. R., de Oe Silva, A. M., Pinto, J. R., Mancini, D. A. (2009) Free phenolic acids from the seaweed *Halimeda monile* with anti-oxidant effect protecting against liver injury. *Z. Naturforsch. C.* 64, 657-663.

Mansouri, A., Embarek, G., Kokkalou, E., Kefalas, P. (2005) Phenolic profile and anti-oxidant activity of the Algerian ripe date palm fruit (*Phoenix dactylifera*). *Food Chem.* 89, 411-420.

Martínez-Tomé, M., Murcia, M., Frega, N., Ruggieri, S., Jiménez, A. M., Roses, F., Parras, P. (2004) Evaluation of anti-oxidant capacity of cereal brans. *J. Agr. Food. Chem.* 52, 4690-4699.

McClintock, J. B. Baker, B. J. (2001) *Mar. Chem. Ecol. CRC Marine Science Series*. CRC Press, pp. 609 ISBN 0-8493-9064-8.

Nagai, T., Yukimoto, T. (2003) Preparation and functional properties of beverages made from sea algae. *Food Chem.* 81, 327-332.

Nagayama, K., Iwamura, Y., Shibata, T., Hirayama, I., Nakamura, T. (2002) Bactericidal activity of phlorotannins from the brown alga *Ecklonia kurome*. J. *Antimicrob. Chemother.* 50, 889-893.

Nair, S. C., Panikka,r B., Akamanchi, K. G., Panikkar, K. R. (1991) Inhibitory effects of Ixora javanica extract on skin chemical carcinogenesis in mice and its antitumour activity. Cancer. Lett. 60, 253-258.

Nakai, M., Kageyama, N., Nakahara, K., Miki, W. (2006) Phlorotannins as radical scavengers from the extract of *Sargassum ringgoldianum*. *Mar. Biotechnol.* 8, 409-414.

Nakamura, T., Nagayama, K., Uchida, K., and Tanaka, R. (1996). Anti-oxidant activity of phlorotannins isolated from the brown alaga *Eisenia bicyclis*. *Fish. Sci.* 62, 923-926.

Nakayama, R., Tamura, Y., Kikuzaki, H. and Nakatani, N. (1999) Anti-oxidant effect of the constituents of susabinori (*Porphyra yezoensis*). *J. Am. Oil Chem, Soc.* 76, 649-653.

Nisizawa, K. (2002) Seaweeds Kaiso: bountiful harvest from the seas. *Japan Seaweed Association*, Kochi, 106 pp.

Norris, J. N., Fenical. W. (1982) Chemical defense in tropical marine algae. In: nutraceuticals and functional foods. Rutzler, K. (Ed), Boca Raton: Taylor and Francis, pp. 259–296.

Novoa A V; Elma R. S. Andrade-Wartha; Linares A F; de O. e Silva A M; Genovese M I; González A E B.; Vuorela P; Costa A; Mancini-Filho J. (2011) Anti-oxidant activity and possible bioactive components in hydrophilic and lipophilic fractions from the seaweed *Halimeda incrassate*. Braz. J. Pharmacog. 21, 53-57.

O'Sullivan, A.M., O'Callaghan, Y.C., O'Grady, M.N. (2011) *In vitro* and cellular anti-oxidant activities of seaweed extracts prepared from five brown seaweeds harvested in spring from the west coast of Ireland. *Food Chem.* 126, 1064-1070.

Ohigashi, H., Sakai, Y., Yamaguchi, K., Umezaki, I., Koshimizu, K. (1992) Possible anti-tumor promoting properties of marine algae and *in vivo* activity of wakame seaweed extract. *Biosci. Biotechnol. Biochem.* 56, 994-995.

Ohshima, H., Joshie, Y., Auriol, S., Gilbert, Isabelle. (1998) *Free Radic Biol Med,* 25, 1057-1065.

Okada, Y., Ishimaru, A., Suzuki, R., Okuyama, T. (2004) A new phloroglucinol derivative from the brown alga *Eisenia bicyclis*: potential for the effective treatment of diabetic complications. *J. Nat. Prod.* 67, 103-105.

Olsen, E. K., Hansen, E., Isaksson, J.,Andersen, J. H. Cellular anti-oxidant effect of four bromophenols from the red algae, *Vertebrata lanosa*. *Mar. Drugs*, 11, 2769-2784.

Ou, B., Huang, D., Hampsch-Woodill, M., Flanagan, J. A., Deemer, E. K., (2002) Analysis of anti-oxidant activities of common vegetables employing oxygen radical absorbance capacity (ORAC) and ferric reducing anti-oxidant power (FRAP) assays: A comparative study. *J. Agric. Food Chem.* 50, 3122-3128.

Parys, S., Rosenbaum, A., Kehraus, S., Reher, G., Glombitza, K. W., König, G. M. (2007) Evaluation of quantitative methods for the determination of polyphenols in algal extracts. *J. Nat. Prod.* 70, 1865-1870.

Pavia, H., Toth, G. B. (2000b) Influence of light and nitrogen on the phlorotannin content of the brown seaweeds *Ascophyllum nodosum* and *Fucus vesiculosus. Hydrobiol.* 440, 299-305.

Pedersen, M., Collen, J., Abrahamsson, K., Ekdahl, A. (1996) Production of halocarbons from seaweeds: an oxidative stress reaction? *Sci. Mar.* 60, 257–263.

Qian, J.Y., D. Mayer., M. Kuhn (1999) Flavonoids in fine buckwheat (*Fagopyrum esculentum* Moench) flour and their free radical scavenging activities. *Dtsch Lebensm Rundsch* 95: 343-349.

Ragan, M. A., Glombitza, K. W. (1986) Phlorotannins, brown algal polyphenols. In *Progress in Phycological Research*. Round, F. E., Chapman, D.J. (Eds.), Biopress Ltd., pp. 129-241.

Ratty, A. K. (1988) Effects of flavonoids on nonenzymatic lipid peroxidation: structure activity relationship. *Biochem. Med. Metabol. Biol.* 39, 67-79.

Rhatigan, P. (2009) Irish seaweed kitchen: the comprehensive guide to healthy everyday cooking with seaweeds. Book-link, Holywood, LA, 288 pp.

Rice-Evans, C. A., Miller, N. J., Paganga, G. (1996) Structure–anti-oxidant activity relationships of flavonoids and phenolic acids. *Free Rad. Biol. Med.* 20, 933-956.

Rice-Evans, C. A., Miller, N. J., Paganga, G. (1997) Anti-oxidant properties of phenolic compounds. *Trends Plant Sci.* 2: 152-159.

Rioux, L. E., Turgeon, S. L., Beaulieu, M. (2010) Structural characterization of laminaran and galactofucan extracted from the brown seaweed *Saccharina longicruris. Phytochem.* 71, 1586-1595.

Rutzler, K., Macintyre, I. G. (ed.), The Atlantic Barrier Reef ecosystem at Carrie Bow Cay, Belize 1: structure and communities. *Smithsonian Contr. mar. Sci.* 12, pp. 417-431.

Saito, T., Ando, Y. (1955) Bromine compounds in seaweeds. I. On a bromophenolic compound obtained from the red algae, *Polysiphonia Morrowii Harv. Nippon Kagaku Zassi.*76, 478-479.

Santoso, J. Y., Yoshie, T., Suzuki, T. (2002) The distribution and profile of nutritions and catechins of some Indonesian seaweed. *Fish. Sci.* 68, 1647-1648.

Santoso, J., Yumiko, Y. and Takeshi, S. (2004) Anti-oxidant activity of methanol extracts from Indonesian seaweeds in an oil emulsion model. *Fish. Sci.* 70, 183-188.

Senevirathne, M., S. Kim, N. Siriwardhana, J. Ha, K. Lee., Y. Jeon, (2006) Anti-oxidant potential of *Ecklonia cava* on reactive oxygen species scavenging, metal chelating, reducing power and lipid peroxidation inhibition. *Food Sci. Technol.* Int. 12, 27-38.

Shahidi, F., Naczk, M. (2004) Anti-oxidant properties of food phenolics. In: Phenolics in food and nutraceuticals. *Boca Raton*: CRC Press, pp. 403–442.

Shibata, T., Ishimaru, K., Kawaguchi, S., Yoshikawa, H., and Hama, Y. (2008) Anti-oxidant activities of phlorotannins isolated from Japanese *Laminariaceae. J. Appl. Phycol. 20, 705-711.*

Shibata, T., Kawaguchi, S., Hama, Y., Inagaki, M., Yamaguchi, K., Nakamura, T. (2004). Local and chemical distribution of phlorotannins in brown algae. *J. Appl. Phycol.* 16, 291-296.

Shin, H. C., Hwang, H. J., Kang, K. J., Lee, B. H. (2006) An antioxidative and anti-inflammatory agent for potential treatment of osteoarthritis from *Ecklonia cava*. *Arch. Pharm. Res.* 29,165-171.

Singh, I. P., Bharate, S. B. (2006). Phloroglucinol compounds of natural origin. *Nat. Prod. Rep.* 23, 558-591.

Siriwardhana, N., Lee, K.W., Kim, S. H., Ha, W. J., Jeon, Y. J. (2003). Anti-oxidant activity of Hizikia fusiformis on reactiveoxygen species scavenging and lipid peroxidation inhibition. *Food Sci. Technol. Int.* 9, 339-346

Sousa, W. R., Rocha, C., Cardoso, C. L., Silva, D. H. S., Zanoni, M. V. B. (2004) Determination of the relative contribution of phenolic anti-oxidants in orange juice by voltammetric methods. *J. Food Compos. Anal.* 17, 619-633.

Stanner., S. A, Hughes, J., Kelly, C. N., Buttriss, J.A. (2004). Review of the epidemiological evidence for the anti-oxidant hypothesis'. *Publ. Health Nutr.*7, 407-422.

Tapas, A. R., Sakarkar, D. M.., Kakde, R. B. (2008) Flavonoids As Nutraceuticals - A Review. *Trop. J Pharm. Res.* 7, 1089-1099.

Targett, N. M., Arnold, T. M. (1998) Predicting the effects of brown algal phlorotannins on marine herbivores in tropical and temperate oceans. *J. Phycol.* 34, 195-205.

Tomas-Barberan, F., Espin, J. C. (2001) Phenolic compounds and relatedenzymes as determinants of quality of fruits and vegetables. *J. Sci.Food Agric.* 81, 853-876.

Valentine, I. K., Maria, V. K., Bruno, B. (2003) Phenolic cycle in plants andenvironment. *J. Mol. Cell Biol.* 2, 13-18.

Valko, M., Leibfritz, D., Moncol, J., Cronin, M. T. D., Mazur, M. and Telser, J. (2007) Free radicals and anti-oxidants in normal physiological functions and human disease. *Int.* J. *Biochem. Cell Biol. 39:* 44-84.

Van Alstyne, K. L., McCarthy, J. J., Hustead, C. L., Duggins, D. O. (1999a) Geographic variation in polyphenolic levels of Northeastern Pacific kelps and rockweeds. *Mar. Biol.* 133, 371-379.

Varnavas, A., Nisi, C., Lassiani, L., Sava, G., Perissin, L., Boccu, E. (1991) Synthesis and antitumor activity of hydrosoluble analogs of p-(3,3-dimethyl-1-triazeno)benzoic acid potassium salt. *Arzneim. Forsch.* 41, 1168-1172.

Vinayak, R. C., Sabu A.S., Chatterji, A. **(2011a) B**io-prospecting of a few brown seaweeds for their cytotoxic and anti-oxidant activities, *Evid. based compl. alt. med.*: Vol. 2011, Article ID 673083, 9 pages, doi: 10.1093/ecam/neq024

Vinayak, R. C., Sabu A.S., Chatterji, A. (2011b) Bio-screening of a few green seaweeds from India for their cytotoxic and anti-oxidant potential. *J. Sci. Food Agr.* 91, 2471–2476.

Wallace, G., Fry, S. C. (1994) Phenolic components of the plant cell wall. *Int. Rev. Cytol.*151, 229-267.

Wang, J., Zhang, Q., Zhang, Z., Li, Z. (2008) Anti-oxidant activity of sulfated polysaccharide fractions extracted from *Laminaria japonica*. *Int. J. Bio. Macromol.* 42, 127–132.

Wang, T., Jónsdóttir, R., Ólafsdóttir, G. (2009) Total phenolic compounds, radical scavenging and metal chelation of extracts from Icelandic Seaweeds. *Food Chem.* 116, 240-248.

Whitfield, F. B., Helidoniotis, F., Shaw, K. J., Svoronos, D. (1999) Distribution of Bromophenols in species of marine alage from Eastern Australia. *J. Agri. Food. Chem.* 47, 2367-2373.

Wijesekara, I., Yoon, N.Y., Kim, S.K. (2010) Phlorotannins from *Ecklonia cava* (Phaeophyceae): biological activities and potential health benefits. *BioFactors.* 36, 408-414.

Wu, D. F., Cederbaum, A. I. (2003) Alcohol, oxidative stress and free radical damage. *Alcohol Res. Health.* 27, 277-284.

Yan, X. J, Chuda, Y., Suzuki, M., Nagata, T. (1999) Fucoxanthin as the major anti-oxidant in *Hizikia fusiformis*, a common edible seaweed. *Biosci. Biotechnol. Biochem.* 63, 605-607.

Yan, X. J., Chuda, Y., Suzuki, M. Nagata, T. (1999) Fucoxanthan as the major anti-oxidant in *Hijikia fusiformis*, a common edible seaweed. *Biosci. Biotechnol. Biochem.* 63, 605-607.

Yan, *X.*, Li, *X.*, Zhou, *C.*, Fan X. (1996) Prevention of fish oil rancidity by phlorotannins from *Sargassum kellmanianum. J. Appl. Phycol.* 8, 201-203.

Yan, X., Nagata, T. and Fan, X. (1998) Antioxidative activities in some common seaweeds. *Plant Foods Hum. Nutr.* 52, 253-262.

Yoshie, Y., Wang, W., Petillo, D., Suzuki, T. (2000) Distribution of catechins in Japanese seaweeds. *Fish. Sci.* 66, 998-1000.

Yoshie-Stark, Y. P., Hsieh, T., Suzuki. T. (2003) Distribution of flavonoids and related compounds from seaweeds in Japan. *J. Tokyo Univ. Fish.* 89, 1-6.

Yuan, Y. V., Bone, D. E. and Carrington, M. F. (2005) Anti-oxidant activity of dulse (*Palmaria palmata*) extract evaluated *in vitro*. *Food Chem.* 91, 485-494.

Yuan, Y. V., Carrington, M. F., Walsh, N. A. (2005b) Extracts from dulse (*Palmaria palmata*) are effective antioxi dants and inhibitors of cell proliferation *in vitro*. *Food Chem. Toxicol.* 43, 1073-1081.

Yuan, Y. V., Walsh, N. A. (2006) Anti-oxidant and antiproliferative activities of extracts from a variety of edible seaweeds. *Food Chem. Toxicol.* 44, 1144-1150.

Yuan, Y.V. (2008) Marine Algal Constituents. In: Marine Nutraceuticals and Functional Foods. Barrow, C., Shahidi, F. (Eds.), CRC Taylor and Francis Press Inc., pp. 259-296.

Zhang, J., Tiller, C., Shen, J., Wang, C., Girouard, G. S., Dennis, D., Barrow, C. J., Miao, M., Ewart, H. S. (2007a) Antidiabetic properties of polysaccharide- and polyphenolic enriched fractions from the brown seaweed *Ascophyllum nodosum. Can. J. Physiol. Pharmacol.* 85, 1116-1123.

Zhao, J., Fan, X., Wang, S., Li, S., Shang, S., Yang, Y., Xu, N., Lü, Y., Shi, J. (2004) Bromophenol derivatives from the red alga *Rhodomela confervoides. J. Nat. Prod.* 67, 1032-1035.

Zhao, X., Xue, C. H., Li, Z. J., Cai, Y. P., Liu, H.Y., Qi, H. T (2004) Anti-oxidant and hepatoprotective activities of low molecular weight sulfated polysaccharide from *Laminaria japonica. J. Appl. Phycol.* 16, 111-115.

Zubia, M., Fabre, M.S.; Kerjean, V.; Deslandes, E. Anti-oxidant and cytotoxic activities of some red algae (Rhodophyta) from Brittany coasts (France). *Bot. Mar.* 2009, *52*, 268-277.

In: Seaweeds
Editor: Vitor H. Pomin

ISBN: 978-1-63117-571-8

Chapter 9

ANTIOXIDANT SULFATED POLYSACCHARIDES FROM SEAWEED

Mariana Santana Santos Pereira Costa[1], Gabriel Pereira Fidelis[2], Hugo Alexandre Oliveira Rocha[2] and Leandro Silva Costa[1]

[1]Instituto Federal de Educação,
Ciência e Tecnologia do Rio Grande do Norte – IFRN, Brazil
[2]Departamento de Bioquimica,
Universidade Federal do Rio Grande do Norte – UFRN, Brazil

ABSTRACT

Seaweeds belong to a group of marine organisms that are frequently subjected to ultraviolet light, high oxygen content, and saline and hydric stress. These conditions produce continuous damage to important biomolecules in these organisms, such as proteins, lipids and DNA, in addition to the high production of free radicals that can destroy the cells. Due to the consequences of this damage, seaweeds have many endogenous mechanisms to prevent the destruction of cells, especially the production of antioxidant molecules. As a result, seaweeds have been studied as a potential source for the extraction of antioxidant compounds, and these molecules might be used in various antioxidant therapies. Among the antioxidant compounds present in the seaweeds, sulfated polysaccharides are the most studied. Each seaweed species produces its own sulfated polysaccharide, which makes sulfated polysaccharide research in algae a vast field of study. Brown seaweeds synthesize polysaccharides rich in sulfated α-L-fucose and are called fucans; red seaweeds synthesize polysaccharides rich in galactose and are called named galactans; while green seaweeds synthesize more heterogeneous sulfated polysaccharides. Many studies have shown that the antioxidant capacity of these polysaccharides depend on structural characteristic, such as sulfation and molecular weight. Moreover, studies have shown that sulfated polysaccharides can act to prevent the production of free radicals or neutralize already existing free radicals. The present study reports the progress of research in antioxidant activity from sulfated polysaccharides obtained from seaweeds, the principal methods used to evaluate this activity, and the principal applications of antioxidant sulfated polysaccharides in several human activities.

Keywords: Fucan, fucoidan, sulfated galactan

1. Introduction

Sulfated polysaccharides are molecules widely spread in nature, with occurrence in a great variety of organisms, such as mammals and invertebrates (Pomim; Mourão, 2008). The most important source of sulfated polysaccharides of non-animal origin is seaweeds. The sulfated polysaccharides come from seaweeds have a quite different structure, varying from species to species (Dietrich et al., 1995), so each new sulfated polysaccharide purified is a new compound with unique structure and consequently with potential new biological activities.

Many pharmacological activities are attributed to the sulfated polysaccharides of seaweeds (Rocha et al., 2006). In the last years one of the pharmacological activities of these polymers that have been put on the spotlight is the antioxidant activity (Shao et al., 2013). Therefore, this paper summarizes the progress in the investigation of antioxidant activity of sulfated polysaccharides obtained from seaweeds.

2. Sulfated Polysaccharides of Seaweeds

The seaweeds are several sources of biopolymers, among these, there are the polysaccharides. (Cardozo et al., 2007; O'Sullivan et al., 2010). A special type of polysaccharide that has aroused the interest of researchers in recent years are the sulfated polysaccharides, such as fucans, fucoidans and different types of sulfated heteropolysaccharide, because they show a great economic potential due to the high number of pharmacological activities that they have (Pomim; Mourão, 2008; Li et al., 2008; Costa et al., 2010).

In seaweed sulfated polysaccharides are located in the mucilage matrix and their biological function, in such beings, is still not well understood, but due to their highly hygroscopic character, it is believed that they protect the algae from dehydration when it is subjected to long periods of exposure to sunlight during low tides (Costa et al., 2012). The mucilaginous nature of these compounds also seems to contribute to making the algae flexible enough to grow in liquid environment and rigid enough to remain extended, and, thus, to capture the light and nutrients available in a better way (O'Sullivan et al., 2010).

The algal sulfated polysaccharides show very different structures, varying from species to species, and sometimes in different parts of the same alga (Dietrich et al., 1995). The sulfated polysaccharides found in Phaeophyta (brown algae) are fucans (Magalhães et al., 2011), which have as their main characteristic the presence of sulfated L-fucose in their structure (Rocha et al., 2006). In the algae, the fucans are found both in homo and heteropolysaccharides form (Berteau & Mulloy, 2003). Two exceptions described until now are the brown algae *Sargassum fusiforme* and *Undaria pinnitafida*. The alga *S. fusiforme* synthesizes sulfated polysaccharides consisting of glucuronic acid, arabinose, glucose, galactose, mannose and xylose (Zhou et al., 2008); on the other hand, the polysaccharides of

U. pinnitafida consist of ramonose, uronic acid, galactose and xylose (Hu et al., 2010), the polysaccharides of both algae do not have fucose in their constitution.

Rhodophyta (red algae) shows the sulfated galactans as a characteristic polysaccharide (Zong et al., 2012). However, recently, in the past 5 years, several studies have reported the presence of other types of sulfated polysaccharides in red algae such as a xilomanana containing various sulfated mannose units in position 2 which was purified from seaweed *Scinaia hatei* (Mandal et al., 2008) and a heteropolysaccharide mainly composed by D-fucose, D-glucose, D-mannose and D-galactose, some fucose residues are shown sulfated in C2 or C4, extracted from seaweed *Gloiopeltis tenax* (Lim; Ryu, 2009). Chlorophyta (green algae) shows more heterogeneous sulfated polysaccharides, which are rich in galactose, mannose, xylose, arabinose and/or uronic acids, among others (Hayakawa et al., 2000, Mao et al., 2006). However, homopolysaccharides have been described as arabinans (Hayakawa et al., 2000; Siddhanta et al., 1999) and galactans (Farias et al., 2008).

The sulfated polysaccharides of algae have been described as having many pharmacological activities (Costa et al., 2010; Li et al., 2008). Among them, the antioxidant activity of sulfated polysaccharides have been the subject of great interest in many research groups in recent years.

3. Reactive Oxygen Species (ROS) x Antioxidants

The term free radical is considered outdated by chemists, but in the biological, pharmaceutical, medical sciences, among others, this is still a term widely used. Therefore, in this chapter the term "free radical" will be used to define any chemical species (atom, ion or molecule) that contains an unpaired electron in its valence layer (top layer). In the case of biological systems, the most common source of free radicals is oxygen (Wickens, 2001).

The molecular oxygen (O_2) is essential for cell metabolism and energy production. However, the degradation of oxygen also produces highly reactive agents that can cause damage to biological tissues, these free radicals are called reactive oxygen species (ROS) (Ferreira; Matsubara, 1997). Among these there are hydroxyl radical ($OH^{\bullet}$), hydrogen peroxide (H_2O_2) and e superoxide anion ($O_2^{-\bullet}$).

Besides being products of the metabolic process, ROS can also be produced by other factors, such as mediators of carcinogenesis, inflammatory injury (Banerjee; Dasguptan; De; 2005) and environmental sources (ultraviolet light, ionizing radiation and pollutants such as paraquat, a toxic substance used as a herbicide) (Wickens, 2001).

The group of substances that neutralize the damaging effects of ROS is the antioxidant defense system, i.e., antioxidants are substances which counteract or prevent oxidation of oxidizable substrates (Somogyi et al., 2007). The exacerbated ROS formation may cause serious damage to biomolecules. In order to counter the reactive species formed, the organisms have defenses that neutralize the presence of ROS (antioxidant enzymes and molecules involved in ROS scavenging) and which prevent the formation of ROS (antioxidant enzymes and chelating compounds of metal ions) (Somogyi et al., 2007). In recent years, sulfated polysaccharides from seaweed have been described as having antioxidant activity both by the mechanism of free radical scavenging and by chelation of metal ions (Rupérez; Ahrazem; Leal, 2002; Zhang et al., 2003; Souza et al., 2007; Wang et al., 2008).

4. Methods for the Determination of Antioxidant Activity

In order to determine the antioxidant activity of sulfated polysaccharides, several methods have been employed, mostly involving *in vitro* tests, which are usually classified according to the chemical characteristics of their reactions (Huang et al., 2005; Selvakumar et al., 2011). It's worth mentioning that some tests measure the same chemical property of the antioxidant, but as the tests are performed under different conditions, it is supposed that they evaluate distinct antioxidant properties. Therefore, a classification based on the features of each test becomes more interesting, by easing the planning of research to assist in the initial selection of antioxidants and/or determination of their possible mechanisms of action. As a suggestion, the antioxidant tests used in research with PS can be classified into four major groups: I) assessing the ability of reducing reactive molecules; II) determining the ability of scavenging free radicals; III) evaluation of the ability to chelate metal ions; IV) determination of the potential of lipid peroxidation (table 1).

The methods of group I are widely used as indicators of antioxidant potential in natural compounds, with the main advantage the fact they are extremely simple, fast and highly reproducible, which makes development of research projects aimed at bioprospecting of antioxidant molecules more economic. There may be mentioned as the most widely used tests for total antioxidant ability (CAT), reduction of the radical 2,2-diphenyl-1-picryl-hydrazyl (DPPH), the reducing power test, reduction of the radical 2,2 - azinobis (3-ethylbenzothiazoline-6-sulfonic acid) (ABTS) and the FRAP (Ferric Reducing Antioxidant Power) method. In recent studies it was shown that natural compounds, including PS, which showed antioxidant potential, especially in CAT and DPPH tests; they also showed significant antioxidant activity through a variety of different mechanisms (Costa et al., 2010; Zhang et al., 2004). These tests are based on the ability of oxidizing radicals react with electron donor molecules, and then reduced, and the product detected by their intense absorption in the visible region. As an example of this type of reaction, CAT detects the reduction of $Molybdenum^{+6}$ to $Molybdenum^{+5}$ and for subsequent formation of a complex greenish $phosphate/molybdenum^{+5}$ in acid pH (Prieto et al., 1999), while the DPPH test determines the ability of this radical being reduced to hydrazine, which now starts to acquire a yellow coloration.

The group II tests are based on the same chemical principles of oxidation-reduction of the tests of group I, but here they are classified in a distinct group because they assess the antioxidant activity of a compound in relation to a specific ROS, such as hydroxyl radicals ($OH^{\bullet}$), and superoxide anion ($O_2^{-\bullet}$), or to nitric oxide (ON), which in high concentrations can react with superoxide anion producing the potentially damaging peroxynitrite, and increasing the oxidative stress state (Peluffo; Radi, 2007). In these tests, free radicals are generated via different chemical reactions, and these periods after incubation with the test sample have their concentration monitored. The production of hydroxyl radicals ($OH^{\bullet}$) is based on Fenton reaction ($Fe^{2+} + H_2O \rightarrow Fe3^{+} + OH^{-} + OH^{-}$) (Smirnoff; Cumbes, 1989), while the superoxide radicals ($O_2^{-\bullet}$) can be generated from different chemical reactions, with the most usual ones those which comprise the systems known as riboflavin-light- Nitro blue tetrazolium (Beauchamp; Fridovich, 1971; Dasguptaa; DE, 2007) and xanthine oxidase (Paya et al., 1992). In turn, nitric oxide is measured by the Greiss method, which has radical-forming system with nitrite, sodium nitroprusside (Radi; Pellufo, 2001).

Table 1. Antioxidant tests used in research with PS

Classification	Principle	Tests	Reference
Group I	Assessment of the ability of reducing reactive molecules	CAT; DPPH; ABTS; reducing power, FRAP	A, B
Group II	Determination of the ability of scavenging free radicals	Hydroxyl radical and superoxide radical scavenging	C, D, E
Group II	Assessment of the ability to prevent the formation of free radicals	Chelation of iron ions, chelation of copper ion; scavenging of hydrogen peroxide (H_2O_2)	F, G
Group IV	Determination of the potential for inhibition of lipid peroxidation	Test 2-thiobarbituric acid (TBA); Kreis test; Carbonyl index; Inhibition of erythrocyte hemolysis	H, I, J

A - Costa et al., 2010;B - Zhang et al., 2004. C - Smirnoff; Cumbes, 1989. D - Beauchamp; Fridovich, 1971. E - Dasguptaa; De, 2007. F - CuzZocrea, 2001. G - Borg, 1993. H - Silva et al., 2012; I - Josephine et al., 2008. J - Ebrahimzadeh et al., 2010.

Within the group III the tests that assess the ability to prevent the formation of free radicals such as tests of copper and iron ions and hydrogen peroxide (H_2O_2) chelating scavenging are highlighted. These tests have become standard tools and extremely necessary to assess the activity of compounds isolated from natural products such as PS and obtained from synthetic sources. This is justified by the fact that ROS, especially superoxide and hydroxyl radicals, requiring few milliseconds to react with biological molecules, making it nearly impossible to combat these after its formation by the body, so for antioxidant therapy is more interesting counteract the formation these radicals than just scavenging them (Cuzzocrea, 2001).

Metal ions in reactions involving H_2O_2 may generate a large amount of ROS through Fenton and Haber-Weiss reactions. In the Fenton reaction hydroxyl radical (HO•) is produced from the reaction of iron ion (Fe^{2+}) with hydrogen peroxide (H_2O_2) while the reaction known as Haber-Weiss is catalyzed by transition metals such as iron and copper also leading to a large formation of hydroxyl radicals (BORG, 1993). Therefore, the binding of metal ions and hydrogen peroxide by antioxidant PS can provide to these applications in antioxidant therapies, both for drug use and for use in cosmetics and food industries, since they act at an early stage of formation of these radicals.

Tests of Group IV are used to assess lipid peroxidation, whose principle is to determine the degree of inhibition of the oxidation of molecules such as lipids or lipoproteins. Lipid peroxidation appears as one of the most aggressive events, triggering a string of cellular injuries that directly affect phospholipids of the plasma membrane, promoting changes in ion flux, selectivity loss and disruption of membrane. Moreover, these peroxided lipids, which are highly reactive, cause a number of deleterious effects, such as mutations in the genetic material and affecting the extracellular matrix components (proteoglycans, collagen and elastin). The methods for analysis of lipid peroxidation are varied, since this process occurs in several steps, involving initiation, propagation and termination of the peroxidation reaction (Gutteridge, 1995), which can be evaluated in different ways, through techniques of chromatography, spectroscopy and immunochemistry. These last ones are the most commonly used due to their good reproducibility and are low cost. They are based on the monitoring of the concentration of unsaturated fatty acids, especially fatty acid malondialdehyde (MDA), which is by far the most used peroxidation marker for assessment of the antioxidant power of sulfated polysaccharides (Silva et al., 2012; Josephine et al., 2008). MDA, which is formed as end-product of lipid peroxidation by test of 2-thiobarbituric acid (TBA), is reacted with this acid to generate a product that absorbs energy at a wavelength of 532 nm. At a lower frequency, tests using other peroxidation markers are also used to determine the antioxidant potential of PS, such as the carbonyl index test, which corresponds to the assessment of carbonyl compounds which are formed during the oxidation process and Kreis test, which is based on the reaction, in acid medium, of phloroglucinol with the primary products of peroxidation.

Also, as a consequence of the susceptibility to lipid peroxidation, some biomembranes are used as oxidation markers, for example to test inhibition of rat erythrocyte hemolysis. In this test, the erythrocytes extracted are subjected to oxidative stress by the system H_2O_2/ascorbate/Fe_2^+ and the lysed cells are quantified (Ebrahimzadeh et al., 2010).

Finally, this suggested classification becomes more interesting when we find that, in general, the studies on the antioxidant activity of sulfated polysaccharides from seaweed follow a standardized sequence that takes into consideration the purpose of each method: PS are subjected to methods of group I, which are tests only used as indicators of antioxidant potential, and then are evaluated by the methods of groups II, III and/or IV, which eases the determination of their possible action mechanisms.

5. Antioxidant Sulfated Polysaccharides from Seaweed

Seaweeds are exposed to ultraviolet light (UV) and high oxygen concentrations; in addition, they live in an environment of constant saline and hydrolytic stress, and these factors lead to formation of free radicals and other strong oxidizing agents, however, the algae cannot bear any oxidation and photodynamic damage processes during the course of their lives. This leads to a consensus that the algal cells have very competent antioxidant mechanisms and they probably should synthesize several compounds with antioxidant action such as phenolic compounds and vitamins (Matsukawa et al., 1997; Hoyer; Karsten; Wiencke, 2002; Rocha et al., 2007). For this reason, the seaweed is one of the potential renewable resources providers of compounds that can be used by humans as antioxidants.

The amount of work related to the antioxidant potential of PS from seaweed has grown considerably in recent years and the results indicate that these compounds have significant antioxidant potential mediated usually through different mechanisms of action, which provides a more efficient defense before free radicals. These results are summarized in Table II.

Table 2. Major marine algae that synthesize antioxidant polysaccharides and the methods by which they were evaluated

	Seaweed	CAT	Reducing power	DPPH	Hydroxyl radical scavenging $OH^•$	NO radical scavenging	superoxide radical scavenging $O_2^{-•}$	Chelation of iron ions	H_2O_2 radical scavenging	Lipid peroxidation	Reference
Phaeophyta	*Laminaria japonica*	nd	+	+	+	nd	+	+	nd	+	A, B
	Sargassum fulvellum	nd	nd	+	+	+	+	nd	+	nd	C
	Sargassum pallidum	nd	nd	+	nd	nd	nd	nd	nd	nd	D
	Sargassum plagiophyllum	nd	nd	+	+	nd	+	nd	nd	nd	E
	Sargassum filipendula	+	+	nd	-	nd	-	+	nd	nd	F
	Sargassum fusiforme	nd	nd	nd	+	nd	+	nd	nd	+	G
	Sargassum tenerrimum	+	+	+	nd	nd	nd	nd	+	nd	H
	Sargassum vulgare	nd	nd	+	nd	nd	nd	nd	nd	nd	I
	Sargassum henslouianum	nd	nd	+	+	nd	+	nd	nd	nd	J
	Fucus vesiculosus	nd	+	nd	+	nd	+	nd	nd	+	K, L
	Padina gymnospora	nd	nd	nd	+	nd	+	nd	nd	+	L
	Padina tetrastromatica	nd	+	+	+	nd	+	nd	nd	+	M
	Undaria pinnitafida	nd	nd	+	+	nd	+	+	nd	nd	N
	Lobophora. variegata	+	+	nd	+	nd	+	nd	nd	nd	O

Table 2. (Continued)

	Seaweed	CAT	Reducing power	DPPH	Hydroxyl radical scavenging $OH^{\bullet}$	NO radical scavenging	superoxide radical scavenging $O_2^{-\bullet}$	Chelation of iron ions	H_2O_2 radical scavenging	Lipid peroxidation	Reference
	Turbinaria ornata	nd	nd	+	nd	+	nd	nd	nd	nd	P
	Dictyota cervicornis	+	+	nd	-	nd	+	+	nd	nd	F
	Dictyota menstrualis	+	+	nd	+	nd	+	+	nd	nd	F
	Dictyota mertensii	+	+	nd	+	nd	+	+	nd	nd	F
	Dictyopteris delicatula	+	+	nd	+	nd	+	+	nd	nd	F,Q
	Spatoglossum schöederi	+	+	nd	-	nd	-	+	nd	nd	F
Rhodophyta	*Porphyra haitanensis*	nd	-	+	+	nd	nd	+	nd	+	R, S
	Porphyra yezoensis	nd	+	nd	+	nd	+	+	nd	nd	T, U
	Eucheuma spinosa	nd	nd	nd	+	nd	+	nd	nd	+	L
	Eucheuma cotonni	nd	nd	nd	+	nd	+	nd	nd	+	L
	Gigartina acicularis	nd	nd	nd	+	nd	+	nd	nd	+	L
	Gigartina pisillata	nd	nd	nd	+	nd	+	nd	nd	+	L
	Gracilaria caudata	+	+	nd	+	nd	-	+	nd	nd	F
	Gracilaria birdiae	nd	nd	+	+	nd	nd	nd	nd	nd	V

	Seaweed	CAT	Reducing power	DPPH	Hydroxyl radical scavenging $OH^{\bullet}$	NO radical scavenging	superoxide radical scavenging $O_2^{-\bullet}$	Chelation of iron ions	H_2O_2 radical scavenging	Lipid peroxidation	Reference
	Gloiopeltis furcata	nd	nd	+	+	nd	+	nd	nd	nd	J
	Gloiopeltis tenax	nd	nd	+	nd	nd	nd	nd	nd	nd	W
	Corallina officinallis	nd	+	+	nd	nd	+	nd	nd	nd	Z
Chlorophyta	*Ulva pertusa*	nd	+	nd	+	nd	+	+	nd	nd	X
	Ulva fasciata	nd	+	+	+	nd	+	nd	nd	nd	J, Y
	Caulerpa cupressoides	+	+	nd	-	nd	-	+	nd	nd	F
	Caulerpa prolifera	+	+	nd	-	nd	-	+	nd	nd	F
	Caulerpa sertularioides	+	+	nd	+	nd	+	+	nd	nd	F
	Codium isthmocladum	+	+	nd	-	nd	-	-	nd	nd	F

A - Wang et al., 2008. B - Wang et al.,2010. C – Kim et al., 2007. D - Ye et al., 2008. E - Suresh et al., 2013. F – Costa et al., 2010. G - Zhou et al., 2008. H - Vijayabaskar et al., 2012. I - Dore et al., 2013. J – Shao et al., 2013a. K - Rupérez; Ahrazem; Leal, 2002. L - Souza et al., 2007. M - Sulaiman et al., 2013. N - Hu et al., 2010. O - Paiva et al., 2011. P - Ananthi et al., 2010. Q - Magalhaes et al., 2011. R - Zhang et al., 2003. S – Zhang et al., 2009. T - Liu et al., 2008. U - Zhou et al., 2008. V - Souza et al., 2012. W - Lim; Ryu, 2009. X - Qi et al., 2005a. Y – Shao et al., 2013b. Z - Yang et al., 2011. +: It shows activity, -: It did not have activity, nd: not determined.

5.1. Sulfated Polysaccharides (SP) of Brown Algae

SP extracted from brown algae are by far the most studied ones, presenting a potential antioxidant primarily mediated through the ability of scavenging free radicals, especially the inhibition of superoxide and hydroxyl radicals and the metal ion chelation (Zhang et al., 2003b, Heo et al., 2005, Wang et al., 2008, Magalhães et al., 2011).

Among the antioxidant PS extracted from species of brown algae, fucoidan from brown alga Laminaria japonica is the most widely evaluated PS. The most systematic studies with this molecule began in 2001, when Xue et al. (2001) analyzed fucoidans from L. japonica as antioxidant agents, using for that the model of lipid peroxidation (TBA test) to quantify the oxidation of low density lipoproteins (LDL);the polysaccharides tested were strongly active in the protection of LDL oxidation induced by AAPH (2,2'-azobis (2-amidinopropane) dihydrochloride), but not to that induced by Cu^{2+}. Subsequently, low molecular weight fucoidans of this same alga obtained by acid hydrolysis demonstrated to have the ability to scavenge the radicals superoxide and hydroxyl (Zhao et al., 2004).

In order to get more information about the antioxidant activity of fucoidan from *L. japonica*, Zhao and colleagues (2008) obtained from the crude sulfated polysaccharides of *L. japonica*, several fractions of fucoidan by gel filtration chromatography (F-A, F-B), by gel filtration chromatography followed by acid hydrolysis (F-A1, F-A2, F-B1, F-B2) and degradation of radical process (LM1, LM2), and found that the low molecular weight fucoidan (F-A2) showed potent scavenging action of hydroxyl and superoxide radicals, superior even to high molecular weight fucoidans FA and FB. Furthermore, the authors found that more sulfated low molecular weight fucoidans showed lower antioxidant activity than those which are less sulfated, these results indicate that the sulfate group of the low molecular weight fractions represents a physical block for the reaction with oxygen radicals. These authors also observed that the presence of more than 5% glucuronic acid in the composition of the polysaccharide abolished its superoxide radical scavenging activity.

Another study also published in 2008 showed that fucoidans from *L. japonica* obtained by fractionation on ion exchange chromatography also showed a high antioxidant activity by scavenging mechanism of superoxide and hydroxyl radicals and moderate ferric chelating ability (Wang et al., 2008). Later, this group also showed that these fucoidans also had antioxidant activity when assessed by DPPH scavenging test and reducing power (Wang et al., 2010). It was also seen that fucoidans from *L. japonica* modified by adding functional groups showed excellent antioxidant activity, mainly the benzoylated (Wang et al., 2009a) and phosphorylated (Wang et al., 2009b).

More recently, other genera of brown algae are evaluated as sources of antioxidant sulfated polysaccharides and one of the genera that has been most studied in this regard is *Sargassum.*

One of the most important early works was published in 2007. In this was demonstrated that a heterofucan from *S. fulvellum* showed excellent antioxidant activity by DPPH test, and scavenging activity of hydrogen peroxide, however, it showed low scavenging activity for hydroxyl and superoxide radicals (Kim et al., 2007). In the following year, it was demonstrated that sulfated heteropolysaccharides with different uronic acid content and different molecular weights were obtained from *S. fusiforme*, these polysaccharides, unlike fucans from *S. fulvellum*, showed scavenging activity of superoxide and hydroxyl radicals, and it is the polysaccharide with higher content of uronic acid and lower molecular mass

(Zhou et al., 2008). SP extracted from *S. pallidum* were also assessed by the DPPH test; however, differently from the polysaccharides mentioned here until now, the SP of *S. pallidum* showed a low antioxidant activity (Ye et al., 2008).

Later it was shown that heterofucans from *S. filipendula* (Costa et al., 2011) and *S. vulgare* (Dore et al., 2013) had lower scavenging activity of hydroxyl and superoxide radicals. Despite the limited data, there appears to be a consensus that this is not the main mechanism of antioxidant action of fucans from algae of the genus *Sargassum*. On the other hand, it was found that a heterofucan from *S. filipendula* showed strong activity in reducing power test that was similar to that found for vitamin C (Costa et al., 2011). As also found that heterofucan from *S. vulgare* showed a good antioxidant activity in the DPPH test (Dore et al., 2013). In other words, the data so far are indicating that the fucans of *Sargassum* genus has antioxidant mechanism as the main fact of acting as reducing agents.

A fucan from the brown alga *Fucus vesiculosus* is easily obtained, because it is sold commercially. Therefore, they were one of the first fucans analyzed as antioxidants. Already in 2002 it was shown that the fucan from this algae had a high antioxidant power measured by FRAP method (Rupérez; Ahrazem; Leal, 2002). Subsequently, it was found that it also had scavenging activity of superoxide and hydroxyl radicals (Souza et al., 2007). These data show that the mechanism of action of the fucan from *F. vesiculosus* is far more complex than the fucans from *Sargassum*. However, this conclusion should be viewed with caution, because several studies show that fucan obtained commercially from *F. vesiculosus* is actually a mixture of different fucans and that the proportion between them varies according to the batch.

Samples with different polysaccharides extracted from the same alga were also evaluated by other authors. As an example, Ananthi et al. (2010) showed that samples rich in heterofucans extracted from algae *Turbinaria ornata* have antioxidant activity in the tests of DPPH scavenging, nitric oxide scavenging and inhibition of lipid peroxidation (Ananthi et al., 2010). When samples rich in heterofucans from other 5 species of algae of the order Dictyotales were assessed by Costa and colleagues (2010): *Dictyota cervicornis, Dictyota menstrualis, Dictyota mertensii, Dictyopteris delicatula* and *Spatoglossum schöederi*, it was noted that only those originated from *D. menstrualis* and *D. mertensii* had the ability to scavenge the hydroxyl radicals, however, this activity was low, and PS have the ability to scavenge less than 8.7% of the radicals formed. As for the scavenging test of superoxide radical, they all showed activity except *S. schröederi*. For the other tests (reducing power, total antioxidant capacity and ferric chelation) the five algae showed moderate to high activity (Costa et al., 2010).

Fucans purified from algal order *Dictyotales* have also been evaluated as antioxidants. Heterofucans from *Padina gymnospora* showed antioxidant activity in the scavenging test of the hydroxyl radical and a fucan extracted from seaweed *Lobophora variegata*, which is called F1, showed activity by scavenging radicals and superoxide hydroxyl and showing an excellent reducing activity (Souza et al., 2007). More recently, five heterofucans isolated from seaweed *Canistrocarpus cervicornis* showed antioxidant activity through the following mechanisms: superoxide and hydroxyl radicals scavenging, reducing capacity and especially chelating ability of iron ions (Camara et al., 2011).

The brown seaweed *Undaria pinnitafida* synthesizes sulfated heteroramnans and these SP showed excellent scavenging activity of superoxide hydroxyl and DPPH radicals, besides having a moderate chelating ability of iron ion (Hu et al., 2010). However, no other studies were identified that evaluated the antioxidant activity of other heteroramnans. So it is not possible to affirm if that this is the mechanisms of antioxidant action of ramnans or if they may act by other mechanisms.

It can be seen from the work based on fucans, that these have a significant antioxidant potential. In addition, each fucan can act by different mechanisms of action of other fucans, which reflects the fact that each alga synthesizes a different kind of fucan that is not found in other alga. On the other hand, when samples (extract) rich in different fucans are evaluated, it is noted that these act by different antioxidant mechanisms. However, these extracts also consist of other molecules that may be toxic, which makes their use by humans difficult. An alternative would be to associate different pure fucans, which present different mechanisms of action and thus have samples with a complex antioxidant properties, which protect more effectively living organisms, including humans, and many industry products such as foods, paints and pharmaceutics.

5.2. Sulfated Polysaccharides from Red Algae

The antioxidant activity mediated through several mechanisms is not exclusive of sulfated polysaccharides from brown algae. Zhang et al. (2003a) isolated three galactans, called F1, F2 and F3, from the red alga *Phorphyra haitanesis* and observed that they had a high power to scavenge superoxide radical and a weaker effect to scavenge hydroxyl radicals, besides inhibiting significantly the lipid peroxidation and partially inhibit hemolysis of mice erythrocytes induced by hydrogen peroxide. Subsequently, this same research group observed that intra-peritoneal administration of PS F1 and F2 of *P. haitanensis* increased the antioxidant capacity of the blood serum and homogenates of several organs in aged mice when they were assessed by CAT test, as well as increased activity of antioxidant enzymes (superoxide dismutase and glutathione peroxidase) (Zhang et al., 2003b; Zhang et al., 2004). It was also observed that the antioxidant activity of sulfated polysaccharides of this alga is dependent on the molecular weight of the polymer, with the most active ones with lower molecular mass (Zhao et al., 2006). When changes were made in chemical polysaccharides from *P. haitanensis* (oversulfation, acetylation, benzoylation and phosphorylation), it was found that these derivatives were more potent than the native polysaccharide (Zhang et al., 2009).

Antioxidant polysaccharides were also extracted from *Porphyra yezoensis*, and these polymers showed significant antioxidant activity by scavenging tests of hydroxyl and superoxide radicals. They also showed activity by DPPH test, however this activity was less pronounced. In addition, these polysaccharides inhibited erythrocyte hemolysis induced by H_2O_2 and the lipid peroxidation in microsomal cells determined by the TBA test (Liu et al., 2007; 2008). Subsequently, Zhou et al. (2008) carried out ultrasonic PS degradation of *P. yezoensis* and tested the antioxidant activity (superoxide and hydroxyl radical scavenging, ferric chelating and reducing power) of the degradation products of these PS. The analysis of the data led the authors to propose that the degraded polysaccharide has higher antioxidant activity than the native polysaccharide, suggesting that the decrease in molecular weight of the polymer is vital to increase its antioxidant activity (Zhou et al., 2008).

Kappa and iota carrageenan extracted from seaweeds *Eucheuma spinosa* and *Eucheuma cottonii,* respectively, and lambda carrageenan obtained from algae (*Gigartina acicularis* and

Gigartina pisillata) showed good activity regarding the superoxide and hydroxyl radicals scavenging (Souza et al., 2007). Sulfated polysaccharides from the edible red alga *Gracilaria caudata* showed excellent activity for the reducing power test, metal chelating activity and activity in total antioxidant capacity test, but they have not had the superoxide radical scavenging activity, or significant activity to scavenge s hydroxyl radicals (Costa, 2010). Sulfated polysaccharides *Gloiopeltis tenax* also have excellent activity for the DPPH test, rather than ascorbic acid (Lim; Ryu, 2009).

Fl and F2, two xilogalactans isolated from the red alga *Corallina officinalis* showed significant antioxidant activity when they were subjected to the scavenging tests of superoxide and hydroxyl radicals, reducing activity and total activity in the DPPH test (2,2-diphenyl-1-picrylhydrazyl). Still, Fl and F2, when they were desulfated, dramatically reduced their antioxidant potential, confirming the importance of sulfate to the antioxidant activity of sulfated polysaccharides (Yang et al., 2011).

Despite the small number of studies carried out with antioxidant SP from red algae, these indicate that the superoxide and hydroxyl radicals scavenging seem to be the main mechanism of action of sulfated galactans from red algae. The only exception is the galactan from seaweed *Gracilaria caudata*, which showed no radical scavenging capacity, which has as the main antioxidant action the chelation of iron ions, which is probably a unique feature of the galactans extracted from this alga.

Thus, it is observed that, as fucans and fucoidans, the SP extracted from red algae can act both preventing the ROS formation, as well as act fighting the ROS already formed. Moreover, it is noted that SP with lower molecular weight and the most sulfated ones have better antioxidant activity.

5.3. Sulfated Polysaccharides from Green Algae

When it comes to green algae, there are few studies on antioxidant sulfated polysaccharides. A study by Qi and colleages (2005a) clearly demonstrated that natural ulvans, sulfated polysaccharides of green alga *Ulva pertusa* had antioxidant activities (hydroxyl radical scavenging, superoxide, reducing power and ferric chelate). In addition, they also showed that when the ulvans are oversulfated they are more potent than natural ulvan (Qi et al., 2005a).

A natural ulvan (157.1 kDa) was degraded into polysaccharides of lower molecular masses (28.5, 64.5 and 50 kDa) and the activities mentioned above were evaluated. The results indicated that the molecular weight had a significant effect on the antioxidant activity of SP from *U. pertusa*, since ulvans of lower molecular weight have more potent antioxidant activity (Qi et al., 2005b). It was also seen that ulvans modified by the addition of acetyl and benzoyl groups showed excellent antioxidant activity, mainly for the test of superoxide radicals, even higher than vitamin C (Qi et al., 2006).

Sulfated polysaccharides from four other species of green algae (*Caulerpa cupressoides*, *Caulerpa prolifera*, *Caulerpa sertularioides* and *Codium isthmocladum*) were also evaluated for their antioxidant potential. From these, only the ones from alga *C. sertularioides* showed activity by scavenging mechanism of hydroxyl radicals and superoxide. In addition, all the polysaccharides showed total antioxidant capacity, reducing power and ferric chelating ability (except for *C. isthmocladum*) (Costa et al., 2010).

Later, this same research group obtained antioxidant galactose-rich polysaccharide fractions from *C. cupressoides* by acetone fractionation followed by gel filtration. These fractions rich in SP had antioxidant test by CAT activity and significant iron chelation, but they showed no scavenging activity of hydroxyl radicals and superoxide (Costa et al., 2012).

The few studies developed with sulfated polysaccharides extracted from green algae do not allow drawing many conclusions about their mechanisms of action, however, it is realized that these molecules appear as antioxidant potential, always presenting significant activity when evaluated through the tests of group I. Still, it is clear that the main mechanism of action of galactans extracted from algae of the genus *Caulerpa* seems to be the ferric chelation, which was significant for the galactans from *C. cupresoides*, *C. sertularioides* and *C. prolifera*.

6. Applications of Antioxidant Sulfated Polysaccharides

Antioxidant compounds have many applications, and the synthetic ones are the most used on the market, mainly due to their low production cost. Among the synthetic antioxidants there are included phenolic compounds such as butylated hydroxyanisole (BHA), butylated hydroxytoluene (BHT), t-butylhydroquinone (TBHQ), and esters of gallic acid (Search for recent reviews). On the other hand, for few decades the safety concerning consumers has increased considerably, which led to a regulation of the use of these compounds, many times their use is prohibited, since studies show the potential toxic effects of these biomolecules (Pacifico et al., 2008).

In this perspective, and considering that natural compounds have low risks for the consumer, the antioxidants of natural origin have been intensively studied as antioxidant potential compounds, which can open up a very wide field of application for these compounds. In this part of the review, these possible applications for the antioxidant PS derived from seaweed will be addressed.

A recent and very promising field of application of antioxidants is the cosmetic industry. Studies show a direct relationship between the presence of antioxidants and anti-aging effect, especially for the protection given to the components of the stratum corneum of the skin, which counteracts the effects of ultraviolet radiation (Beatty et al., 2013; Cestari et al., 2012; Calabrese et al., 2000).

Seaweed extracts and complexes consisting primarily of fucoidan extracted from algae *F.vesiculosus* and *U. pinatifida* now are often used in cosmetic formulation, which in turn creates an interesting market for use of other seaweed. The use of fucoidan is justified by its supposed benefits to health of the skin, including the inhibition of oxidative degeneration of molecules such as collagen and hyaluronic acid and consequently the anti-aging action (Fujirama et al., 2000, 2001, 2002).

The use of extracts rich in fucoidan is not only exclusivity in the cosmetic industry. A brief search on the internet allows you to find tens of dietary supplements that have the fucoidan in their composition. The properties attributed to these compounds have a wide variety, highlighting the high antioxidant power, which in turn is associated with the prevention of chronic diseases and muscle fatigue in athletes, as well as aid in the treatment of burns, clinical picture in which there an overproduction of free radicals (Horton, 2003).

Still, this strong commercial appeal associated to fucoidan is directly linked to the consumption of fresh brown algae, particularly in Asian cultures. Seaweed has a high nutritional value, are rich in carotenoids, proteins, dietary fiber, essential fatty acids, vitamins (B complex and vitamins C, D, E and K) and minerals (calcium, phosphorus, sodium, magnesium, iron, copper, manganese, potassium, vanadium, iodine), and SP can act synergistically with these elements in a series of physiological events, including the most effective defense against the damaging effects of free radicals.

Until this moment, applications for the antioxidant sulfated polysaccharides are restricted to those mentioned above, however, due to the growing demand for natural products in the market and the significant antioxidant potential reported for some SP, it is natural that they can be applied in several areas, competing with synthetic products. As an example of these possible applications there are included: canned food preservation and sausages rich in fatty acid, whereas the oxidation of this type of fat is primarily responsible for the presence of odors and strange taste to the food (especially the rancid) (Sousa et al., 2007; Jardini; Mancini-Filho, 2007); in retarding the oxidation of biodiesel in which phenolic synthetic compounds are used, thus increasing the lifetime of these fuels and consequently industrial machines, and as adjuvants in medicines, in order to maintain the stability of the components present, especially as antioxidants and chelators of metals; and as drugs in antioxidant therapies, in the prevention and treatment of chronic diseases.

Conclusion

The evidences shown above suggest that, among many antioxidant compounds extracted from natural sources, sulfated polysaccharides from seaweed are promising candidates as antioxidants, by operating both in preventing ROS production, as well as scavenging these radicals when they are already formed. The antioxidant activity of these polymers can be easily accessed by effective, simple and low cost methods. Furthermore, it is observed that the antioxidant potential of PS is dependent on several structural parameters such as degree of sulfation, molecular weight and type of monosaccharide.

The data reported in these studies suggest that future seaweeds may be used as a rich source of antioxidant sulfated polysaccharides to be used as antioxidants in the cosmetic, food and pharmaceutical industry.

References

Ananthia, S; Raghavendrana, HRB; Sunilb, AG; Ramakrishnana, VGG; Vasanthi, HR. In vitro antioxidant and in vivo anti-inflammatory potential of crude polysaccharide from Turbinaria ornata (Marine Brown Alga). *Food Chemistry Toxicology*, 2010, (48), 187–192

Banerjee, A; Dasgupta, N; De, B. In vitro study of antioxidant activity of Syzygium cumini fruit. *Food Chemistry*, 2005, (90), 727–733

Beauchamp, C; Fridovich, I. Superoxide dismutase: Improved assays and an assay applicable to acrylamide gels. *Analytical Biochemistry*, 1971, (44), 276-287

Berteau, O; mulloy, B. Sulfated fucans, fresh perspectives: structures, functions, and biological properties of sulfated fucans and an overview of enzymes active toward this class of polysaccharide. *Glycobiology*, 2003, (13), 29 - 40

Borg, DC. Oxygen free radicals and tissue injury. In M Tarr, and Samson, F. (Eds.). Oxygen free radicals in tissue damage. Birkhauser, Boston, 1993, 12-55

Camara, RBG; Costa, LS; Fidelis, GP; Nobre, LTDB; Dantas-Santos, N; Cordeiro, SL; Costa, MSSP; Alves, LG ; Rocha, HAO. Heterofucans from the Brown Seaweed Canistrocarpus cervicornis with Anticoagulant and Antioxidant Activities. *Marine Drugs*, 2011, (9), 124-138

Cardozo, KH; Guaratini, T; Barros, MP; Falcão, VR; Tonon, AP; Lopes, NP; Campos, S; Torres, MA; Souza, AO; Colepicolo, P; Pinto, E. Metabolites from algae with economical impact. Comparative Biochemistry and Physiology Part C: *Toxicology & Pharmacology*, 2007, 60-78

Costa, LS; Fidelis, GP; Cordeiro, SL; Oliveira, RM; Sabry, DA; Câmara, RBG; Nobre, LTDB; Costa, MSSP; Almeida-Lima, J; Farias, EHC; Leite, EL; Rocha, HAO. Biological activities of sulfated polysaccharides from tropical seaweeds. *Biomedicine & Pharmacotherapy*, 2010, (64), 21–28.

Costa, LS; Fidelis, GP; Telles, CB; Dantas-Santos, N; Camara, RBG; Cordeiro, SL; Costa, MSSP; Almeida-Lima, J; Melo-Silveira, RF; Oliveira, RM; Albuquerque, IRL; Andrade, GPV; Rocha, HAO. Antioxidant and Antiproliferative Activities of Heterofucans from the Seaweed Sargassum filipendula. *Marine Drugs*, 2011, 952-966

Costa, MSSPC; Costa, LS; Cordeiro, SL; Almeida-Lima, J; Dantas-Santos, N; Magalhães, KD; Sabry, DA; Albuquerque, IRL; Pereira, MR; Leite, EL; Rocha, HA. Evaluating the possible anticoagulant and antioxidant effects of sulfated polysaccharides from the tropical green alga Caulerpa cupressoides var. flabellate. *Journal of Applied Phycology*, 2012, 1159-1167

Cuzzocrea, S; Riley, DP; Caputi, AP; Salvemini, D. Antioxidant therapy: a new pharmacological approach in shock, inflammation, and ischemia/reperfusion injury. *Pharmacological Reviews*, 2001, 53(1), 135-59.

Dasgupta, N; De, B. Antioxidant activity of some leafy vegetables of India: A comparative study. *Food Chemistry*, 2007, (101), 471–474

Dietrich, CP; Farias, GGM; Abreu, LRD; Leite, EL; Silva, LF; Nader, HBA. A new aproach for characterization of polysaccharides from algae: presence of four main acidic pollysaccharides in three specie of the class Phaeophycea. 1995. *Plant Science*, 108, 143-53.

Dore, CM; Alves, MGCF; Will, LS; Costa, TG; Sabry, DA; Rêgo, LAS; Accardo, CM; Rocha, HA; Filgueira, LG; Leite, EL. A sulfated polysaccharide, fucans, isolated from brown algae Sargassum vulgare with anticoagulant, antithrombotic, antioxidant and anti-inflammatory effects. *Carbohydrate Polymers*, 2013, 467-475

Ebrahimzadeh, MA; Nabavi, SF; Nabavi, SM; Eslami, B. Antihemolytic and antioxidant activities of Allium paradoxum. *Central European Journal of Biology*, 2010, (5), 338-345

Farias, EHC; Pomin, VH; Valente, AP; Nader, HB; Rocha, HAO; Mourão, PAS. A preponderantly 4-sulfated, 3-linked galactan from the green alga Codium isthmocladum. *Glycobiology*, 2008, (18), 250–259

Ferreira, ALA; Matsubara, LS. Radicais livres: conceitos, doenças relacionadas, sistema de defesa e estresse oxidativo. *Revista da Associação Médica Brasileira*, 1997, (43) 61-68.

Gutteridge, JM. Lipid peroxidation and antioxidants as biomarkers of tissue damage. *Clinical Chemistry*, 1995, (41), 1819-1828.

Hayakawa, Y; Hayashi, T; Lee, JB; Srisomporn, P; Maeda, M; Ozawa, T; Sakuragawa, N. Inhibition of thrombin by sulfated polysaccharides isolated from green algae. *Biochimica et Biophysica acta*, 2000, (1543), 86-94.

Heo, SJ; Park, EJ; Lee, KW; Jeon, YJ. Antioxidant activity of enzymatic extracts from brown seaweeds. *Bioresearch Technology*, 2005, (96), 1613-1623.

Horton, JW. Free radicals and lipid peroxidation mediated injury in burn trauma: the role of antioxidant therapy. *Toxicology*, 2003, (189), 75-88.

Hoyer, K; Karsten, U; Wiencke, C. Induction of sunscreen compounds in Antarctic macroalgae by different radiation conditions, *Marine Biology*, 2002, (141).

Huang, D; Ou, B; Prior, RL. The chemistry behind antioxidant capacity assays. *J. Agric. Food Chemistry*, 2005, (53), 1841-1856.

Hu, T; Liu, D; Chen, Y; Wu, J; Wang, S. Antioxidant activity of sulfated polysaccharide fractions extracted from Undaria pinnitafida in vitro. *International Journal of Biological Macromolecules*, 2010, (46), 193–198.

Josephine, A; Nithya, K; Amudha, G; Veena, CK; Preetha, SP; Varalakshmi, P. Role of sulphated polysaccharides from Sargassum Wightiiin Cyclosporine A-induced oxidative liver injury in rats. *BMC Pharmacology*, 2008, 1 – 9.

Kim, AH; Chiu, DS; Athukorala, Y; Jeon, YJ; Senevirathne, M; Rha, CK. Antioxidant Activity of Sulfated Polysaccharides Isolated from Sergassum fulvellum. *Journal of Food Science and Nutrition*, 2007, 65 – 73.

Li, B; Lu, F; Wei, X; Zhao, R. Fucoidan: structure and bioactivity. *Molecules*, 2008, (13), 1671–1695.

Lim, BL; Ryu IH. Purification, structural characterization, and antioxidant activity of antioxidant substance from the red seaweed Gloiopeltis tenax. *Journal of Medicinal Food*, 2009, (12), 442-451

Liu, F; Liu, J; Gu, J; Zhang, L; Shen, W; Guo, T; Liu, C; He, P. Ex vivo Anti-Oxidation Activity of Polysaccharides from Red Alga Porphyra yezoensis. *Research Journal of Pharmacology*, 2007, 61 – 66.

Liu, F; Liu, J; Gu, J; Zhang, L; Shen, W; Guo, T; Liu, C; He, P. Antioxidation Activity of Polysaccharides from the Red Alga Porphyra yezoensis. *Ciencias Marinas*, 2008, 253-261.

Magalhães, KD ; Costa, LS; Fidelis, GP; Oliveira, RM ; Nobre, LTDB; Dantas-Santos, N; Camara, RBG ; Albuquerque, IRL; Cordeiro, SL; Sabry, DA; Costa, MSSP; Alves, LG; Rocha, HAO. Anticoagulant, Antioxidant and Antitumor Activities of Heterofucans from the Seaweed Dictyopteris delicatula. *International Journal of Molecular Sciences*, 2011, (12), 3352-3365

Mandal, P; Pujol, CA; Carlucci, MJ; Chattopadhyay, K; Damonte, EB; Ray, B. Anti-herpetic activity of a sulfated xylomannan from Scinaia hatei. *Phytochemistry*, 2008, (69), 2193–2199

Mao, W; Zang, X; Li, Y; Zhang, H. Sulfated polysaccharides from marine green algae Ulva conglobata and their anticoagulant activity. *Journal of Applied Phycology*, 2006, 9-14

Matsukawa, R; Dubinsky, Z; Kishimoto, E; Masakki, K; Masuda, Y; Takeuchi, TA. Comparison of screening methods for antioxidant activity in seaweeds. *Journal of Applied Phycology*, 1997, (9), 29-35

Mohsin, S; Kurup, GM; Mahadevan, R. Effect of Ascophyllan from Brown Algae Padina tetrastromatica on Inflammation and Oxidative Stress in Carrageenan-Induced Rats. *Inflamation*, 2013.

O'Sullivan, L; Murphy, B; McLoughlin, P; Duggan, P; Lawlor, PG; Hughes, H; Gardiner, GE. Prebiotics from Marine Macroalgae for Human and Animal Health Applications. *Marine Drugs*, 2010, 8(7), 2038–2064.

Pacifico, S; Abrosca, B; Golino, A; Mastellone, C; Piccolella, S; Fiorentino, A; Monaco, P. Antioxidant evaluation of polyhydroxylated nerolidols from redroot pigweed (Amaranthus retrofexus) leaves. *Food Science Technology*, 2008, (41), 1665-1671

Peluffo, G; Radi, R. Biochemistry of protein tyrosine nitration in cardiovascular pathology. *Cardiovascular Research*, 2007, (75), 291–302

Pomin, VH; Mourão, PA. Structure, biology, evolution, and medical importance of sulfated fucans and galactans. *Glycobiology*, 2008, 1016-1027.

Prieto, P; Pineda, M; Aguilar, M. Spectrophotometric quantitation of antioxidant capacity through the formation of a Phosphomolybdenum Complex: Specific application to the determination of vitamin E. *Analytical Biochemistry*, 1999, (269), 337-341

Qi, H; Zhang, Q; Zhao, T; Chen, R; Zhang, H; Niu, X; Li, Z. Antioxidant activity of different sulfate content derivatives of polysaccharide extracted from Ulva pertusa (Chlorophyta) in vitro. *International Journal of Biological Macromolecules*, 2005a, (37), 195–199

Qi, H; Zhao, T; Zhang, Q; Li, Z; Zhao, Z. Antioxidant activity of different molecular weight sulfated polysaccharides from Ulva pertusa Kjellm (Chlorophyta). *Journal of Applied Phycology*, 2005b, 527–534.

Qi, H; Zhang, Q; Zhao, T; Hu, R; Zhang, K; Li, Z. In vitro antioxidant activity of acetylated and benzoylated derivatives of polysaccharide extracted from Ulva pertusa (Chlorophyta). *Bioorganic & Medicinal Chemistry Letters*, 2006, 2441–2445

Radi, R; Peluffo, G. Unraveling peroxynitrite formation in biological systems. *Free Radical Biology and Medicine*, 2001, (30), 463-488

Rocha, FD; Pereira, RC; Kaplan, MAC; Teixeira, VL. Natural products from marine seaweeds and their antioxidant potential. *Revista Brasileira de Farmacognosia*, 2007, (17), 631-639.

Selvakumar, K; Madhan, R; Srinivasan, G; Baskar, V. Antioxidant Assays in Pharmacological Research. *Asian Journal of Pharmacy and Technology*, 2011, (1), 99-103

Silva, RO; Santana, APM; Carvalho, NS; Bezerra, TS; Oliveira, CB, Damasceno, SRB; Chaves, LS; Freitas, ALP; Soares, PMG; Souza, MHLP; Barbosa, ALR; Medeiros, JR. A Sulfated-Polysaccharide Fraction from Seaweed Gracilaria birdiae Prevents Naproxen-Induced Gastrointestinal Damage in Rats. *Marine Drugs*, 2012, 2618-2633

Smirnoff, N; Cumbes, QJ. Hydroxyl radical scavenging activity of compatible solutes. *Phytochemistry*, 1989, (28), 1057-1060.

Somogyi, A; Rosta, K; Pusztai, P; Zsolt, T; Nagy, G. Antioxidant measurements. *Physiological measurement*, 2007, (28), 41 - 55

Paya, M; Halliwel, B; Hoult, JR. Interactions of a series of coumarins with reactive oxygen species: scavenging of superoxide, hypochlorous acid and hydroxyl radicals. *Biochemical pharmacology*, 1992, (44), 205-214.

Rocha, HAO; Leite, EL; Medeiros, VP; Lopes, CC; Nascimento, FD; Tersariol, ILS; Nader, HB. Natural sulfated polysaccharides as antithrombotic compounds. Structural characteristcs and effects on the coagulation cascade. In: *Verli H, editor. Carbohydrate Structure and Biological Function*. Kerala: Transworld research network, 2006, 51 - 67.

Ruperez, P; Ahrazem, O; Leal, JA. Potential antioxidant capacity of sulfated polysaccharides from the edible marine brown seaweed Fucus vesiculosus. *Journal of Agricultural and Food Chemistry*, 2002, (4), 840-845

Siddhanta, AK; Shanmugam, M; Mody, KH; Goswami, AM; Ramavat, BK. Sulphated polysaccharides of Codium dwarkense Boergs. from the west coast of India: chemical composition and blood anticoagulant activity. *International journal of biological macromolecules*, 1999, (26), 151-154.

Souza, MCR; Marques, CT; Dore, CMG; Silva, FRF; Rocha, HAO; Leite, EL. Antioxidant activities of sulfated polysaccharides from brown and red seaweeds; *Journal of Applied Phycology*, 2007, (19), 153–160

Souza, BWS; Cerqueira, MA; Bourbon, AI; Pinheiro, AC; Martins, JT; Teixeira, JA; Coimbra, MA; Vicente, AA. Chemical characterization and antioxidant activity of sulfated polysaccharide from the red seaweed Gracilaria birdiae. *Food Hydrocolloids*, 2012, (27), 287–292.

Shao, P; Chen, X; Sun, P. In vitro antioxidant and antitumor activities of different sulfated polysaccharides isolated from three algae. *International Journal of Biological Macromolecules*, 2013, 155 - 161.

Shao, P; Chen, X; Sun, P. In vitro antioxidant and antitumor activities of different sulfated polysaccharides isolated from three algae. *International Journal of Biological Macromolecules*, 2013, (62), 155–161

Suresh, V; Senthilkumar, N; Thangam, R; Rajkumar, M; Anbazhagan, C; Rengasamy, R; Gunasekaranc, P; Kannan, S; Palani, P. Separation, purification and preliminary characterization of sulfated polysaccharides from Sargassum plagiophyllum and its in vitro anticancer and antioxidant activity. *Process Biochemistry*, 2013, 364 - 373.

Vijayabaskar, P; Vaseela, N; Thirumaran, G. Potential antibacterial and antioxidant properties of a sulfated polysaccharide from the brown marine algae Sargassum swartzii. *Chinese Journal of Natural Medicines*, 2012, 421 - 428

Wang, J; Zhang, Q; Zhang, Z; Li, Z. Antioxidant activity of sulfated polysaccharide fractions extracted from Laminaria japonica. *International Journal of Biological Macromolecules*, 2008, 42, 127–132.

Wang, J; Wang, F; Zhang, Q Zhang, Z; Shi, X; Li P. Synthesized different derivatives of low molecular fucoidan extracted from Laminaria japonica and their potential antioxidant activity in vitro. *International Journal of Biological Macromolecules*, 2009a, 379-84.

Wang, J; Zhang, Q; Zhang, Z; Zhang, J; Li, P. Synthesized phosphorylated and aminated derivatives of fucoidan and their potential antioxidant activity in vitro. *International Journal of Biological Macromolecules*, 2009b, 44(2), 170-174.

Wang, J; Zhang, Q; Zhang, Z, Song, H; Li, P. Potential antioxidant and anticoagulant capacity of low molecular weight fucoidan fractions extracted from Laminaria japonica. *International Journal of Biological Macromolecules*, 2010, (46), 6-12.

Wickens, AP. Aging and the free radical theory. Respiration Physiology. 2001. (128) 379–91

Xue, CH; Fang, Y; Lin, H; Chen, L; Li, ZJ; Deng, D; Lu, CX. Chemical characters and antioxidative properties of sulfated polysaccharides from Laminaria japonica. *Journal of Applied Phycology*, 2001, (13), 67-70.

Yang, Y; Liu, D; Wu, J; Chen, Y; Wang, S. In vitro antioxidant activities of sulfated polysaccharide fractions extracted from Corallina officinalis. *International Journal of Biological Macromolecules*, 2011, (49), 1031–1037.

Ye, H; Wang, KQ; Zhou, CH; Liu, J; Zeng, XX. Purification, antitumor and antioxidant activities in vitro of polysaccharides from the brown seaweed Sargassum pallidum. *Food Chemistry*, 2008, (111), 428-432.

Zhang, Q; Yu, P; Li, Z; Zhang, H; Xu, Z; Li, P. Antioxidant activities of sulfated polysaccharide fractions from Porphyra haitanesis. *Journal of Applied Phycology*, 2003a, 305-310

Zhang, QB; Li, N; Zhou, GF; Lu, XL; Xu, ZH; Li, Z. In vivo antioxidant activity of polysaccharide fraction from Porphyra haitanesis (Rhodephyta) in aging mice. *Pharmacological Research*, 2003b, 48(2), 151–155.

Zhang, QB; Li, N; Liu, X; Zhao, Z; Li, Z; Xu, Z. The structure of a sulfated galactan from Porphyra haitanensis and its in vivo antioxidant activity. *Carbohydrate Research*, 2004, (399), 105 – 111.

Zhang, Z; Zhang, Q; Wang, J; Zhang, H; Niu, X; Li, P. Preparation of the different derivatives of the low-molecular-weight porphyran from Porphyra haitanensis and their antioxidant activities in vitro. *International Journal of Biological Macromolecules*, 2009, (45), 22-26.

Zhao, T; Zhang, Q; Qi, H; Zhang, H; Niu, X; Xu, Z; Li, Z. Degradation of porphyran from Porphyra haitanensis and the antioxidant activities of the degraded porphyrans with different molecular weight. *International Journal of Biological Macromolecules*, 2006, 45 – 50.

Zhao, X; Xue, CH; Li, BF. Study of antioxidant activities of sulfated polysaccharides from Laminaria japonica. *Journal of Applied Phycology*, 2008, 431-436.

Zhou, J; Hu, N; Wu, YL; Pan, YJ; Sun, CR. Preliminary studies on the chemical characterization and antioxidant properties of acidic polysaccharides from Sargassum fusiforme. *Journal of Zhejiang University SCIENCE B*, 2008, (9), 721–727.

Zhou, C; Wang, Y; Ma, H; He, R. Effect of Ultrasonic Degradation on In Vitro Antioxidant Activity of Polysacharides from Porphyra yezoensis (Rhodophyta). *Food Science and Technology International*, 2008.

Zong, A; Cao, H; Wang, F. Anticancer polysaccharides from natural resources: A review of recent research. *Carbohydrate Polymers*, 2012, (90), 1395–1410.

EDITOR'S CONTACT INFORMATION

Dr. Vitor H. Pomin
Professor of Biochemistry and Chemical Biology
Institute of Medical Biochemistry
Federal University of Rio de Janeiro
Rio de Janeiro, Brazil
vhpomin@gmail.com

INDEX

B

C

F

G

H

I

J

K

L

M

N

O

P

Q

R

S

T

U

V

W

X

Y

Z